10699673

Electromyography in Ergonomics

Born in India, Professor **Shrawan Kumar** holds an MSc (Zoology) from the University of Allahabad, India and a PhD (Human Biology) and DSc (Science) from the University of Surrey, UK. Professor Kumar held academic positions at the University of Dublin, Ireland, in Engineering Sciences and the University of Toronto, Canada in Physical Medicine and Rehabilitation. He joined the University of Alberta, where he continues to work, as Assistant Professor in Physical Therapy in 1977 and rose to become Associate and Full Professor in 1979 and 1982 respectively.

Anil Mital is Professor of Industrial Engineering and Physical Medicine and Rehabilitation at the University of Cincinnati. He is also the founding director of the Ergonomics and Engineering Controls Research Laboratory and former Director of Industrial Engineering at the University of Cincinnati.

He holds a BE degree in Mechanical Engineering from the University of Allahabad, India, and an MS and PhD in Industrial Engineering from Kansas State University and Texas Tech University, respectively. He is editor-in-chief of the *International Journal of Industrial Ergonomics* and the *International Journal of Industrial Engineering*, and author/coauthor/editor of nearly 350 technical publications including 15 books. His current research interests are in the areas of human integration in advanced manufacturing technology, rehabilitation ergonomics, application of systems methodologies to ergonomics, metal cutting, economic justification and traditional industrial engineering.

Professor Mital is the founder of the International Society (formerly Foundation) of Occupational Ergonomics and Safety and winner of its first Distinguished Accomplishment Award (1993). He is also a Fellow of the Human Factors and Ergonomics Society (1992) and winner of the Liberty Mutual Insurance Company Best Paper Award (1994). He is a senior member of the Institute of Industrial Engineers and past director of its ergonomics division.

Electromyography in Ergonomics

EDITED BY

SHRAWAN KUMAR

AND

ANIL MITAL

UK Taylor & Francis Ltd, One Gunpowder Square, London EC4A 3DE
USA Taylor & Francis Inc., 1900 Frost Road, Suite 101, Bristol, PA 19007

A catalogue record for this book is available from the British Library
ISBN 0 7484 0130 X

Printed in Great Britain by T J Press, Padstow.

Contents

Preface

Electromyography has been used in ergonomics for many years. Due to the multifaceted nature of ergonomics, none of the techniques employed can ever assume a dominating role. Different approaches and different techniques are employed at different times to solve different problems. Though ergonomics is a rigorous science as any other, such a seemingly shifting emphasis may create a perception of lack of focus. Ergonomics has evolved out of biological, behavioral, and engineering roots. Rarely people have thorough understanding of all issues involved. The vastness of the field coupled with seeming lack of focus has hindered the growth, development, and establishment of our discipline. It is, therefore, of considerable significance to take an extensive look at what is done in ergonomic context with ergonomic relevance. It has been our intent to try and put electromyography in an ergonomic perspective. Such an endeavor will demystify, rationalize, and exemplify one of the frequently used techniques in ergonomic application. A thorough discussion and presentation of such a subject will help reduce possible errors in ergonomic endeavors. It is hoped this book will help to achieve this goal even ever so slightly.

To accomplish the above-mentioned goal, the book has been organized in a sequence of logical progression. The first chapter by Kumar deals with the overview of electromyography in ergonomics. In this chapter, the necessity, the uses, and the limitations of electromyography with particular reference to ergonomics has been described. Particular care has been taken to highlight the ergonomic perspectives and present them with ergonomic issues. It is hoped that such a scientific and technical platform will help launch beginners or unsure readers with a better understanding of the issues. The second chapter on the principles and concepts by Luttmann elaborates on the biological and physiological principles which govern the physiology of muscle contraction. It will be of particular interest to know what happens, but more interestingly why and how the muscle contracts. Such understanding begins to provide an intuitive feel for the behavior of muscle. The third chapter by Ortengren addresses the aspects of noise in signal. What may be considered noise and why? How does it creep into the signals and how can one discriminate. It is of particular significance for novices to remember that EMG is only an electrical signal and electrical signals can be generated from sources other than muscles. Therefore, it is of vital importance to be able to discriminate between signal and noise. The fourth chapter on interpretation of the EMG signals by Winter deals with various aspects of processing and interpretation of EMG. The fifth chapter by Moritani on muscle energetics and electromyography introduces less commonly addressed subject areas of energy metabolism and its relationship with electromyography. The uniqueness of this approach lies in linking more generalized systemic stress with those of local biomechanical and neuromotor loads. The aspects of skill acquisition and muscle soreness, of considerable significance to ergonomics, are also addressed. In Chapter 6, Hagg and Kadefors discuss aspects of fatigue and their significance to ergonomics.

Chapter 7, 8, and 9 deal with the application of electromyographic technology in ergonomics. In Chapter 7, Strasser presents an extensive account of ergonomic application through the use of electromyography. In particular, evaluation of repetitive tasks, splitting of EMG into static and dynamic, and an ergonomic assessment of tools will be of considerable interest to ergonomists of all shades. Westgaard *et al.* in Chapter 8 discuss the use of surface electromyography to assess the risk of musculoskeletal symptoms. They present the use and limitation of such an approach in particular for the trapezius muscle. Roy and De Luca describe their experiences with surface electromyographic assessment of low-back pain in Chapter 9. The prevalence of low back pain and its economic impact in addition to human suffering make it particularly significant. Finally, Feldman discusses a clinical perspective of EMG in ergonomics.

SHRAWAN KUMAR
ANIL MITAL

CHAPTER ONE

Electromyography in ergonomics

SHRAWAN KUMAR

University of Alberta, Edmonton, Canada

1.1 INTRODUCTION

Conceptually, biological and socioeconomic evolution have occurred concurrently. However, their rates have been differential. Within a minuscule fraction of time socioeconomic evolution has occurred at an explosive rate. The latter was considerably aided by the Industrial Revolution of the New World. So much so, this has revolutionized the human obsession with wealth far beyond the biological necessity of survival. The progressive development of technology has also continued to change rapidly the nature of work and the demand on the worker. Nonetheless, most occupations require varying amounts of perceptual, cognitive, and motor efforts to execute any task. Since the media of action related to task execution changed from psychological to physical to environmental realms, the limitations of and optimal interfacing with them have become critical for safety, comfort, and efficiency. Ergonomics is a science which deals with characteristics of each of these realms in the context of work and endeavors to optimize the compatibility between them.

1.2 THE PHYSIOLOGICAL BASIS OF ELECTROMYOGRAPHIC TECHNIQUE

The technique of electromyography is based on the phenomenon of electromechanical coupling in muscle. Electrical signals generated in the muscle eventually lead to the phenomenon of muscle contraction through intermediate processes. Briefly, as a single or a train of action potentials sweep the muscle membrane (sarcolemma), these electrical potential differences travel deep into the muscle cells through t-tubules. The t-tubules are invaginations of the muscle membrane inside the muscle cells. Such invaginations are numerous and occur at the junctions of the light and dark bands of the myofibrils where they surround it as a ring on a finger. These rings are interconnected with the rings of neighboring myofibrils making an extensive system of tubules. Such structural organization allows the electrical potential to travel to the deepest parts of the muscle almost instantly as it sweeps the surface. These action potentials trigger the release of Ca^{2+} ions from the sarcoplasmic reticulum into the muscle cytoplasm. These calcium ions are responsible for facilitating the muscle contraction which in turn manifests itself in the motion of body members and generation of force. Thus, though there is an electromechanical coupling in the muscle, it is mediated through biochemical means. The latter, therefore, does modify this coupling differently in different conditions, such as, motion, duration, geometrical configuration, and fatigue, etc. These considerations will be discussed later in the section on 'analysis and interpretation' (see Winter, Chapter 4).

1.3 THE RECORDING METHODOLOGY

The sweep of action potentials over the muscle membrane can be likened to waves. Information about the electrical activity in the muscle can be represented as the height of the wave at a given point with respect to the other, and also the density of the waves. Such information represented graphically or pictorially is called an electromyogram (EMG) and its recording is electromyographic recording.

Since the electromyographic recording is most commonly done to pick up the electrical potential difference between two points, a bipolar electrode configuration is used. These bipolar electrodes can be needle or wire electrodes, but surface electrodes are most commonly used in ergonomic applications. The recording methodology for an electromyogram can vary greatly depending upon the purpose, application, desired signal, available equipment, and analytical methodology. For more complete discussion of all these aspects the readers are directed to Basmajian and DeLuca (1985), Loeb and Gans (1986), and others. In most common use, two silver–silver chloride surface electrodes are placed on skin overlying the muscle of interest to line up with the predominant fibre direction closer to the motor point. The subject is grounded by placing an electrode at an inactive region of the body and connected to the chassis of the pre-amplifier. At the pick-up site the magnitude of the EMG signals are very small (in the order of microvolts). These small signals can be drowned in the noise of the system or electromagnetic interference. Therefore, these signals are pre-amplified close to the source of generation before noise has a chance to contaminate the signals irrevocably. Such pre-amplified signals are then fed to an amplifier through coaxial cables. Between the pre-amplifier and amplifier the signal can be amplified by a factor of several thousands. Most amplifiers do signal conditioning and may be able to filter undesirable frequency components. Such signals can then be either registered directly on a recording medium such as an oscilloscope, ink jet recorders, a light pen recorder, or magnetic FM tape recorders, or they can be fed to a computer through an A to D converter. The choice of the recording medium will be dependent on the selected subsequent analysis. The advantage of 'on line' recording is the instant availability of various types of processing.

1.4 USES AND LIMITATIONS OF ELECTROMYOGRAPHY

The technique of electromyography can be a powerful ergonomic tool. However, due to its sophistication it is highly selective and requires a good understanding, thorough consideration, and considerable discretion in application. Electromyography can answer some questions decisively, some ambiguously, and others not at all. It can also provide some quantitative answers but some remain qualitative or at best semi-quantitative. Therefore, an ergonomist must know precisely the question to be answered and choose the relevant technique. One must know the muscles to be tested, apply electrodes appropriately, make a suitable recording, and analyze and interpret the data correctly to obtain a valid answer. Some of these aspects are addressed later in this chapter and other chapters in this book. The technique of electromyography is used in ergonomics to obtain various types of answers. These may be as follows.

1 Whether the muscle in question is active or inactive in a given task?
2 When does the muscle turn on and off?
3 What is the phasic relationship between the muscles of interest?
4 Is the muscle activation pattern efficient to indicate skill acquisition?
5 Is the magnitude of the electromyographic activity greater to imply higher stress task as measured by one or more of the following semi-quantitative variables?
   - (i) Magnitude of raw signals.
   - (ii) Number of zero crossing.
   - (iii) Number and/or magnitude of spikes.
6 The magnitude of the electromyographic activity based on quantitative measures listed below.
   - (i) Voltage.
   - (ii) Wave rectification and envelope detection.
   - (iii) Integration.
   - (iv) Root mean square.
   - (v) Normalization and force calculation from EMG signal.
7 Is the muscle fatigued?

1.4.1 Presence of Muscle Activity

The primary information of value to an ergonomist, at times, is simply to establish the active or inactive state of the muscle in question. Frequently this information may be gleaned through consideration of the functional anatomy of the area in relation to the task being performed. However, an electromyographic recording will confirm this information and provide a better understanding and an intuitive feel of the muscle involvement. The muscle activity recorded and seen is the end-product of the net activity of excitatory and inhibitory neural input to the motor neuronal pool in the central nervous system. This centrally produced nervous signal travels along the nerve and terminates at the motor end plate, where it triggers a chain of electrochemical and mechanical events leading to muscle contraction (see Luttmann, Chapter 2). This phenomenon is a frequency-modulated rate of action potentials in each of the motor neurons innervating the muscle in question. Thus, each of the motor neurons through their train of action potentials contributes to the muscle contraction and generation of muscle tension. The technique of electromyography endeavors to pick up the electrical phenomena associated with the muscle contraction.

Since the muscle environment is not electrically compartmentalized and insulated the electrical signals generated from one motor neuron upon reaching that of another sum up. Thus, these volume-conducted signals, a small distance away from the source, become compound action potentials. The farther the electrode from the source the greater the number of action potentials total. With increasing distance between the site of action potential and the recording site, the resistance of the intervening tissue progressively attenuates the signals. In ergonomic applications surface electrodes are commonly used. These have an area of pick-up in the shape of a broad inverted cone with electrodes making the apex. Thus the signals obtained and recorded are summed into an amplitude-modulated envelope of AC signals with a fairly wide bandwidth (0–500 Hz). A visual record of these signals is called an electromyogram and the pattern of the trace recorded is called an interference pattern. Since this electrical output is a consequence of the centrally generated command signal to execute the task being performed, it is a reliable index of muscle involvement in the activity. If the information required was simply to establish activity/inactivity of the muscle(s) concerned, the objective will be achieved by such recordings. However, one cannot help but noticing the general magnitude of activity to classify, subjectively as none, slight, moderate, strong, and explosive.

Carlsöö (1961) determined the presence of muscle activity and thereby the load in different work positions. He employed surface electrodes as well as needle electrodes to determine muscle activity/inactivity of the sacrospinalis, neck musculature, gluteus maximus, gluteus medius, tensor fasciae latae, quadriceps femoris, sartorius, iliopsoas, biceps femoris, semimembranosus, semitendinosus, gastrocnemius, soleus, flexor hallucis longus, peroneus longus, peroneus brevis, tibialis anterior, rectus abdominis, the external and internal obliques, and the transverse abdominal muscles. Carlsöö (1961) carefully described the postures as the symmetric standing at rest position, forward leaning position, backward leaning, slight and pronounced stooping postures, fully bent postures, backward bending posture, tiptoe position, knee-bending posture, asymmetric standing posture, rotated trunk posture, and lateral bending posture. In all these static postures the electromyographic activities were recorded and classified symbolically as follows.

−none, + slight, (+) less than slight, + (+) more than slight, + + strong

These recordings and scores were interpreted in the light of posture and balance to understand the muscle load.

1.4.2 Onset, Offset, and Synergies

Onset of muscle activity is important information for an ergonomist. It allows one to recognize and understand the role of the muscle in relation to the task, and determine agonistic and antagonistic activity for defining the biomechanical forces (stresses) in the region. The onset–offset information also allows a quantitative assessment of the duration of muscle load and its temporal relationship with the posture and/or task cycle. The ergonomic significance of such information is obvious. It allows an ergonomist to focus attention on the relevant mechanisms and factors. One of the prime examples of demonstration of interactions of mechanics and muscle activity is the phenomenon of flexion relaxation of the erector spinae in full flexion. In spite of a claim by Fick (1911) that the erector spinae undergo relaxation in full flexion, the phenomenon remained unproven until it was demonstrated by Floyd and Silver (1950) using electromyography. They placed electrodes over the erector spinae at T_{10}, T_{12}, L_2 and L_4 levels in addition to using needle electrodes for confirmation. Floyd and Silver (1950) showed that during the initial phases of the trunk flexion, the motion was controlled by the erector spinae. In a follow-up study, Floyd and Silver (1955) studied function of trunk muscles in greater detail. With respect to the function of the erector spinae they reported that the erector spinae quietened in advanced flexion. Even during heavy lifting from a fully flexed posture, the erector spinae remained silent until the weight had reached knee level height. This phenomenon of flexion relaxation of the erector spinae has a major bearing on our understanding of load sharing among active and passive elements of trunk structures during stressful maneuvers. Such information can play a useful role in task design as well as safety training for workers.

With respect to the motion of the elbow and the function of muscles operating this joint, Bechtel and Caldwell (1994) investigated the effect of changing the internal mechanical variables and task demands on muscle activity and torque production during strong isometric contractions of the elbow flexors. Through surface electromyography of the biceps brachii, brachio-radialis, and triceps brachii they studied the effect of superimposing a 50% MVC of supination on a maximal isometric elbow flexion at several angles between 30–110° of elbow flexion. The magnitude of bicep brachii

EMG progressively increased with increasing flexion angle. It also increased in magnitude or remained unchanged by superimposition of supination at all angles. As a result of supination the magnitude of brachioradialis EMG decreased and so did the torque at most angles of elbow flexion. Bechtel and Caldwell (1994) concluded that the reduction in torque was attributable to the reduction in brachioradialis contribution. These authors suggested that the manner in which the central nervous system distributes activities to the synergists is determined by the mechanical nature of the tasks in question. Since the task-specific muscle activity determines the joint torque, it may be possible to reduce fatigue-mediated work injuries by modifying the job and also facilitate post-injury rehabilitation by use of this knowledge. It may be pointed out that muscle synergism can be affected significantly by a change in the characteristics of the task being performed. An elbow in pure active flexion will have the brachioradialis actively participating in the task but a superimposition of supination will cause an offset in its activity. In ergonomic investigations, therefore, a consideration of exact job requirements may be a critical factor to consider.

1.4.3 Phasic Interrelationship, Agonism, and Antagonism

All joints of the body are operated by a variable number of muscles but never by a single muscle. The greater the degrees of freedom of motion, the larger the number of muscles operating the joint. Since the muscles are soft contractile tissues with a point of origin and a point of insertion, during their contraction they are capable of generating tension only along the direction of predominant fiber orientation. Complex motions or motions non-coincidental with the vector force of a single muscle are assisted by others. The muscle responsible for creating a motion primarily in the direction of its vector force is the prime mover and called the agonist muscle. Those muscles which assist the agonist muscles in their function by contributing additional force through their contraction are synergistic muscles. Within the normal range of motion of every joint there is little passive resistance.

It is essential to have muscles which oppose the force and motion of the agonist and synergistic muscles for two reasons. First, the activity of agonistic and synergistic muscles has to be controlled otherwise every motion will be sudden and jerky at the least, if not explosive. The braking processes of limb movements were studied by Cnockaert (1978) and Lestienne (1979). They reported that the viscoelastic forces and the pattern of EMG activity played a significant role in braking. Maton *et al.* (1980) showed that the elbow extensors carried out the task of braking during elbow flexion. The activity of the extensors occurred in two bursts. The first occurred during the onset of acceleration which was a weak activity. Evidently this corresponded to the occurrence of smaller forces of the joint to stabilize the articulation. This was found to be simultaneous for all elbow extensors and occurred sooner and

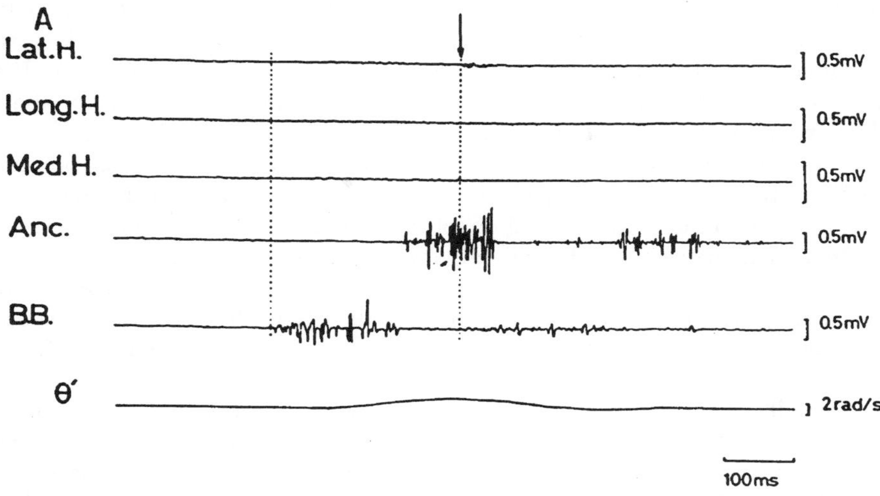

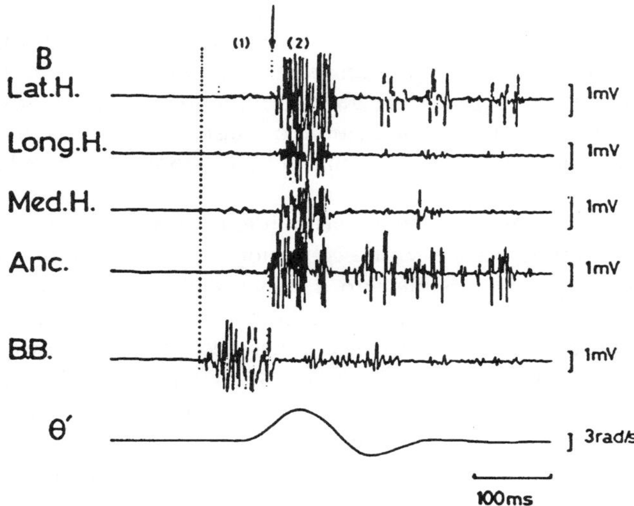

Figure 1.1 Agonism and antagonism among muscles of elbow. Typical records of flexion movements performed against one inertial load (1 = 0.021 kg m^2). (a) Slow movement. (b) Fast movement. Lat. H: surface EMG of lateral head of the triceps muscle. Long. H: surface EMG of long head of the triceps muscle. Med. H: surface EMG of medial head of the triceps muscle. Anc.: surface EMG of anconeus muscle. B.B.: surface EMG of biceps brachii. O': angular velocity. The arrow represents the onset of activity of the lateral head. (1) First burst. (2) Second burst. Reprinted with permission from Maton, B., Le Bozec, S. & Cnockaert, J.C. (1980) The synergy of elbow extensor muscles during dynamic work in man: II. Braking elbow flexion. *European Journal of Applied Physiology*, **44**, 271–78.

sooner with the increase in the velocity of flexion. The second burst of activity corresponded more to the braking of the flexion (Figure 1.1). For an ergonomist it is important to recognize the totality of the situation. For example, if one were to consider antagonism of only one muscle in elbow flexion one would overestimate the contribution of viscoelastic resistance in the system. The latter is because the activities of the anconeus precede significantly that of the triceps in slow motion. If the viscoelastic resistance was to have a limited mechanical effect it will not be true of its physiological activity. The most valuable role of the viscoelastic resistance lies in the regulation of the precision of the movement regardless of the generated force. Thus, the role of the anconeus in braking the elbow flexion in slow speed motions is important but loses its significance, at least to some extent, in faster activities.

Secondly, having reached the extreme range of motion, the joint cannot return to the original position. The group of muscles which oppose the agonist and synergist muscles is called the antagonist group of muscles. Generally each muscle provides a limited but requisite amount of force and motion. The combination of joint geometry, connective tissue support system, and muscle contraction of all types of muscles determine the full range of function of every joint. In order to assess the contribution of muscles in the range of functional activity determination of their temporal interrelationship in a mechanical (or spatial) context becomes valuable for ergonomic application. This understanding generally assists optimization of tasks as well as the physical activity for efficiency and safety among normal people and people with disabilities.

One of the earliest examples of such study was in human gait. Locomotion is the primary activity in getting a variety of occupational activities done. In bipedal walking the gait cycle is divisible into a stance and a swing phase. Most of the joints of the human body are involved in locomotion to a varying degree, therefore, the muscles operating them are also recruited. The lower extremities are primarily responsible for locomotion, the upper body playing a compensatory role. Therefore, a brief consideration of lower extremities during human gait is in order. The literature on gait is vast and it is not intended to give a comprehensive picture here, but rather an example of the interaction of muscles, joints, and mechanics. In a pioneering study, Radcliffe (1962) reported on the interaction between the knee and ankle joints and the phasic activity of muscle groups of the lower extremity (Figure 1.2). At the start of the gait cycle during heel strike, the hamstrings and the pretibial group of muscles have their peak activity. The quadriceps group of muscles follow soon reaching a peak while maintaining the stability of the knee. At heel-off, the calf group of muscles begin to increase in activity and become silent at toe-off. Immediately before the toe-off, the quadriceps group of muscle begin to fire reaching another but smaller peak at approximately 60% of the gait cycle. The pretibial group of muscles also reach another smaller peak around toe-off (60% of the gait cycle). In addition, the pretibial

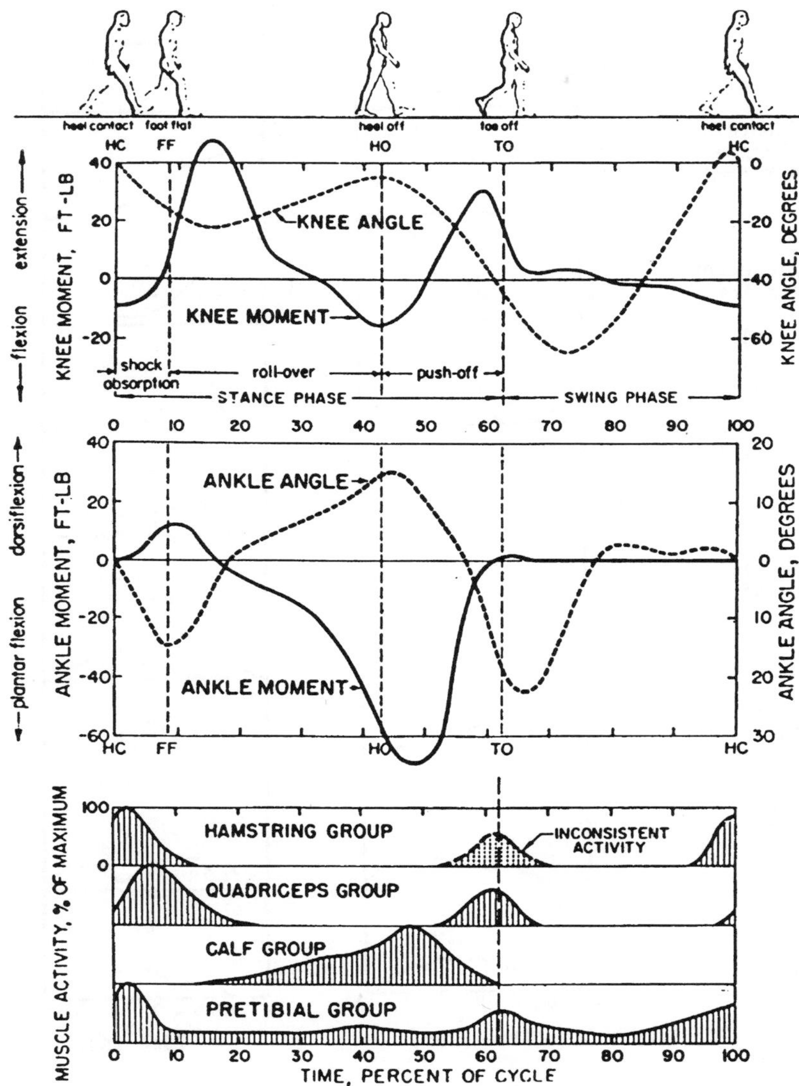

Figure 1.2 The phasic interrelationship of the regional muscles. Normal walking: knee and ankle moments (in foot pounds) compared with muscular activity during one cycle of walking (right heel to right heel contact) on level ground. Reprinted with permission from Radcliffe, C.W. © (1962) The biomechanics of below knee prostheses in normal, level, bipedal walking. *Artificial Limbs*, 6, 16–24.

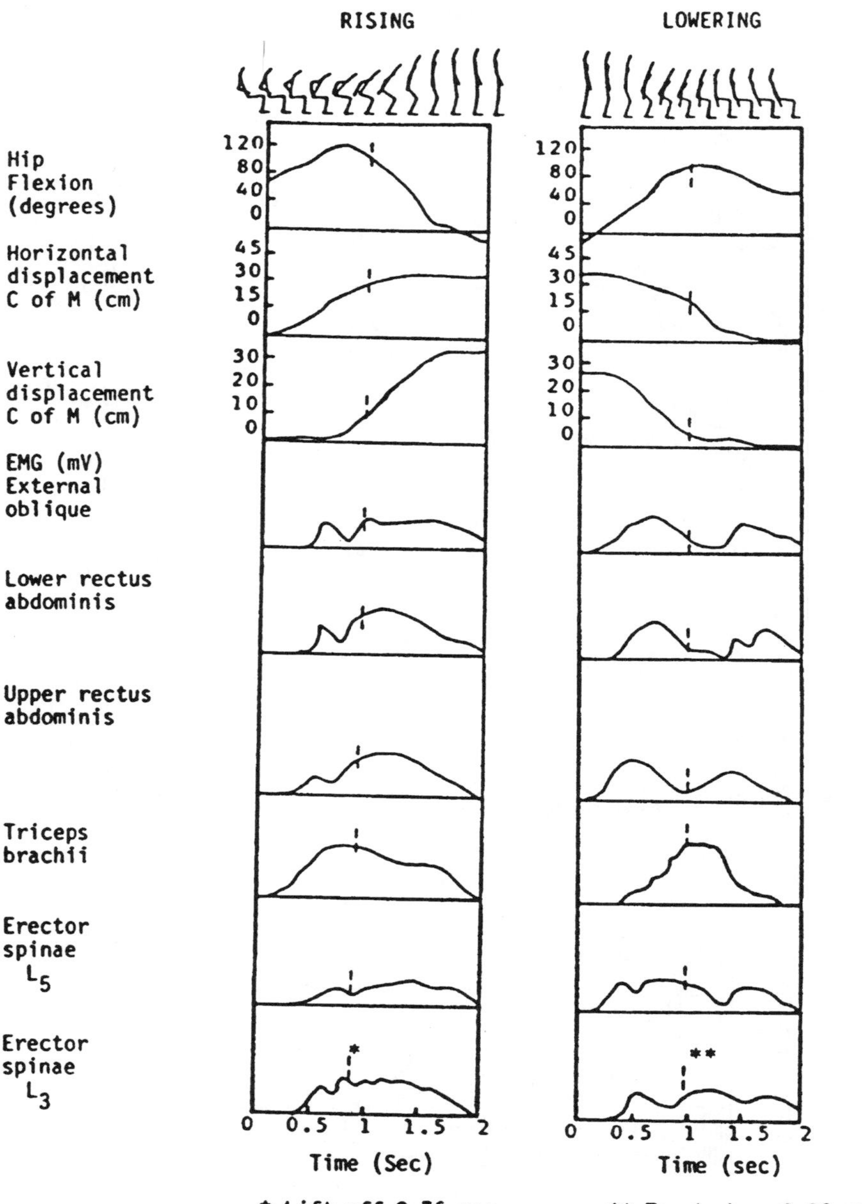

Figure 1.3 The phasic interrelationship between muscles and temporal relationship between posture and the activities. Subject 1 rising with armrest hand support using seat height position 4 (10 cm lower than position 1) and subject 2 lowering with no hand support using seat height position 1 (knee angle 90°). Reprinted with permission from Robertshaw, C., Kumar, S. & Quinney, H.A. (1986) Electromyographic and cinematographic analysis of rising and lowering between sitting and standing. *Trends in Ergonomics/Human Factors III* (Ed. W. Karwowski) pp. 739–49. Elsevier Science Publishers B.V. (North-Holland).

group of muscles continue to maintain some degree of activity throughout the cycle. During these activities the knee and ankle undergo flexion–extension and dorsiflexion–plantarflexion, respectively. Also evident from the figure is the variation in the moment as the function of different muscle groups and the velocity of motion. This is a good example of force, motion, and muscle activity interaction. In all occupational activities these variables are equally involved.

In a study of the phasic relationship between posture and muscle activity during rising and lowering between sitting and standing Robertshaw *et al.* (1986) concluded that lower seat heights were detrimental to patients with low-back pain. In order to obtain this information, these authors studied the kinematics and the electromyograms of spinal and triceps brachii muscles during rising and lowering between sitting and standing with varying seat heights and varying hand support. The four seat heights were subjects' under knee height, 5 cm less, 5 cm more, and 10 cm more than the first height. For hand support subjects had standard arm support, support of the seat pan, and no arm support (arms dangling in air). Using Beckman bipolar miniature silver–silver chloride surface electrodes an EMG was recorded from the erector spinae at the L_3 and L_5 levels, rectus abdominis above and below the umbilicus, external oblique, and the triceps brachii. Such hard wired subjects were marked for their joint centers at the wrist, elbow, shoulder, vertex, hip, knee, ankle, and the little toe. While these subjects rose from sitting or sat down from standing they were cinematographed at 50 frames per second, and the EMG was recorded from the muscles stated above. The phasic relationship between muscles, and the temporal relationship between posture and the activities are presented in Figure 1.3.

In a study of unresisted normal velocity-free axial rotation from a neutral position to the extreme left followed by the extreme right and finally to the neutral position, Kumar, Narayan and Zedka (1995) reported the phasic interrelationship between eight muscles of the trunk and angular rotation (Figure 1.4). In order to produce a repeatable axial rotation, the subjects were stabilized in an axial rotation tester (Kumar, 1994) establishing a fixed axis of rotation. The objectives of the study were to determine the mechanisms of initiation, sustenance, and execution of axial rotation. Kumar, Narayan and Zedka (1995) studied 50 normal young subjects through the full cycle of axial rotation. He studied the EMG of the external obliques, internal obliques, rectus abdominis, pectoralis major, erector spinae at T_{10} and L_3, and latissimus dorsi bilaterally simultaneously with the trunk rotation. He recorded the timing and relative magnitude of the EMG traces and correlated the EMG with angular displacement and the regression between them. He reported a variable pattern of muscle activation. However, the contralateral external obliques, ipsilateral erector spinae, and latissimus dorsi were established to be the initiators of axial rotation. These agonists and synergists were responsible for 65% of the total EMG activity whereas the antagonists and stabilizers contributed approximately 35% of the total EMG output. Their relative contribution is presented in

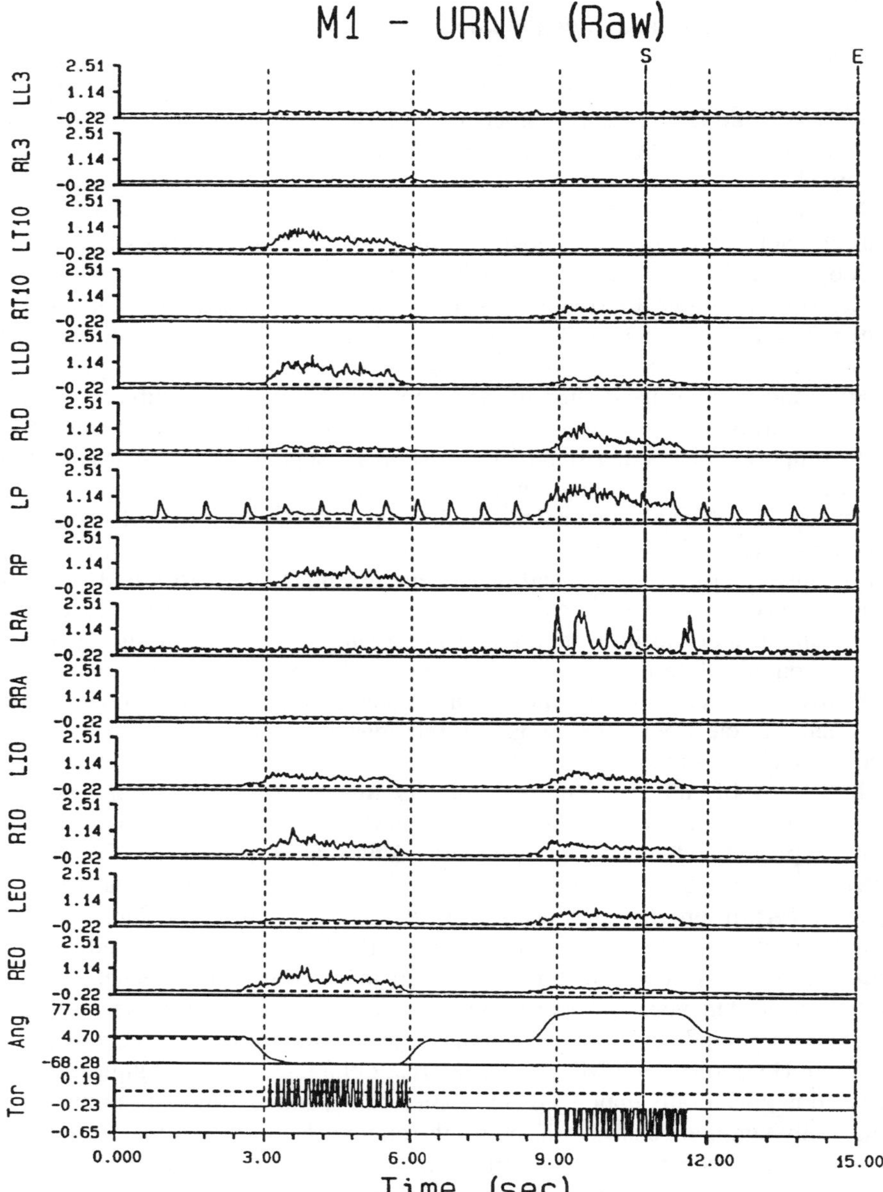

Figure 1.4 The phasic interrelationship between the trunk muscles during axial rotation. Reprinted with permission from Kumar, S., Narayan, Y. and Zedka, M. (1995) An electromyographic study of unresisted trunk rotation with normal velocity among healthy subjects. *Spine* (in press).

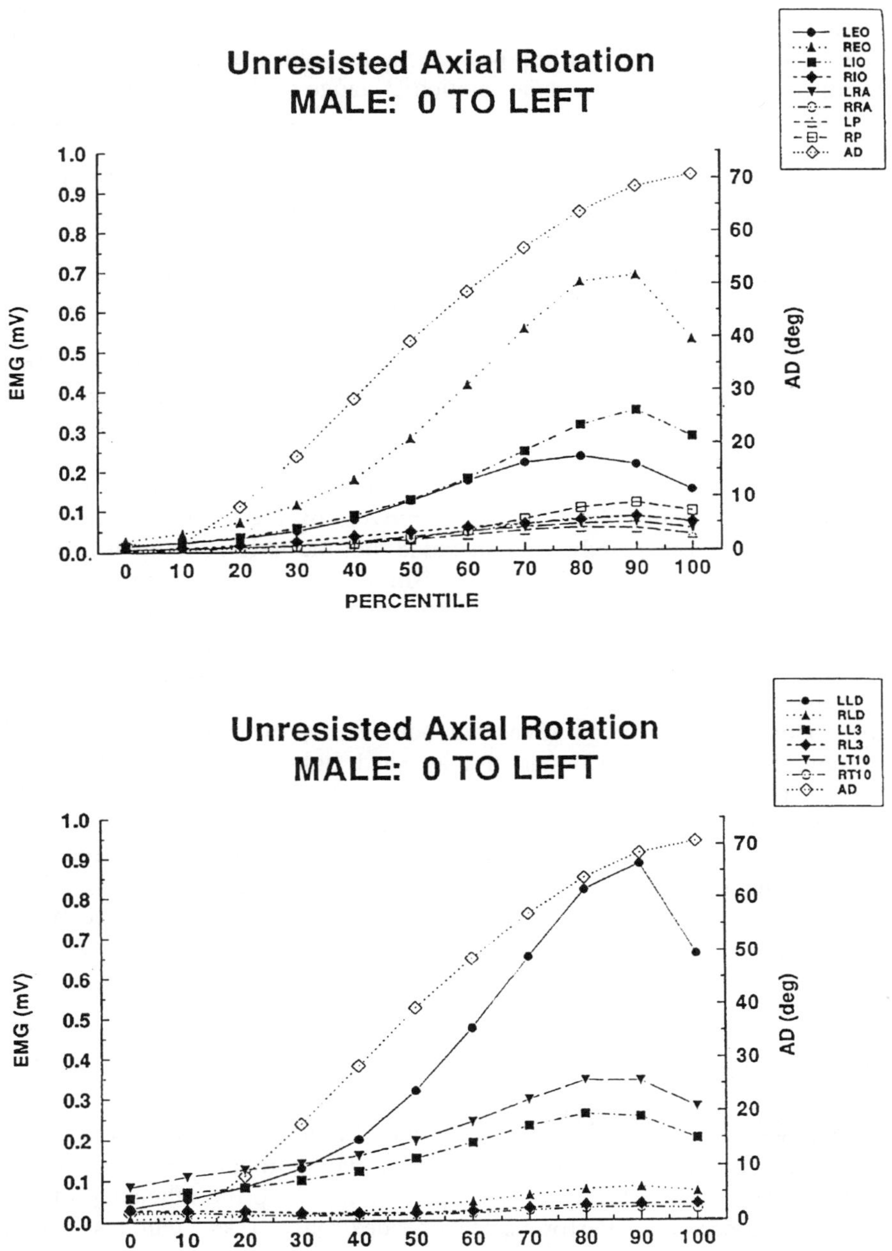

Figure 1.5 The relative contribution of trunk muscles in axial rotation from neutral posture. Reprinted with permission from Kumar, S., Narayan, Y. and Zedka, M. (1995) An electromyographic study of unresisted trunk rotation with normal velocity among healthy subjects. *Spine* (in press).

Figure 1.5. The return to neutral position from a rotated posture was largely achieved by elastic recoil of the passive tissues with the initial phase being controlled by the agonists of axial rotation. A range of approximately 10–15° on either side of the anatomical mid-sagittal plane involved very little muscle effort but beyond this region the osteoligamentous structures became stiff requiring increasing effort to execute axial rotation (Figure 1.6). The foregoing information now clearly establishes that the muscles which initiate also sustain and allow reversal of the rotary process. These muscles are, therefore, under continuous load in an activity which requires rotation. Perhaps, it is due to this prolonged load that 60% of back injuries are associated with trunk rotation (Manning *et al.*, 1984).

Kumar (1993) reported the phasic interrelationship between intra-abdominal pressure and electromyographic activity of the erector spinae at T_{12} and L_3, external oblique, and rectus abdominis during lifting tasks. On the basis of timing of onset of these variables he reported the onset of intra-abdominal pressure invariably delayed (250–750 ms) by variable time in relation to the erector spinae, the prime movers of the trunk, during lifting activity. Based on this study, he concluded that the intra-abdominal pressure is not a primary or protective mechanism for the human trunk in lifting activities. It is simply a by-product of the physiological and mechanical phenomena. The presence of raised intra-abdominal pressure, however, may add to the spinal stability during motion of the spine under stress. This phasic interrelationship information then negates the clinical wisdom where raised intra-abdominal pressure is advocated both as a preventative and remedial measure. Such phasic information may shed light on some complex and obscure phenomena. The work of Maton *et al.* (1980) illustrated the usefulness of such an approach. It shed light on the role of passive tissue resistance due to its viscoelastic property. In addition, the velocity-dependent contributions made by the smaller anconeus and larger triceps in elbow flexion is a valuable information. Furthermore, the coordination between these agonists and antagonists allows us to assess the total muscle load and joint stress.

1.4.4 Skill Acquisition

Effectiveness and efficiency in the work-place is one of the primary goals of ergonomics. In many jobs, the acquisition of the appropriate level of skill is essential even for the safety of the operator let alone the efficiency of the operation concerned. Performance of dextrous tasks requires motor coordination, appropriate timing, and targeting of the motion. Economy of motion saves production time as well as physiological energy. A cumulative effect of uncoordinated motion can raise a significant problem over time.

Neurophysiologically it has been shown that the strategy for recruitment of individual motor units does not alter with training (De Luca *et al.*, 1982). It is,

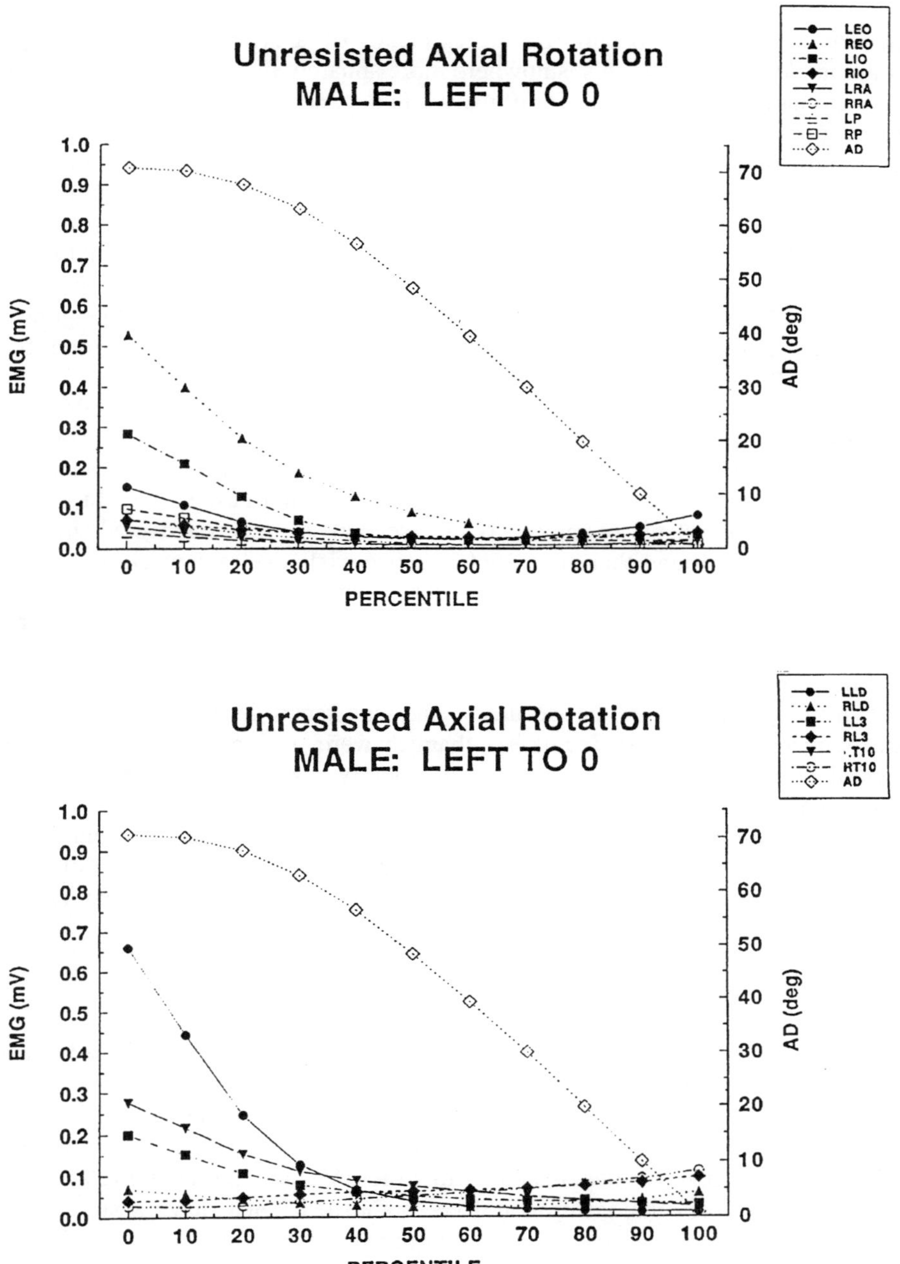

Figure 1.6 The relative contribution of trunk muscles in return to neutral position from axially rotated posture. Reprinted with permission from Kumar, S., Narayan, Y. and Zedka, M. (1995) An electromyographic study of unresisted trunk rotation with normal velocity among healthy subjects. *Spine* (in press).

therefore, concluded that the timing and coordination of their recruitment must be able to be altered to optimize the efficiency of a given operation through training. An example of this is nowhere more apparent than in maturing infants as they learn to walk and perform many complex activities. Even for an adult an unfamiliar activity results in exaggerated motion and wasted energy. However, practice makes such activities smooth and efficient. Person (1958) reported that subjects chopping and filing indulged in wasteful physiological activities of the biceps brachii and triceps brachii. These activities involved a high level of activity of the antagonists which was reduced significantly by repetition. Similar findings have been reported by many authors (Kamon and Gormley, 1968; Lloyd and Voer, 1973; Hobart *et al.* 1975). Antagonist inhibition also allows accentuation of torque development as reported by Ciriello (1982).

Moritani (Chapter 5) presents an excellent review of work which deals with the effect of skill acquisition in motion targeting. His discussion in the context of Fitts Law describes the motor unit activity and involvement in 'speed–accuracy trade-off' as described by Pachella and Pew (1968). In works reported by Moritani (Chapter 5) of several studies conducted in his laboratory, they found that after 1500 practice trials over a 1 week period of producing shots of force-varying isometric contractions corresponding to 20 and 60% of maximal voluntary contraction of the bicep brachii muscle there was a significant learning effect which could be seen through the motor output parameters. Specifically, significant improvements were recorded subsequent to the training session in the accuracy and constancy in the rate of force development. Accompanying these changes were a significant reduction in neural output variability. The latter will clearly result in repeatable smooth motion of the limb segment.

An electromyographic recording through the training procedure could then continually indicate the gain in efficiency and its consistency through different stages of skill acquisition. Such information can be a powerful ergonomic tool.

1.4.5 Quantification

Since all physical activities are performed by muscle contraction, the harder the work the harder the contraction. Electromyographic recordings have also shown that the frequency of firing as well as the magnitude of these EMG signals increase with increased force application. Such an observation allows one to have an intuitive understanding that the EMG and torque are related. If the nature of this relationship could be established it will have a significant advantage in predicting the forces exerted in hard-to-measure activities without interfering with them. Therefore, a variety of methods and techniques have been developed and employed in interpreting task demands through the use of EMG techniques. Broadly they fall into two categories: semi-quantitative and quantitative.

1.4.5.1 *Magnitude of Raw Signals*

The method of measuring the magnitude of raw EMG signals has been frequently used in earlier studies (Kamon, 1966; Jonsson, 1970; Grieve and Pheasant, 1976; Kumar and Davis, 1983; Kumar, 1988; and others). It was based on the observation that stronger contractions generating higher forces produced larger signals. Since the neuromuscular drive could not be measured precisely and accurately due to the nature of the phenomenon and limitation of the technique, the magnitude of excursion provided a reliable semi-quantitative method of indicating the intensity of the task. This comparison generally has been performed for either isometric activities in the same channel (e.g. Kumar and Davis, 1983) or the identical phases of activity of the same muscle recorded on the same channel (Kumar, 1988). These magnitudes have simply been measured either in divisions of excursion (Kumar and Davis, 1983; Kumar, 1988) or millimeters of deflection or arbitrary units assigned by workers (Grieve and Phesant, 1976). These excursions have traditionally been eye balled to determine the peak and average values. Once these activities were identified they were assigned values according to the established scale. These could then be correlated and regressed with the load to determine a qualitative relationship. However, one must bear in mind due to a varying relation and ratio between the deflection and load a quantitative assessment of the load would never be possible. However, a qualitative comparison between two loads or tasks with respect to their severity could be performed.

Using this methodology Kamon (1966) studied the muscle activity during static and dynamic postures in some sporting physical activities. He used surface electrodes on gymnasts during an exercise sequence on a pommelled horse. Gymnast's bodies were supported on their arms in a static position before executing a sideways swing. Kamon (1966) studied the trapezius, serratus anterior, latissimus dorsi, teres major, infraspinatus, pectoralis major, deltoideus, biceps brachii, triceps brachii, flexor carpi radialis, extensor carpi radialis brevis, rectus abdominis, external obliques, and erectores spinae. Through the rhythmical interplay of the right and left group muscles and the marked sudden bursts of activity of specific duration and sequence he indicated the coordination and skill required to accomplish the movements. In terms of quantitative analysis, Kamon (1966) used a common gain on all channels without varying it throughout the recording and scored the muscle activities in terms of millimeters of pen deflection.These scores were then compared between the exercises establishing the roles of the muscles of the upper extremities and trunk in this exercise.

Jonsson (1970) reported the functions of individual muscles in the lumbar region of the erectores spinae, more specifically the multifidi, longissimus, and iliocostalis at different lumbar vertebral levels. He inserted wire electrodes guided by TV fluoroscopy to place them in the pick-up area accurately. He also used a common gain for all channels for studying static contractions in prone, standing and sitting postures. He graded all EMG recordings on a three-point scale: 0 = no activity, 1 = slight activity (individual action potentials were perceptible and the baseline was not obscured), and 2 = marked activity (the base line was obscured by the action potentials). He only measured the mean activity. Using this simple scoring system, Jonsson (1970) was able to demonstrate that there was no significant difference between muscle loads to left and right sides of the body in symmetrical postures. Furthermore, he reported no difference between the medial or lateral placement of electrodes in the longissimus. He also concluded that there were differences in function between the multifidi, longissimus, and iliocostalis muscles even at the same level; sometimes differences in function occur between different vertebral levels even in the lumbar region.

Grieve and Pheasant (1976) investigated the effect of change of the muscle length on partially summated twitches of the soleus and tibialis anticus muscle using surface electrodes. The amplified, full-wave-rectified, and smoothed signals were assigned scores according to an arbitrary scale. Using these scores they developed log–linear regression equations to estimate EMG levels that corresponded to 25, 50, and 75% of the maximum possible torque.

For studying dynamic loading in a static stooping trunk posture and its comparison with stoop lifting of an equivalent and smaller load, Kumar and Davis (1983) recorded the EMG of the erectores spinae at the T_{12} and L_3 vertebral levels in addition to intra-abdominal pressure. They used silver–silver chloride surface electrodes and differential channel gains but compared the results within the channel. For dynamic loading and unloading a steady flow of 25 kg of water into or out of a plastic tub held in the hand while maintaining a stooping posture was used. The subjects also performed stoop lifting of weights of 15 and 25 kg. The raw EMG signals recorded were assigned the score of divisions of deflection on the chart. These were quantitatively compared for peak and average activities during lifting and lowering and dynamic loading and unloading. Based on their observations, Kumar and Davis (1983) reported that the mean electromyographic activities of the erectores spinae at the T_{12} level were approximately one-half that of the L_3 level. At the L_3 level they reported that the magnitude of the EMG activity increased by 50% during unloading. In addition to the temporal analysis describing the pattern of loading during static holds of changing load Kumar and Davis (1983) reported that a comparable stoop lift was at least twice as stressful compared to the former. Kumar and Davis (1983) stated that the dynamic activity requiring a postural change requires not only a biomechanical matching of the physical demand of load, but also the effort required in executing the motion. The latter is performed with a margin of safety requiring stability of the moving body parts. Therefore, muscle activity and, hence, their load is significantly higher compared to a static hold.

Stoop lifts involving a progressively increasing load was studied by Kumar (1988) using electromyography as one of the three techniques used simultaneously. He studied the EMG of the erectores spinae at the L_3 level and the external oblique using silver–silver chloride surface electrodes. The raw EMG signals were amplified and recorded on a paper chart. Throughout the experiment the gains of the channels were kept constant thereby allowing a qualitative within-channel comparability. Using these scores as indicators of the muscle load they reported a correlation matrix indicating a strong correlation between the load, on the one hand and the EMG activity of the erectores spinae and external obliques on the other (Table 1.1). They reported this correlation to range between 0.80 and 0.99 ($p<0.001$). The regressions calculated by them were also invariably significant ($p<0.05$) (Table 1.2).

In this way using qualitative scores of deflection of raw signals several authors have shown the activity/inactivity and pattern of activity (Kamon, 1966; Jonsson, 1970; Kumar and Davis, 1983), physiological function of muscle length change (Grieve and Pheasant, 1976), and qualitative and semi-quantitative relationship between the external load, on the one hand and the muscle load as well as the joint stress, on the other (Kumar, 1988).

Table 1.1 A simple correlation between the weight (Wt) of the lifts and the myoelectric activity of the erector spinae (ES) and the external oblique (EO) for peak, sustained, and averaged values during lift-up and lift-down

		Weight lb	Erector spinae	External oblique
Peak				
Up	Wt	1.000	0.989***	0.955***
	ES		1.000	0.918***
	EO			1.000
Down	Wt	1.000	0.990***	0.943***
	ES		1.000	0.924***
	EO			1.000
Plateau				
Up	Wt	1.000	0.943***	0.933***
	ES		1.000	0.807**
	EO			1.000
Down	Wt	1.000	0.987***	0.967***
	ES		1,000	0.936**
	EO			1.000
Average				
Up	Wt	1.000	0.974***	0.921***
	ES		1,000	0.856**
	EO			1.000
Down	Wt	1.000	0.991***	0.911***
	ES		1.000	0.866***
	EO			1.000

** $p<0.01$, $p<0.001$.

Table 1.2 A regression analysis of EMG values for lifting activities

Intercept, gradient, and variance ratio	Up			Down		
	a	*b*	∫	*a*	*b*	∫
Erectors spinae on weight						
Peak	37.9	0.9	393.2	27.0	1.0	414.7
Plateau	22.5	0.5	64.8	12.9	0.5	305.0
Average	29.6	0.7	148.5	19.7	0.7	461.7
External oblique on weight						
Peak	−0.3	0.5	82.6	1.1	0.4	64.8
Plateau	3.1	0.1	53.6	3.0	0.1	116.0
Average	1.2	0.3	44.9	1.2	0.3	38.9
External oblique on erectors spinae						
Peak	−18.7	0.5	43.0	−8.8	0.3	46.5
Plateau	−1.3	0.2	14.9	−0.3	0.2	56.4
Average	−9.3	0.4	21.5	−5.4	0.3	24.1

a = intercept, *b* = gradient, and ∫ = variance ratio.

1.4.5.2 Zero Crossing

As the name implies the technique requires counting the number of times the electromyographic signal crosses the baseline between the positive and negative phases. It is thought that higher muscle activity implying increased muscle load will be generating more EMG signals. The greater the number of signals the more frequently zero crossing will occur. It is assumed that the effect of summation and occlusion of the signal will be random, and therefore will not affect the general relationship. Theoretically it may be possible to count zero crossing manually if either the signals are sparse or the speed of the sweep is fast enough. However, it is impractical, imprecise, and tedious. Therefore this counting is done exclusively electronically. An electronic circuit designed to do this counting may produce a single digital output much like a ratemeter. A caution for adjusting the threshold level of signal must be exercised to prevent false positives. There are two important limitations with this technique which must be borne in mind. First, the accuracy of this technique is compromised at a higher firing frequency. At approximately 60% of maximal voluntary contraction the zero crossing count levels off (Christensen *et al.* 1984). The latter will introduce significant inaccuracy into the technique. Secondly, if some muscles have a preponderance of homogenous motor units generating same size signals and being recruited at similar levels of effort, the electronic counter can be saturated and not provide a valid index of activity. However, in efforts producing low-frequency or sparse signals the readings are likely to be more accurate and reliable.

Despite the limitation and tedium of the technique several people have used this technique successfully. Hagg *et al.* (1987) and Suurkula and Hagg (1987) have used zero crossings for determining shoulder and neck disorders among assembly line workers. The authors argue its particular usefulness of this technique in ergonomic applications. The investigator may, however, consider their requirements and balance them against the limitations before deriving their conclusions.

1.4.5.3 EMG Spike Frequency and Magnitude

Spikes represent the action potentials of the active muscle. With stronger contraction a larger number of motor units are recruited creating more numerous spikes and generating a larger muscle force. Although the magnitude of the individual spikes remains constant, their frequency as well as the number of sources will increase. Due to the volume-conducted nature of the action potentials, fusion and summation of the signals takes place. The latter phenomenon results in an increased frequency as well as an increased magnitude of the summated spikes. Bergstrom (1959) analyzed a graphical record of EMG signals from a human forefinger abductor muscle. He reported a linear relation between the frequency of spikes and the area under the EMG time record. Close *et al.* (1960) confirmed this using an electronic counter. Kumar and Scaife (1975) investigated the frequency–amplitude relationship of EMG signals in an isometric surface electromyogram. They examined the EMG recordings on either side of the baseline at which individual peaks became just perceptible. They counted the spikes and measured the height above the mask. They found a linear relationship between the combined magnitude of the mask and the spike and the frequency of spikes per second provided all data were with the same mask value. However, when they analyzed the data from several records with the mask values between 2 and 12 mm, a logarithmic relationship was established (Figure 1.7).

Several authors exploited this relationship to gain an understanding of muscle load and task stress in industrial tasks. An example of this was published by Kumar and Scaife (1979) in which the authors investigated the postural stresses of a precision task. Kumar and Scaife (1979) investigated seated women operatives who were using a low-power stereoscopic microscope for threading core memory arrays for digital computers (Figure 1.8). Surface electromyography of the trapezius and erectores spinae in the thoracic and lumbar region was done. They took simultaneous photographs for biomechanical analysis. They adjusted the workstation to the individual's optimum height and tipped it 5 ° forward and backward in addition to studying the task at the level desk. They reported that narrow constraints were placed on the operative due to the demand of the task. Based on the biomechanical and EMG task analysis, Kumar and Scaife (1979) presented a biomechanical model (Figure 1.9). They suggested that a combination of the task factors could account for the neck–shoulder and low-back pain problems in these operatives. They designed and proposed an adjustable set-up for reducing the postural stresses and thereby the occupational health problems.

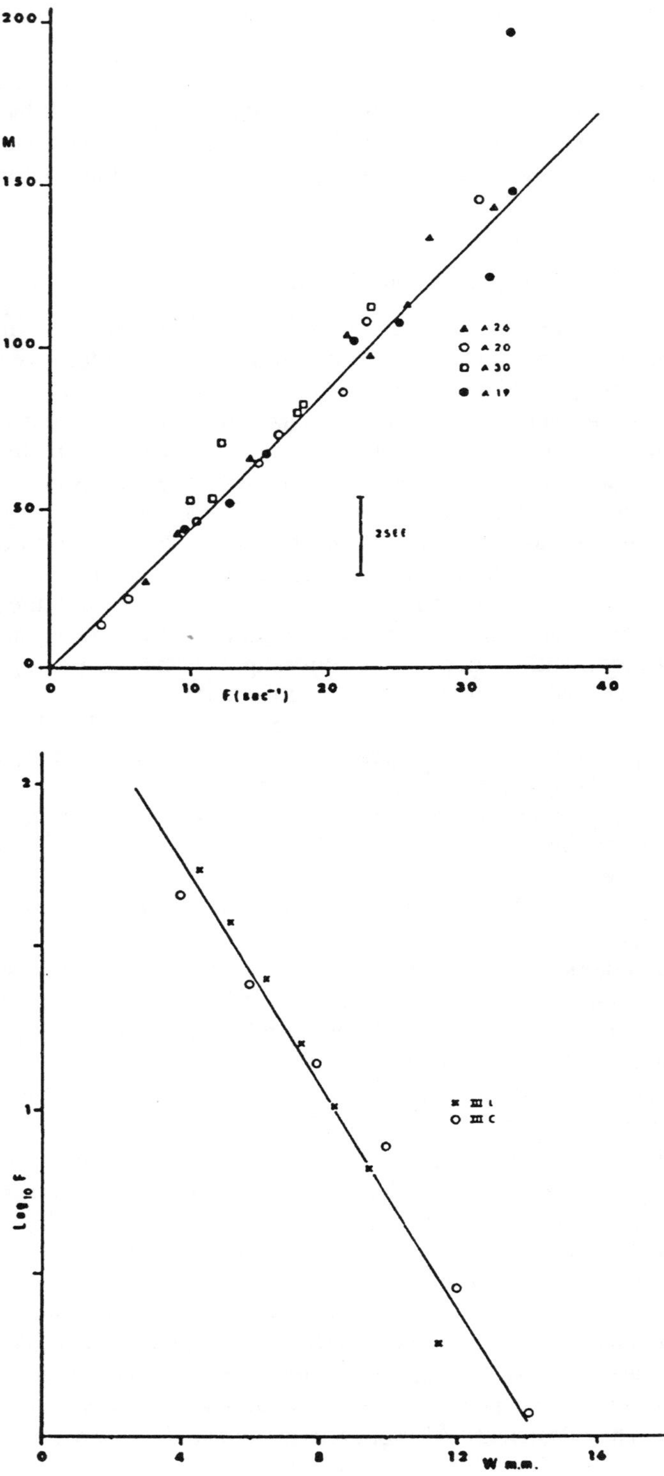

200
M
150
100
50
0
0
10
F(sec⁻¹)
20
30
40
▲ A 26
○ A 20
□ A 30
● A 19
2SEE
2
1
Log10 F
0
4
8
12
W mm.
16
III L
III C

Figure 1.8 The working posture of operatives in the core-memories plant. Side view of subject engaged in threading a core: memory array for a digital computer (laboratory set-up). Reprinted with permission from Kumar, S. & Scaife, W.G.S. (1979) A precision task posture and strain. *Journal of Safety Research*, **11** (1), 28–36.

A variant of this methodology is to determine the area under the curve formed by the total raw signal. Such an approach is based on the premise that no part of the signal is unimportant and can be disregarded. Therefore, a line is drawn around the raw trace covering all signals and the area covered under this line can be measured planimetrically. The magnitude of this area will indicate the muscle load. A simple method for such comparative evaluation is to cut out

---

Figure 1.7 *Opposite* The logarithmic relationship between the magnitude of the signals and their frequency. Reprinted with permission from Kumar, S. & Scaife, W.G.S. (1975) Frequency amplitude relationship of signals in isometric surface electromyograms. *Electromyography in Clinical Neurophysiology*, **15**, 539–44.

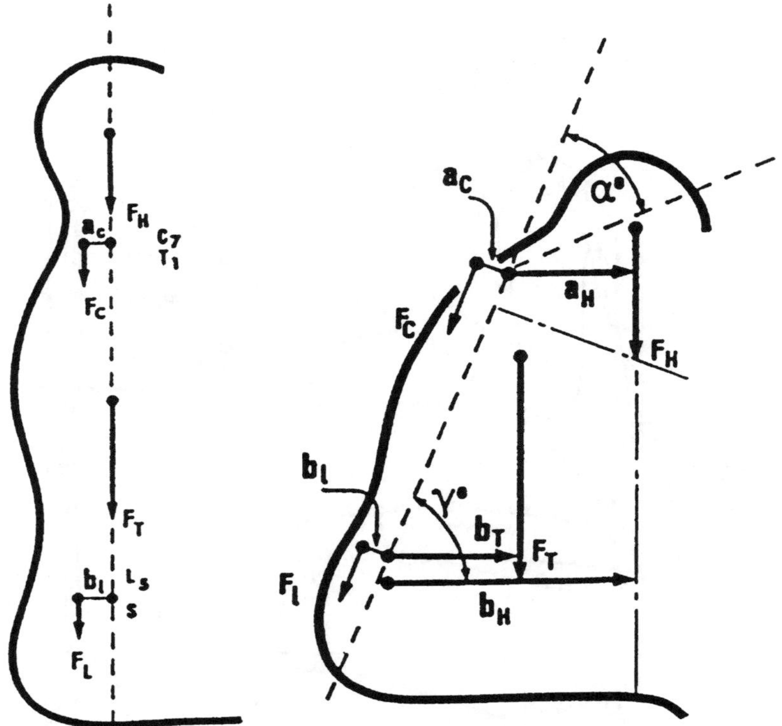

Figure 1.9 A biomechanical model of the task performed in the core-memories plant. Forces acting on the spine in resting and in a forward inclined posture. Reprinted with permission from Kumar, S. & Scaife, W.G.S. (1979) A precision task posture and strain. *Journal of Safety Research*, **11** (1), 28–36.

the area under the curve on the paper and weigh the paper cuttings. These weights also correlate strongly with the muscle loads of given tasks.

1.4.6 Quantitative EMG

For ergonomic applications, the magnitudes of electromyographic records are only of secondary importance. Those ergonomists who are also curiosity-driven physiologists may have a greater interest in the nature of the signal and its physiological implications. However, for an ergonomist or a practitioner the load on a joint or the muscle load is of greater interest as it impacts on safety and productivity. There have been numerous studies to uncover the relationship between EMG signals and muscle tension. To this day this effort continues as some aspects, in particular the length–tension–EMG, continue to elude us. Furthermore, different muscles seem to have different responses. However, a brief consideration of the EMG–torque relationship follows to set the stage for a consideration of quantitative EMG.

Inman *et al.* (1952) and Lippold (1952) reported their findings of investigations of human muscles. Inman *et al.* (1952) measured the muscle tensions and the muscle lengths in amputees having cineplastic muscle tunnels. They used surface as well as inserted wire electrodes into the muscles and activated them by voluntary effort. The amplified raw EMG signals, integrated EMG, and the muscle tension were recorded simultaneously. Inman *et al.* (1952) reported a reliable and repeatable relationship between the tension, spike amplitude of the raw EMG, and the integrated EMG of the human biceps brachii. Based on their observation, they concluded that an integrated EMG output faithfully reflects the tension developed in a given muscle (Figure 1.10). Therefore, they suggested that an integrated EMG could be used as an index of the isometric tension. However, with increasing length of the muscle

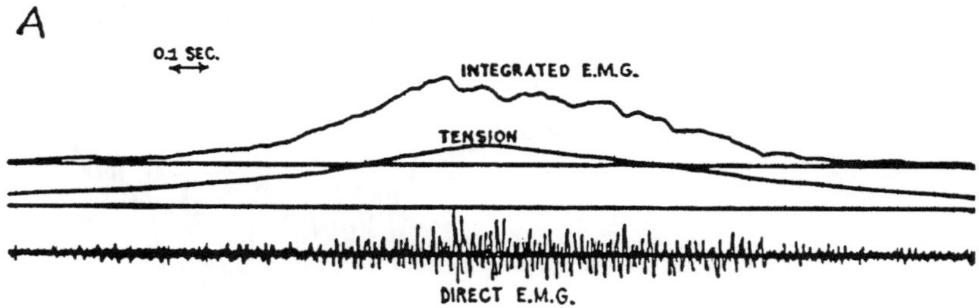

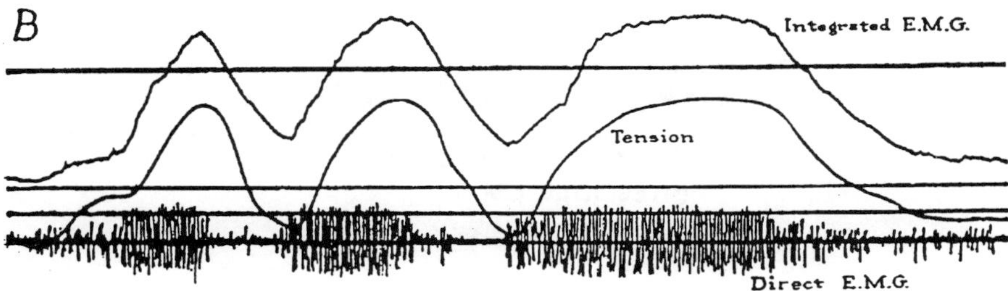

Figure 1.10 The relationship between raw EMG signals, integrated EMG, and muscle tension. (A) Biceps brachii of human cineplastic amputee. Integrated electromyogram, isometric tension and electrical 'spikes' (direct EMG) with varying degrees of effort. Note the parallelism between integrated EMG and tension. (B) Isometric tension, direct EMG, and integrated EMG of the anterior tibial muscle of a normal subject. The force measured was produced by dorsiflexion of the foot, keeping the muscle length constant. Note the parallelism between tension and integrated EMG. Reprinted with permission from Inman, V.T., Ralston, H.J., Saunders, J.B. de C.M., Feinstein, B. & Wright, E.W. (1952). Relation of human electromyogram to muscular tension. *EEG Clinical Neurophysiology*, **4**, 187–94.

the EMG drops off in spite of the tension remaining the same or even increasing (Figure 1.11). Thus Inman *et al.* (1952) reported that because of the length–tension relationship in muscles there is no constant relationship between tension and an EMG.

Lippold (1952) also reported that the relationship between the isometric tension and an EMG is directly linear. He reported a correlation ranging between 0.93 and 0.99. For his experiment Lippold (1952) used the gastrocnemius–soleus muscle group exerting in plantar flexion. He used surface electrodes for recording the EMG. The signals were integrated. Different incremental levels of plantar flexion efforts were used for generating isometric tension. He explained that, though the individual signals do not change in magnitude, the summated electrical activity of all units firing rises linearly with increasing tension.

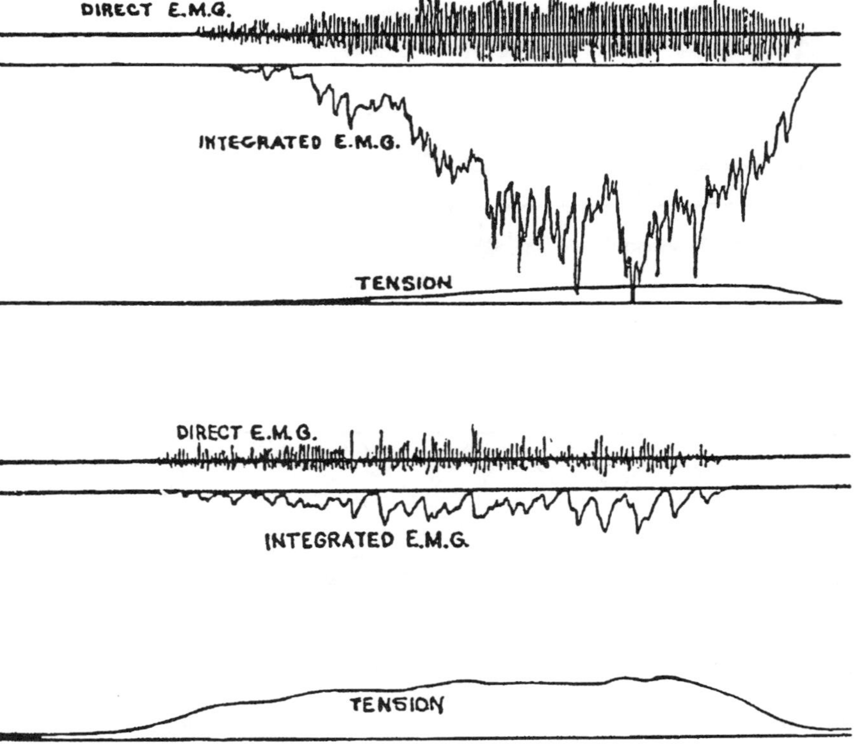

Figure 1.11 The effect of muscle length on raw and integrated EMG in relation to tension. Right triceps of human cineplastic amputee. Above, muscle very short. Below, muscle stretched well beyond rest length. Note the fall in the EMG in the stretched muscle. Maximal effort in both cases. Reprinted with permission from Inman, V.T., Ralston, H.J., Saunders, J.B. de C.M., Feinstein, B. & Wright, E.W. (1952) Relation of human electromyogram to muscular tension. *EEG Clinical Neurophysiology*, **4**, 187–94.

These two pioneering works led to intense activity in this area. The prime motivating factor, of course, is the value of an accurate prediction of force based on an indirect and unobtrusive technique. In a subsequent publication Bigland and Lippold (1954) reported on the relationship between force, velocity of contraction, and the EMG activity. Again using gastroc–soleus group they had their subjects perform an isotonic plantar flexion at a constant velocity with different loads and different velocities at a fixed load in concentric and eccentric modes of contraction. They recorded the EMG using surface electrodes and integrated the signals. At a constant-velocity isotonic contraction, the integrated EMG was linearly related to the load for both concentric and eccentric contractions with correlation coefficients of 0.93 and 0.88, respectively. The slope of the rise for concentric isotonic activity was significantly steeper compared to the eccentric contraction. With respect to the velocity of contraction they found that the EMG activity increased linearly with the velocity of contraction in concentric activity. However, in eccentric contraction the myoelectric activity was considerably lower and independent of the velocity of contraction.

Vredenbregt and Rau (1973) also investigated the torque–EMG relationship of the bicep brachii and reported somewhat different findings. In their experiments, like those of others, the relationship varied with the length of the muscle at which the measurements were made. However, they found that there was a non-linear relationship between the EMG and torque contrary to the findings reported by Inman *et al.* (1952), Lippold (1952), and Bigland and Lippold (1954). They also reported that when the EMG was plotted as a function of percent maximum voluntary contraction (MVC) at the same muscle length all values fell on the same non-linear line without any change in the standard deviation. In Vredenbregt and Rau's (1973) study, the EMG–torque relationship for the brachioradialis and biceps brachii were different for the same forces measured. There was a greater degree of variability among different subjects in the EMG–torque relationship for the brachioradialis. The latter indicates the possibility of not only a different EMG–torque relationship for different muscles but also different variability responses for different muscles. Whether such a response was due to slight differences in effective mechanical variables due to the muscle attachment, muscle role in joint motion, or joint geometry is not entirely clear. Vredenbregt and Rau (1973) also reported that in a prolonged contraction of a constant force the magnitude of the EMG signals increased linearly and, if the magnitude of EMG signals was held constant, the magnitude of the force declined with time. The work of Vredenbregt and Rau (1973) threw considerable light on the EMG–torque relationship confirming the existence of such a relationship. However, by a significant variation from other authors it also cautioned workers to assume a relationship for any predictive purposes.

In a carefully controlled study of 26 human subjects, Moritani and De Vries (1978) reported a linear relationship between the EMG and torque in a unipolar recording system. The same signal when picked up using a bipolar electrode configuration was reported to be non-linear. No explanation or a theoretical model for such a finding was offered in the paper. Hof and van den Berg (1977) also reported a linear rise in the EMG with an increase in the force of exertion in isometric contraction. The authors also described their interesting finding of load sharing among the muscles operating the joint. They found the sum of the individual muscle torques as derived from their respective EMGs was equal to the mechanically measured total torque, though the extent of contribution from different muscles involved was variable.

To add to the complexity of the EMG–torque relationship Metral and Cassar (1981) and Stokes *et al.* (1987) reported a biphasic bilinear relationship between the EMG and torque in the muscles of hand and back, respectively. Metral and Cassar (1981) chose the extensor carpi radialis and studied isometric and anisotonic contractions. They considered the muscle as a system presenting maximum force and the other in which the EMG was linearly related to the submaximal force. Metral and Cassar (1981) accounted for this behaviour by a double exponential function. The authors suggested that there was a change in the slope of the EMG–torque relationship of between 50 and 70% of MVC. The initial linearity has been described by Milner-Brown *et al.* (1973) to be due to the combination of recruitment and rate coding giving a linear relationship between force and EMG. In the latter portion of the response curve, it was suggested that the EMG amplitude of a motor unit with a high threshold was greater than of that with a low threshold. Stokes *et al.* (1987) measured on integrated EMG of the lumbar erectores spinae and the torque produced by pulling against a harness around the shoulders in an isometric extension effort. They reported a non-linear EMG–torque relationship. The EMG response was reported to be biphasic/bilinear with an inflection point in the middle. However, they reported these slopes to be variable with a coefficient of variation of 25%. This large variability may have been affected by the synergism of other muscles. With the back being a complex organ with numerous muscles controlling its movement, measurement of one set of muscle may not tell the complete story.

Finally, the relationship between surface electromyogram and the force generated is not entirely settled. Linear, non-linear, and bilinear relationships have been reported. Clearly, there is a relationship between EMG and force. However, an accurate assessment of the muscles, the mode of activity, the conditions of activity, and the duration of activity all must be considered in arriving at the EMG–force relationship.

1.4.6.1 Voltage Measurement

Conventionally a quantitative measure requires finding the number of units of the variable being measured. In the case of electromyography, ergonomists are interested in determining the absolute units of electromotive force of the motor units action potentials. These are compound action potentials which are volume conducted and become summated. Thus, if one could determine the absolute magnitude of these EMG signals in microvolts or millivolts it might allow determination of the force of contraction. Such a measure will permit a direct comparison between different tasks, postures, stress, etc. However, a voltage measurement does not necessarily allow one to do all of the above. One has to be aware of the uses and the limitations of such a measure.

In order to be able to use a microvolt measure of an EMG signal, the technique must be standardized and the signal should be carefully calibrated before any meaningful conclusion can be drawn. The microvolt measurements do offer an indication of the baseline of the muscle unless related to the maximal voluntary contraction. Since the individual anthropometric characteristics play heavily on the magnitude of the recording due to differences in conditioning, subcutaneous fat, thickness of skin, and oil and/or hair on the surface, the recorded value may not be of much use in a quantitative sense. In addition, slight differences in electrodes, electrode jelly, and lead characteristics can also modify the values recorded significantly due to varying levels of electrode impedance. Therefore, it is imperative that a careful and precise calibration must be conducted in the most sensitive range of the equipment to obtain any meaningful information.

One must also be aware of the characteristics and limitations of technique. The signals represent only the area of pick-up. If the area of pick-up is too small, then the signal is not representative of the entire muscle. On the other hand, if it spans a large area, the electrodes will pick up the co-activity of the neighboring muscles. Since electromyographic signals are volume conducted, their pick-up is considerably affected by the frequency in addition to the magnitude. The frequency of EMG signals in different muscles may be different due to the role of the muscle – operating as a precision control or a power generator. The latter can significantly change the microvolt value even if the number of fibers per unit cross-sectional area are the same in two muscles. Kadefors (1978) pointed out that up to a certain frequency the electrical activities of the same motor unit sum up linearly. The action potentials of different motor units add in square. Therefore, some workers use voltage measurement of the electromyographic activity in conjunction with biomechanical model. The latter strategy must also take into account the geometry of the muscle. The fiber arrangement and orientation may have a significant impact on the relationship between the magnitude of the signals and the force generated.

1.4.6.2 *Rectification, Averaging and Envelope Detection*

For a more detailed account of demodulation and interpretation, see Winter (Chapter 4) in this volume. However, for completion of treatment of the subject matter a brief description of broad principles follows. From previous treatment of the subject, it is obvious that EMG signals are time and force dependent. Their amplitude varies randomly above and below the baseline. Thus, a true average of these signals will be zero representing the DC level.

In order to carry out one of the many analyses, the EMG signals are rectified. Rectification may involve elimination of negative components of the signals (half-wave rectification), or the negative parts of the wave form may be inverted forcing them to fall on the positive side of the baseline (full-wave rectification). The latter retains every part of the signal. Such rectified signals are low-pass filtered for smoothing, a process which smooths out all high frequency components making the trace move up and down gradually with different levels of muscle activity. Thus, such a rectified and smoothed signal will rise above the baseline with increasing muscle activity in proportion to the intensity of effort. When the muscle is relaxed the trace will return to the baseline. Thus, a microvolt reading for peak and average activities will reflect the peak and average forces generated by the muscles concerned. The area under the curve will represent the total muscle activity during this effort.

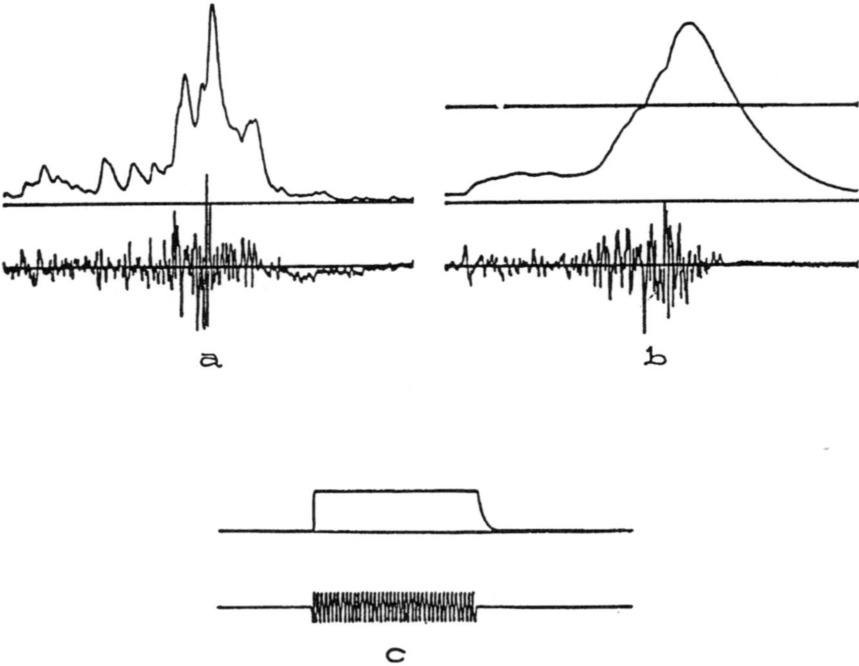

Figure 1.12 The relationship between the raw signals and the rectified linear envelope detected signals with different electronic processors. (a) Electromyogram and integrator trace with filter in the 2 mfd. position. (b) Electromyogram and integrator trace showing undesired decay curve due to excessive discharge time constant. (c) Response of integrator in the 2 mfd. position to a burst of a constant amplitude 100 cs^{-1} signal. Reprinted with permission from Inman, V.T., Ralston, H.J., Saunders, J.B. de C.M., Feinstein, B. & Wright, E.W. (1952) Relation of human electromyogram to muscular tension. *EEG Clinical Neurophysiology*, **4**, 187–94.

Rectification, averaging, and linear envelope detecting causes a time delay in the EMG signals (Figure 1.12). The electronic characteristics of these processes should be considered to determine the extent of asynchrony and choose the best suited specifications.

1.4.6.3 Integration

Integration is a mathematical process achieved electronically which calculates the area under a curve. For integration the EMG signals are full-wave rectified prior to processing. The integration may be achieved electronically or geometrically. For geometrical integration planimetry has been frequently used. However, due to ease of operation electronic integration is increasingly being used. The integrated electromyogram (IEMG) represents the total muscle activity and is a function of amplitude, duration, and frequency. Since the IEMG curve will keep on rising, at a pre-set magnitude or intervals it is reset to zero. Figure 1.13 shows the variously processed EMG signals.

1.4.6.4 Root Mean Square

The root mean square (RMS) value of an AC signal is the total value of the quantity. It measures the electrical power in the signal. Basmajian and DeLuca (1985) recommend this as the most desirable processing method for the analysis of the EMG signal. The RMS value provides the moving average. There have been many studies which have employed this processing methodology.

1.4.6.5 Normalization

Regardless of the quantitative measure (microvolts, full-wave rectified linear envelope, root mean square, integrated value) it is useful to calibrate the EMG signal against a known reference. This process is termed normalization. Normalization allows one to compare quantitatively different activities for the same muscle, different muscles, activities on different days, and different subjects for same or different activities. Since EMG signals are volume conducted and the tissue environment is subject to large variation between subjects an absolute measure of an activity is not feasible. Furthermore, there may be significant differences in pick-up and conductivity characteristics of the electrode and lead system. The latter adds another unknown variable. In addition to the technique-related reasons, there are physiological reasons which do not permit absolute quantitative measure. Among them is the fact that the electrical changes on the muscle membrane may not reflect faithfully the contractile force generated in the muscle fibers due to the physiological status of the muscle membrane such as fatigue or stage of its progression. For these and other reasons the process of normalization allows one to derive meaningful and quantitative conclusions.

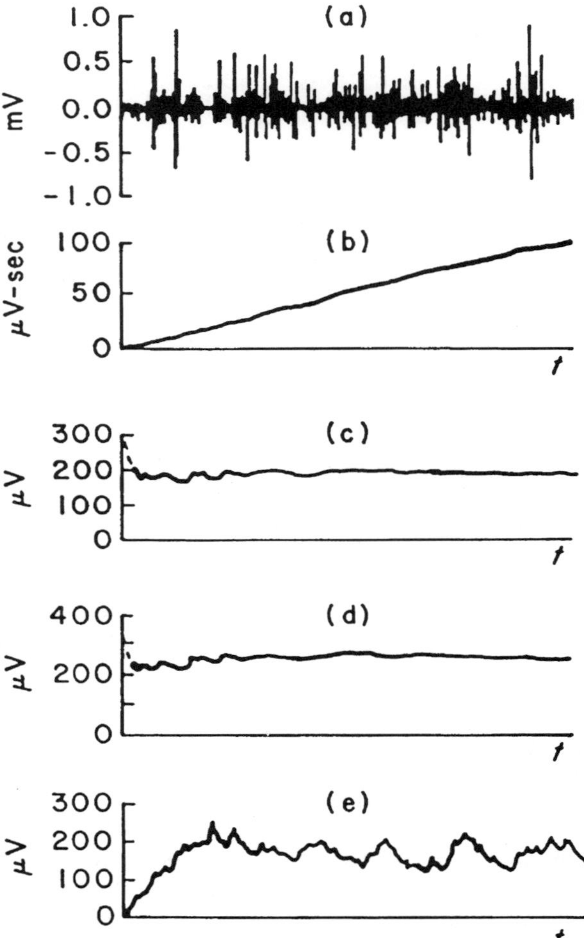

Figure 1.13 A comparison between different processing techniques. Comparison of four data reduction techniques. (a) Raw EMG signal obtained with wire electrodes from the biceps brachii during a constant-force isometric contraction. (b) The integrated rectified signal. (c) The average rectified signal. (d) The root mean square signal. (e) Smoothed rectified signal. In the latter case, the signal was processed by passing the rectified EMG singal through a simple RC filter having a time constant of 25 ms. The time base for each plot is 0.5 ms. Reprinted with permission from Basmajian, J.V. & De Luca, C.J. (1985) *Muscle Alive: Their Functions revealed by Electromyography* (5th edn). Williams & Wilkins, Baltimore, MD.

For normalization, after appropriate standardized electrode placement and hooking up to appropriately stabilized and calibrated recording set-up, well-defined and standardized activities are undertaken by the experimental subjects. These activities are repeated several times after an appropriate rest period to obtain a representative response. The activities chosen by different authors have varied considerably. Some workers have used unresisted normal activity of the joint(s) concerned, others have used a maximal voluntary contraction in a standard activity, yet others have used maximal voluntary contraction in a functional activity, still others have used maximum voluntary isometric contraction in primary functional activities for each of the separate muscles, and, finally, some have employed a peak EMG in a dynamic activity as reference. In principle any standardized and repeatable activity can be used provided the rest of the EMG activity is referenced to it. One must, however, recognize that some normalizations provide better quantitative estimates whereas others are more qualitative.

Although normalization has been widely used as a technique to overcome some of the problems of electromyography in obtaining quantitative information of the body mechanics, there are many limitations which must be recognized. In processes using maximal voluntary isometric contraction, the variability of force generation due to motivational and physiological reasons cannot be discounted. Apart from the foregoing, the relationship between the EMG and torque is not universally reported to be linear. Many authors have reported a non-linear relationship. Therefore, a proportional extrapolation based on the maximum voluntary isometric contraction may have a significant error. The reader is referred to previous discussion on EMG–torque relationship (pp. 24–8). Depending on the behaviour of the muscle in question, the information may be in significant error. The muscle response is clearly based on its innervation and motor unit firing pattern. Generally there is an increase in the number of firing motor units with increment in force, but the firing rate increases only at a higher force level (Milner-Brown and Stein, 1975). Furthermore, slow motor units are recruited much later and they fire only at much higher force level. Thus, if the EMG–torque relationship for the muscle in question is non-linear, the EMG values at lower levels of contraction will overpredict the force.

In contrast, many authors have used submaximal contraction as the normalizing reference value (Perey and Bekey, 1981; Yang and Winter, 1983; and others). In these studies, either one single submaximal activity was chosen or a series submaximal isometric contraction was used. In an environment of linear relationship, this method may be somewhat more accurate, as it will not be affected by the motivational or physiological factors so much. However, with a non-linear behaviour pattern, significant inaccuracies will creep in, underpredicting the force levels. However, for most ergonomic applications where submaximal forces are generated this may be a preferred method.

In yet other studies, either submaximal dynamic activity is chosen as a reference for normalization (Bobet and Norman, 1984;, Kumar, 1994) or an activity out of many being compared to each other (Kumar, 1995). In other experimental circumstances, the highest EMG activity of the task cycle has been chosen as the reference and other phases described as a percentage thereof (Gregor *et al.*, 1985; Kumar, 1988). Such applications may give some information about the relative activity in different phases of the task cycle. However, it must be borne in mind that the EMG significantly changes with the change in length of the muscle (Bigland and Lippold, 1954). The EMG also changes significantly with a change in the velocity of execution of the task altering the relationship (Bigland and Lippold, 1954). Therefore, if meaningful quantitative information is desired, it is essential that the task must be standardized with respect to posture (joint angles), movement, and the velocity of motion.

Mirka (1991), in a systematic study, reported that the error in normalization can be large when dynamic activities are compared with a MVIC at a fixed joint. He had his subjects perform a series of controlled trunk extensions. These consisted of 20 and 40% of upright MVC at varying velocities of 5, 10, 15, and $20^{\circ}\ s^{-1}$. When the EMGs obtained from such a controlled constant force and constant velocities were normalized against MVIC at 0° of trunk flexion, large errors (up to 99%) were observed. The errors were larger at larger angles of trunk flexion. Therefore, caution must be exercised in selection of normalization criteria.

Due to a variety of needs of the investigators and the driving hypotheses of their projects, an infinite number of permutations and combinations have been used in EMG studies employing different measurement variables and normalization techniques. Only a few selected studies are presented here to illustrate some of the techniques employed for ergonomic purposes. First, an account of papers dealing with sitting and seated tasks are described followed by papers which deal with lifting and whole body tasks.

Seats, Sitting, and Office Work Andersson and Ortengren (1974a) measured and reported on back muscle activity during sitting in a chair with varying backrest inclination in an attempt to determine an optimum set-up. They also measured the electromyographic activity of the erectores spinae during the use of an office chair (Andersson and Ortengren, 1974b). In the former study they measured the EMG of the spinal extensor using surface electrodes at the C_4, T_1, T_5, T_8 and L_3 levels, and over the trapezius muscle. For this investigation they studied three support parameters namely, backrest inclination, lumbar support, and thoracic support. In addition to these, they also studied three standing and five unsupported sitting positions. The EMG signals were full-wave rectified and averaged, and their microvolt values determined. These microvolt values were compared against sine waves of known amplitudes to determine the exact magnitude of the signals. These readings were subjected to statistical analysis. Using back muscle activity as the criterion measure the authors concluded that the backrest inclination was the most important variable, the EMG activity decreasing with increasing inclination. The thoracic support had no effect on back muscle activity whereas the lumbar support had only a minor influence only in some settings. Whereas these authors were successful in determining the effect of different conditions on their small subject population, they were unable to compare between subjects as they had not normalized their readings. The purpose of the study must be borne in mind while selecting measurement variables and analysis techniques. In their study of the office chair Andersson and Ortengren, (1974b), used the same variable and technique to study upright sitting, sitting with the back dorsiflexed over the backrest, and four sedentary tasks – writing, typewriting, pedal depression, and lifting a weight of 1.2 kg. The authors reported that the work activity was of greater consequence to the EMG score than the change in table height or location of the backrest. The study confirmed the design principles of the desirability of an adjustable set-up. However, once again lack of normalization did not allow between-subject comparisons.

Kumar (1994) used electromyography in addition to biomechanics and psychophysics in evaluating a bifocal computer desk, which he designed. Through the use of these techniques he determined the optimum set-up of the workstation for workers using bifocal lenses. Since neck and shoulder aches due to excessive head movement for moving the sight line from the monitor to the keyboard have been reported to be a stress factor, Kumar (1994) positioned the monitor such to bring it and the keyboard into a single span of vision. He recorded the EMG signals from the trapezius and sternomastoids bilaterally using silver–silver chloride surface electrodes. He low-pass filtered the EMG signals before rectifying and averaging them over time using a time constant of 25 ms. The linear envelope values were recorded in microvolts and were normalized against the unresisted extension of the subject. Such normalized EMG values were subjected to statistical analysis. Based on the results of the study, Kumar (1994) concluded that a sunken computer monitor tipped backward at a 35° angle was the most desirable set-up for workers using bifocal lens. The latter not only engendered the least electromyographic activity but also brought the monitor and keyboard within a narrow span of vision eliminating the necessity to raise and lower the head. Using normalization allowed him to proceed with quantitative and statistical analysis. The latter would not have been possible if only microvolt values were recorded without a reference.

Lifting, Carrying and Holding The involvement of spinal, paraspinal, and abdominal muscles in lifting activity has long been established (Floyd and Silver, 1955). However, prevalent low-back injury/pain problems in industrial society have necessitated the discovery of any lead which may enable one to control the problem somewhat. In a study reported by Kumar (1988) electromyography of the erectores spinae in addition to the external oblique was done. The EMG from 32 male volunteers during lifting and lowering loads from 10 to 55 kg in incremental steps of 5 kg was recorded. He recorded the EMG using surface electrodes. The microvolt values of peak to peak deflection of raw signals were normalized against the EMG recorded from the largest lift. In addition to describing the pattern of the EMG, Kumar (1971, 1988) reported the results of statistical analysis of the quantitative data. The EMG activity was significantly correlated with the load ($r = 0.92$–0.99; $p < 0.01$) and had a highly significant regression ($p < 0.01$). However, in this study the peak and average values of only raw signals were used. Lack of integration and subsequent normalization rendered the results comparative rather than quantitative. However, using electromyography Kumar (1979, 1980) studied 30 different lifting activities in symmetric and asymmetric planes. These activities consisted of symmetric lifts up and down from the ground to knee, knee to hip, hip to shoulder, ground to hip, and ground to shoulder. In the oblique plane he studied lifts up and down from the ground to knee, ground to hip, and ground to shoulder. Finally several same-level weight transfers were also studied (Figure 1.14). He studied full-wave rectified, averaged and envelope-detected microvolt values of the erectores spinae and external oblique. These EMG scores were normalized against the maximum scores recorded during oblique lifts. The microvolt values recorded are shown in Figure 1.15. The normalized values were subjected to statistical analysis with analysis of variance. The latter revealed that there was no significant difference in the electromyographic activity between different lifting conditions between the ground and knee. However, at other levels, the EMG scores were significantly different between different lifts. He suggested that during the low lifts, the spine remains considerably flexed and the erectores spinae in flexion relaxation preventing significant activity. In this way a quantitative response in individual activities and a qualitative and statistical comparison between activities could be performed (Figure 1.16). Such information, clearly, is of considerable ergonomic value.

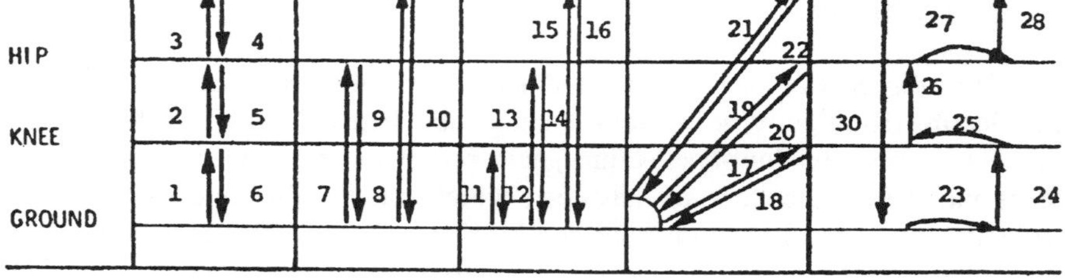

Figure 1.14 The experimental design of lifting tasks tested. Scheme of weight manipulation. The sequences of exercises is numbered. Reprinted with permission from Kumar, S. (1979) Variations in stress and response due to lifting in different planes. In *Science in Weight Lifting* (Ed. J. Terauds), Human Kinetics Pubs, Champaign, Il. pp. 31–41.

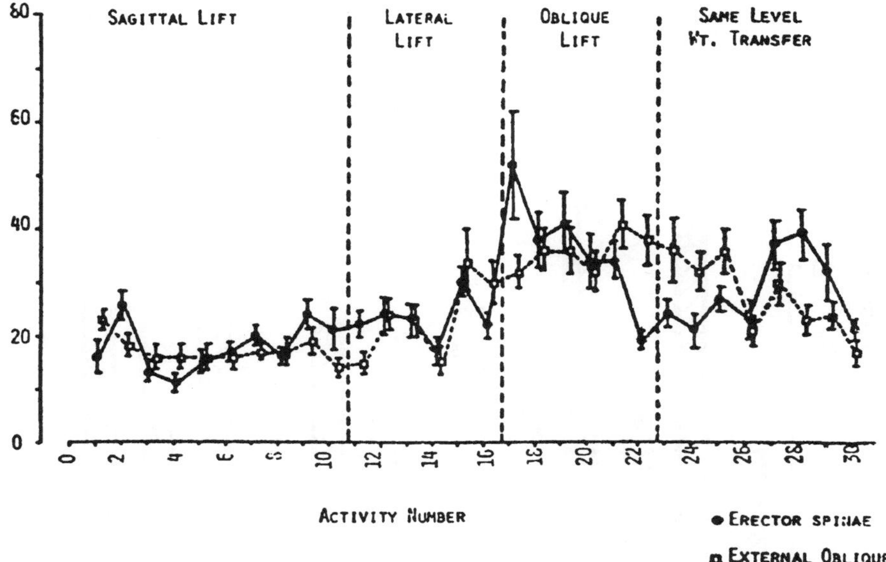

Figure 1.15 A comparative EMG response of the erectores spinae and external obliques in 30 lifting tasks. Means and standard error of means of myoelectric activity of erector spinae and external oblique during 30 activities of weight manipulation. Reprinted with permission from Kumar, S. (1979) Variations in stress and response due to lifting in different planes. In *Science in Weight Lifting* (Ed. J. Terauds), Human Kinetics Pubs, Champaign, Il. pp. 31–41.

Electromyography can be used for optimizing a procedure/technique and even maximizing safety in a given procedure if the risk factors lend themselves to being indexed by this technique. Delitto *et al.* (1987) compared two techniques of squat lifting with a view to optimizing safety. They studied the EMG of the erectores spinae and oblique abdominal muscles to examine the effects of two different alignments of the lumbar spine (bowed-in and bowed-out) and three different loads during squat lifting. The EMG signals were full-wave rectified and linear envelope detected. The authors used MVIC as the normalizing reference. The subjects performed both types of lifts with weights of 6.8, 11.4, and 13.6 kg in a crate weighing 0.9 kg. The quantitative values normalized against the MVIC values were subjected to a three-way ANOVA and demonstrated a significant reduction in the EMG activity of both the erectores spinae and the oblique abdominals in the back bowed-out posture (Figure 1.17). Based on their findings with respect to the recruitment pattern during the phases of the lifts, they concluded that due to the protective contraction of the muscles during the first half of the lift overcoming inertia of the load, a bowed-in back was a safer method to execute the squat lift. This is valuable information which could only be gleaned due to the quantification, normalization, and appropriate statistical analysis. A qualitative approach will not enable one to deduce this information.

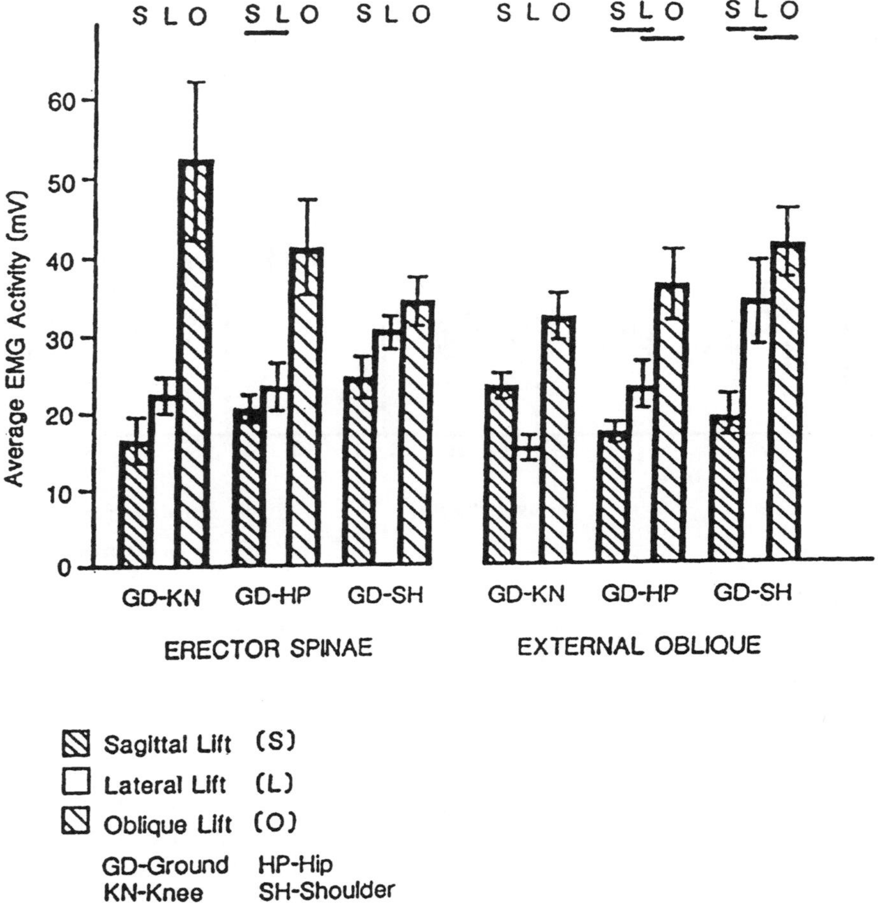

Figure 1.16 Quantitative and statistical analysis of the erector spinae and external oblique in symmetrical and asymmetrical lifting. Bar chart showing the mean and standard error of the mean of the erector spinae and external oblique electromyograph activity during ground to knee, ground to hip and ground to shoulder lifts in sagittal, lateral and oblique planes. The overlap between the means is shown by the line underlining the planes. Reprinted with permission from Kumar, S. (1980) Physiological response to weight lifting in different planes. *Ergonomics*, **23**(10), 987–93.

In an electromyographic study of holding tasks, Schultz *et al.* (1982) obtained an excellent agreement between their biomechanical model and the EMG scores validating the predictive relationship between the the EMG and torque in the confines of their experiment. For the measurement of the EMG, they used silver–silver chloride-recessed miniature surface electrodes. The EMG was recorded from the erectores spinae, at the C_4, T_8, and L_3 levels and the rectus abdominis and external obliques. The amplified, filtered, and root

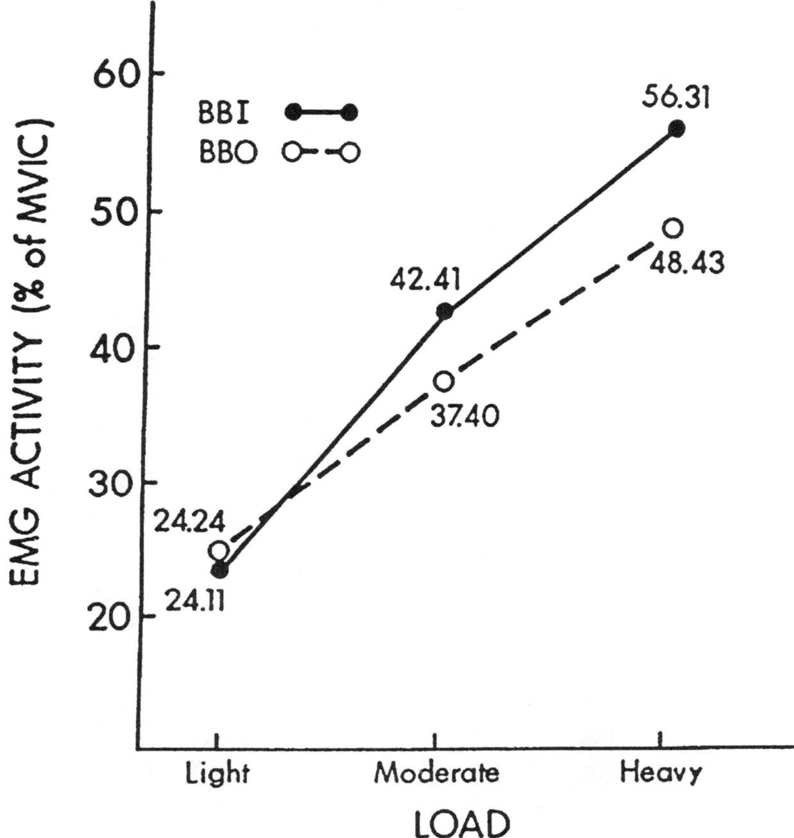

Figure 1.17 An electromyographic comparison of two methods of squat lifting. Interaction between style of lift and load for erector spinae muscle electromyographic activity (AC interaction). Reprinted with permission from Delitto, R.S., Rose, S.J. & Apts, D.W. (1987) Electromyographic analysis of two techniques for squat lifting. *Physical Therapy*, 7(9), 1329–34.

mean squared signals were converted from analog to digital forms. These were then compared with the external loads, the total flexion moment, and the calculated spinal compression in several isometric tasks in upright posture. The tasks they studied were resistance to flexor and extensor torques of 0, 150, and 300 N applied by hanging weights on the end of a cable running over a pulley and attached to a shoulder harness. In addition, they also studied weight holding in an upright and 30° flexed trunk posture. Three external loads (0, 40 and 80 N) were held in the hands close to chest and farthest away from it. The relationship between the EMG-derived muscle contraction forces and the biomechanically calculated spinal compression are shown in Figure 1.18. The muscle force derivations were based on the relationships shown in Figures 1.19

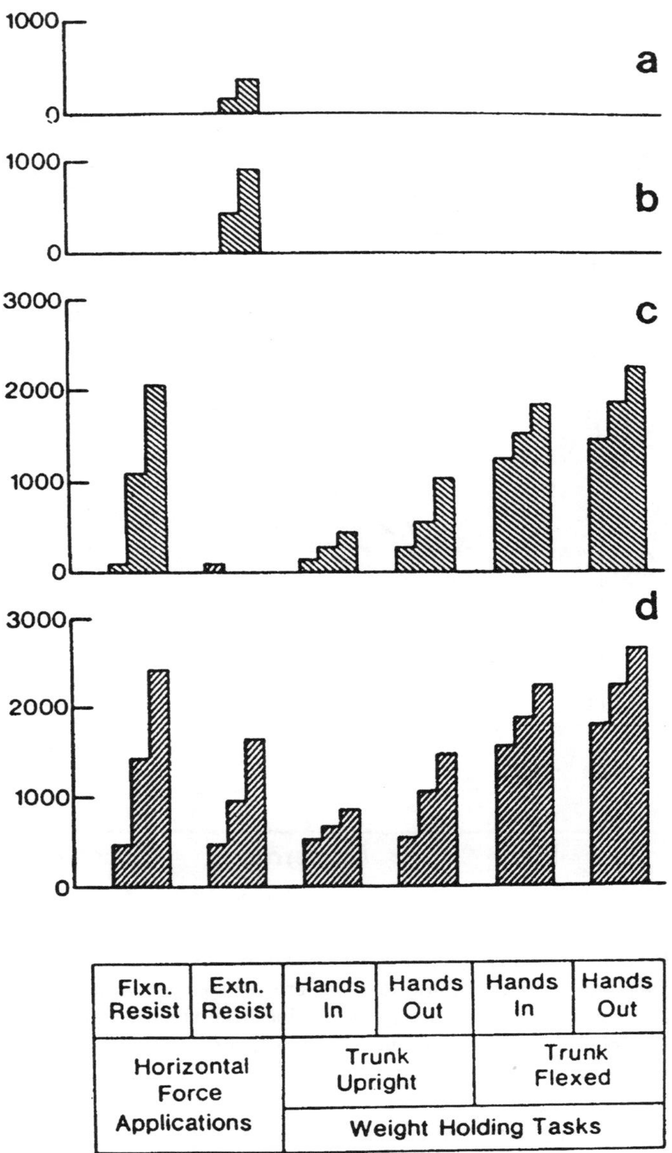

Figure 1.18 A comparison between the myoelectric activity and the spinal compression forces. Muscle contraction forces and spine compression forces predicted by the analysis. For the horizontal applications, bar levels from left to right correspond to external loads of 0, 150, and 300 N. In the weight-holding tasks, the bar levels correspond to held weights of 0, 40, and 80 N. (a) Rectus abdominus muscle tension. (b) Abdominal oblique muscle tension. (c) Posterior back muscle tension. (d) Spine compression. Reprinted with permission from Schultz, A., Andersson, G.B.J., Ortengren, R., Bjork, R. and Nordin, M. (1982) Analysis and quantitative myoelectric measurements of loads on the lumbar spine when holding weights in standing postures. Spine, Vol. 7, No. 4, 390–397.

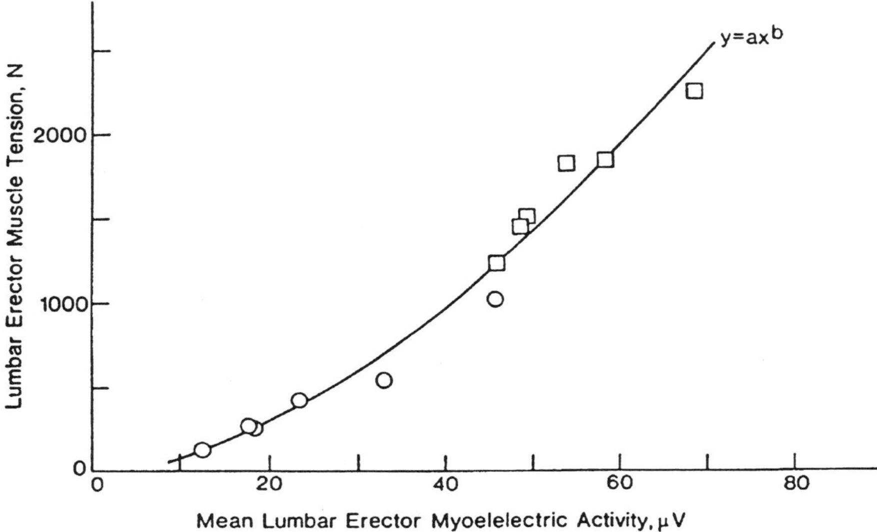

Figure 1.19 The relationship between the muscle tension and the myoelectric activity of the erector spinae muscles. Relationship between predicted muscle contraction forces and myoelectric activities. Each point shows the relationship in one of the weight-holding tasks studied. Circles correspond to upright positions of the trunk and squares to flexed trunk positions. The correlation coefficient is 0.922. Reprinted with permission from Schultz, A., Andersson, G.B.J., Ortengren, R., Bjork, R. & Nordin, M. 61982) Analysis and quantitative myoelectric measurements of loads on the lumbar spine when holding weights in standing postures. *Spine*, **7**(4), 390–7.

and 1.20. A series of studies published by these authors in the early and mid-1980s enhanced the fidelity of biomechanical models and paved the way for EMG-driven modelling activity. Such indirect extrapolation has valuable ergonomic applications. Cheng and Kumar (1991) published a three-dimensional static segmental torso model which validated the external loading through the internal muscle forces derived from EMG measurements. The latter allowed the capability of determining spinal compression at several segmental levels in a variety of industrial tasks. Such a capability has a desirable ergonomic significance. Safe and unsafe tasks can be identified. Task redesign to bring them within the safe limits becomes feasible by use of such ergonomic tools.

Whole Body Tasks Bobet and Norman (1982) looked into the usefulness of the information content of the generalized average electromyogram in whole body load carriage tasks. They investigated the EMG of the tibialis anterior, vastus lateralis, biceps femoris, erectores spinae, and trapezius muscles. These authors studied the effect of carrying three loads (20, 25, and 32 kg) with high and low load placement on the back after 5, 37, and 97 min of

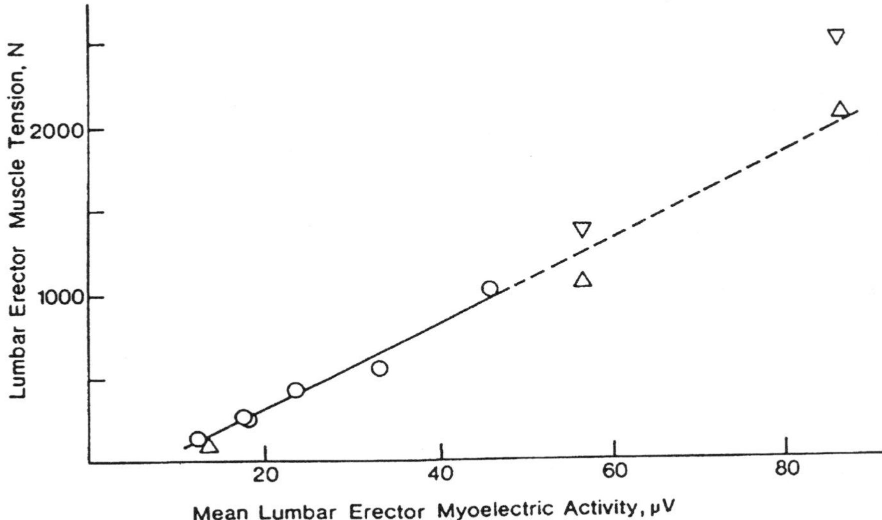

Figure 1.20 The relationship between predicted muscle contraction forces and myoelectric activity of the erector spinae. Relationship between predicted muscle contraction forces and myoelectric activities. Circles correspond to weight-holding tasks executed with the trunk upright and triangles to the flexion-resist tasks with antagonistic activity assumed to be zero. The inverted triangles also correspond to the flexion-resist tasks, but here the computations of erector tension included the additional tensions required to overcome the antagonistic contraction detected in the anterior muscles. The least-squares regression line through the six circled data points is shown extrapolated (broken line) to the higher tensions imposed by the flexion-resist tasks. Reprinted with permission from Schultz, A., Andersson, G.B.J., Ortengren, R., Bjork, R. & Nordin, M. (1982) Analysis and quantitative myoelectric measurements of loads on the lumbar spine when holding weights in standing postures. *Spine*, **7**(4), 390–7.

continuous walking at an average velocity of 5.6 $\mathrm{km\,h^{-1}}$. The signals were full-wave rectified and normalized against 50% MVIC. The normalized magnitude values were subjected to analyses of variance and subsequent *post hoc* analysis. The authors identified the unique combinations of load magnitude, height of load placement, and the duration of exercise. Their investigation revealed some useful information about the task. However, a more useful conclusion was that in a whole body generalized task a whole body EMG profile may be less useful compared to tasks involving specific regions and specific muscles.This highlights the importance of choosing techniques and dependent variables appropriately suited to the questions being investigated.

In contrast to the previous study, Andersson and Ortengren (1984) assessed the load on the back in assembly line work using electromyography. The amplified surface electromyographic signals were root mean squared and quantified by means of an amplitude histogram. They divided the amplitude into five intervals and the time spent in each of these intervals was expressed as the percentage of the task cycle. The magnitude of the myoelectric signal, however, was calibrated against known weights held in the hands in specific postures. Such magnitude calibrations were performed several times during the entire task period for each of the muscles separately. Such a strategy allowed the estimation of the local loads on the workers' backs in quantitative terms of microvolts and the relative time spent under each of these loads by plotting a time–amplitude histogram (Figure 1.21). Such an approach allows a pictorial temporal determination of relative load on the back. By analyzing one task cycle one can, therefore, make a quantitative assessment of the overall load on an assembly line. Using the same methodology it may be possible to assess redesigned tasks for any improvements.

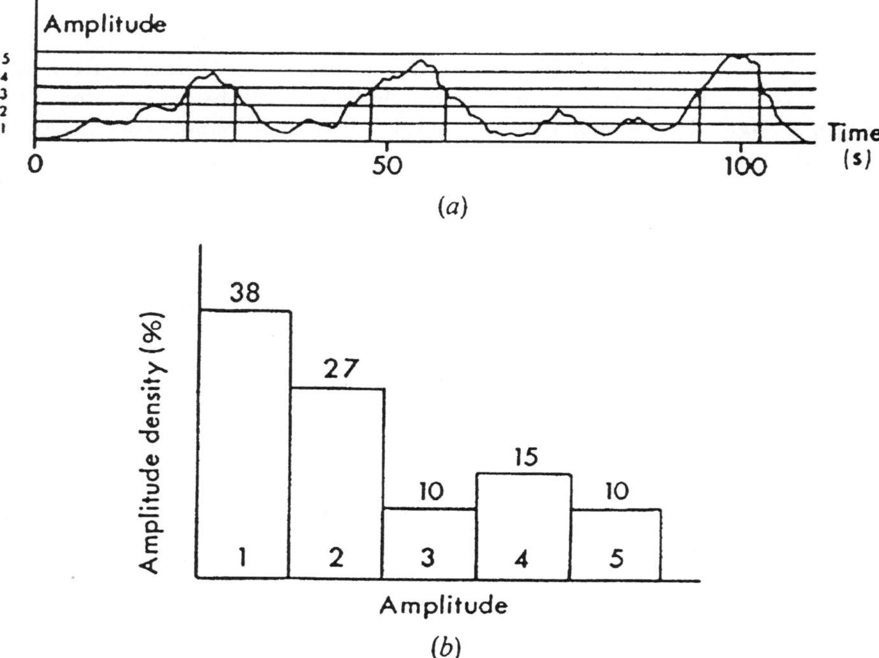

Figure 1.21 A temporal distribution of amplitude histogram. (a) Examples of myoelectric signal amplitude variations during a work spell. The curve is obtained after full-wave rectification and low-pass filtering of the raw myoelectric signal. Five amplitude intervals are marked. (b) Ampltiude histogram showing the total time the amplitude curve in (a) spends in each interval in percent of the total recording time (amplitude density diagram). Reprinted with permission from Andersson, G.B.J. & Ortengren, R. (1984) Assessment of back load in assembly line work using electromyography. *Ergonomics*, **27**(11), 1157–68.

Electromyography can be successfully used in product evaluation as it may impact on the stress on the operator for a given task and as the tasks vary. Kumar (1995) studied the spinal and abdominal electromyograms during garden raking with two different rakes and rake handles. An effective marketing strategy had been used to promote garden rakes and shovels with handles bent such that they allowed an upright posture during task execution. It was claimed by manufacturers that by eliminating the need to bend these devices will reduce the stress on the back. Using surface electrodes Kumar (1995) studied the EMG of the erectores spinae, external obliques, and rectus abdominis bilaterally in raking wet and dry dirt and sand with two rake blades having straight and bent handles. The EMG signals were amplified, low-pass filtered, and linear envelope detected. The millivolt EMG values were subjected to numerical, graphical comparisons, analyses of variance, and *post hoc* analyses. Kumar (1995) found a significant effect of handle and rake types on the electromyographic activity (Table 1.3). He reported that the normal rake generated a marginally higher EMG activity compared to the modified one. However, in pushing activity the modified rake generated a significantly higher amplitude than the normal rake ($p < 0.01$). On the basis of this study it was concluded that the normal rake with a straight handle will be far superior to the modified rake. Kumar and Cheng (1991), in another biomechanical study, established that during pushing the modified rake generated five to eight times greater spinal compression on the operator. One may, therefore, use such electromyographic technique for product evaluation.

Table 1.3 Statistical analysis of electromyographic data in product evaluation

Part of model	*F* ratio	df	*P*
Activity	30.88	1	0.0001
Muscle	6.91	7	0.0001
Medium	6.00	1	0.0280
Water	47.50	1	0.0001
Rake	14.06	1	0.0022
Activity × muscle	10.61	7	0.0001
Activity × rake	29.23	1	0.0001
Muscle × water	3.08	7	0.0057
Muscle × rake	6.94	7	0.0001
Medium × water	9.92	1	0.0071
Activity × muscle × medium	2.05	7	0.0566
Activity × muscle × water	4.33	7	0.0003
Activity × muscle × rake	3.91	7	0.0008
Activity × medium × water	5.99	1	0.0282

1.4.7 Muscle Fatigue

One of the very useful parameters in the industrial application of ergonomics is determination of localized muscle fatigue. Due to its importance to ergonomics, an entire chapter is dedicated to the topic in this book. The readers are referred to Chapter 6 by Hagg and Kadefors in this volume. However, a very brief account with a few examples follows.

Habes (1984) investigated localized muscle fatigue of the erectores spinae of workers in an automobile upholstery plant over the entire workshift. He performed a comparative evaluation of the accumulation of localized fatigue from working for a full day on a job requiring 914 mm reach and another requiring 813 mm reach. He showed that the integrated EMG amplitude over the course of the day increased considerably and significantly more with the 914 mm reach distance. Ortengren *et al.* (1975) studied power spectrum changes in the EMG signals to determine the fatigue in assembly line work. They studied two types of assembly line work considered heavy by workers. EMG signals were recorded from five muscles bilaterally. They reported significant power spectrum changes due to the heavy work. As a result of their observations the authors suggested that such measurements will be a valuable method in determining fatigue for occupational activities.

The power spectrum analyses are achieved through the fast Fourier transform (FFT) technique which changes raw EMG signals from the time domain to frequency domain. Once this has been accomplished various frequency components, the median frequency, and the power contained (area under the curve) can be calculated and studied. Obtaining this electrophysiological data of the muscle allows one to discern physiological changes and progression towards eventual fatigue even though no external or mechanical indicators can be apprehended. The study of muscle fatigue has been an area of vigorous activity to which numerous workers have contributed (see Chapter 6 by Hagg and Kadefors). Here only the most basic information is presented to rationalize its ergonomic application. With the onset of muscle fatigue, an increase in the magnitude of the EMG signal and a frequency shift toward lower frequency components have been reported extensively. Lindstrom (1970) suggested that these two phenomena are related, as during prolonged contraction the low-frequency components of EMG signals increase, more EMG signals will be transmitted through the low-pass filtering effect of the body tissue. There is controversy over the reason for such behaviour. Some advocate that this phenomenon occurs due to recruitment of additional motor units to maintain force output (Edwards and Lippold, 1956; Vredenbregt and Rau, 1973). However, others argue against this suggestion (Milner-Brown *et al.*, 1973; De Luca *et al.*, 1982). Another reason suggested for this behaviour has been cited to be motor unit synchronization (Chaffin, 1973; Bigland-Ritchie *et al.*, 1981). However, the synchronization becomes increasingly pronounced with the progression of contraction and the frequency shift has been shown to be more pronounced at the beginning of the contraction. Thus, there appears to be a discrepancy in such an argument. Whatever the mechanism, the electrophysiological phenomena increased the amplitude of the EMG signals and caused a shift in median frequency toward lower values, which are widely accepted as reliable indicators of localized muscle fatigue.

Dolan *et al.* (1995) used 'frequency banding' of the surface electromyographic signals in assessing fatigue of the erectores spinae. The

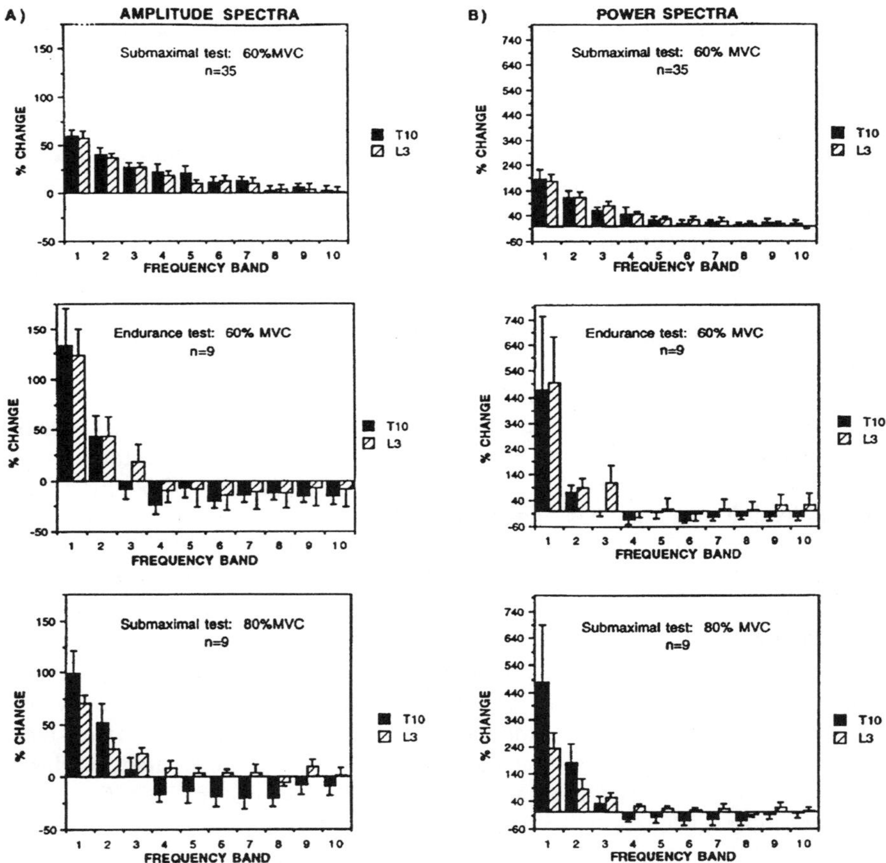

Figure 1.22 A distribution of amplitude and power change in different frequency band due to fatigue. The percentage change in the EMG signal during the sustained isometric contractions is shown for each of ten discrete frequency bands, ranging from ban 1 (5–30 Hz) up to band 10 (270–300 Hz). Results are compared for the three different fatigue tests and are calculated separately from (A) the amplitude spectra and (b) the power spectra. Reprinted with permission from Dolan, P., Mannion, A.F. & Adams, M.A. (1995) Fatigue of the erector spinae muscles – a quantitative assessment using 'frequency banding' of the surface electromyography signal. *Spine*, **20**(2), 149–59.

authors recorded the surface EMG from the erectores spinae at the T_{10} and L_3 levels and performed a fast Fourier transform of the data. The range of frequency from 5 to 300 Hz was studied by authors. They divided the power spectra into ten frequency bands between 5 and 300 Hz. Dolan *et al.* (1995) calculated the median frequency, total power, and the peak amplitude of the spectra. They found that the median frequency decreased steadily during the contractions, whereas the total power and the peak amplitude increased. They found that the most repeatable and linear index of change was the increase in the EMG signal in the 5–30 Hz frequency band (Figure 1.22). They reported that the middle to high frequency components of the EMG signals increased during the early stages of the contractions, but decreased as the endurance limits were approached. The median frequency has received a lot of attention in the literature. However, it appears that the changes in the lower frequency band are more sensitive and a linear index of the muscle fatigue. It may, therefore, be a more convenient measure to record.

References

ANDERSSON, B.J.G. & ORTENGREN, R. (1974a) Myoelectric back muscle activity during sitting. *Scand. J. Rehabil. Med. (Suppl)*, **3**, 73–90.

ANDERSSON, B.J.G. & ORTENGREN, R. (1974b) Lumbar disk pressure and myoelectric back muscle activity during sitting: II. Studies on an office chair. *Scand. J. Rehabil. Med. (Suppl)*, **6**, 115–21.

ANDERSSON, G.B.J. & ORTENGREN, R. (1984) Assessment of back load in assembly line work using electromyography. *Ergonomics*, **27**, 1157–68.

BASMAJIAN, J.V. & DELUCA, C.J. (1985) *Muscles Alive: Their Functions Revealed by Electromyography*, (5th edn). Williams & Wilkins, Baltimore, MD.

BECHTEL, R. & CALDWELL, G.E. (1994) The influence of task and angle on torque production and muscle activity at the elbow. *J. Electromyogr. Kinesiol.*, **4**, 195–204.

BERGSTROM, R.M. (1959) The relation between the number of impulses and the integrated electrical activity in the electromyogram. *Acta Physiol. Scand.*, **45**, 97–101.

BIGLAND, B. & LIPPOLD, O.C.T. (1954) The relationships between forces, velocity and integrated electrical activity in human muscles. *J. Physiol. (London)*, **123**, 214–224.

BIGLAND-RITCHIE, B., DONOVAN, E.F. & ROUSSOS, C.S. (1981) Conduction velocity and EMG power spectrum changes in fatigue of sustained maximal efforts. *J. Appl. Physiol.*, **51**, 1300–5.

BOBET, J. & NORMAN, R.W. (1982) Use of average electromyogram in design evaluation – Investigation of a whole body task. *Ergonomics*, **25**, 1155–63.

BOBET, J. & NORMAN, R.W. (1984) Effects of load placement on back muscle activity in load carriage. *Eur. J. Appl. Physiol.*, **53**, 7–75.

CARLSÖÖ, S. (1961) The static muscle load in different work positions: an electromyographic study. *Ergonomics*, **4**, 193–211.

CHAFFIN, D.B. (1973) Localized muscle fatigue: definition and measurement. *J. Occup. Med.*, **15**, 346–54.

CHENG, R.C.K. & KUMAR, S. (1991) A three dimensional biomechanical model for human back. *Int. J. Indust. Ergonom.* **7**, 327–39.

CHRISTENSEN, H., LO MONACO, M., DAHL, K., FUGLSANG-FREDERIKSEN, A. (1984) Processing of electrical activity in human muscle during a gradual increase in force. *EEG Clin. Neurophysiol.*, **58**, 230–39.

CIRIELLO, V.M. (1982) A longitudinal study of the effects of two training regimens on muscle strength and hypertrophy of fast twitch and slow twitch fibers. PhD dissertation, Boston University, Boston, MA.

CLOSE, J.R., NICKEL, E.D. & TODD, F.N. (1960) Motor-unit action-potential counts. *J. Bone Joint. Surg (Am)*, **42**, 1207–22.

CNOCKAERT, J.C. (1978) Comparison electromyographique du travail en allongement et en raccourcissement au course de mouvement de va-et-vient. *EMG Clin. Neurophysiol.*, **15**, 477–89.

DELITTO, R.S., ROSE, S.J. & APTS, D.W. (1987) Electromyographic analysis of two techniques for squat lifting. *Physical Therapy*, **67**, 1329–34.

DE LUCA, C.J., LEFEVER, R.S., MCCUE, M.P. & XENAKIS, A.P. (1982) Behaviour of human motor units in different muscles during linearly varying contractions. *J. Physiol.*, **329**, 113–28.

DOLAN, P., MANNION, A.F. & ADAMS, M.A. (1995) Fatigue of the erector spinae muscles – a quantitative assessment using 'frequency banding' of the surface electromyography signal. *Spine*, **20**(2), 149–59.

EDWARDS, R.G. & LIPPOLD, O.C.J. (1956) The relationship between force and integrated electrical activity in fatigued muscle. *J. Physiol. (London)*, **132**, 677–81.

FICK, R. (1911) *Handbuch der Anatomie und Mechanik der Gelenke*, Vol 3. Gustav Fischer, Jena, Germany.

FLOYD, W.F. & SILVER, P.H.S. (1950) Electromyographic study of patterns of activity of the anterior abdominal wall muscles in man. *J. Anat.*, **84**, 132–45.

FLOYD, W.F. & SILVER, P.H.S. (1955) The function of the erectores spinae muscles in certain movements and postures in man. *J. Physiol.*, **129**, 184–203.

GREGOR R.J., GAVANAGH, P.R. & LAFORTUNE, M. (1985) Knee flexor moments during propulsion in cycling: a creative solution to Lombard's Paradox. *J. Biomech.*, **18**, 307–16.

GRIEVE, D.W. & PHEASANT, S.T. (1976) Myoelectric activity, posture and isometric torque in man. *EMG Clin. Neurophysiol.*, **16**, 3–21.

HABES, D.J. (1984) Use of EMG in a kinesiological study in industry. *Appl. Ergonom.*, **15**, 297–301.

HAGG, G.M., SUURKULA, J. & LIEW, M. (1987) A worksite method for shoulder muscle fatigue measurements using EMG, test contractions and zero crossing technique. *Ergonomics*, **30**, 1541–51.

HOBART, D.J., KELLEY, D.L. & BRADLEY, L.S. (1975) Modification occurring during acquisition of a novel throwing task. *Am. J. Phys. Med.*, **54**, 1–24.

HOF, A.L. & VAN DEN BERG, J.W. (1977) Linearity between the weighted sum of the EMGs of the human triceps surae and the total torque. *J. Biomech.*, **10**, 529–39.

IINMAN, V.T., RALSTON, J.H., SAUNDERS, J.B. de C.M. *et al.* (1952) Relation of human electromyogram to muscular tension. *EEG Clin. Neurophysiol.*, **4**, 187–94.

JONSSON, B. (1970) The functions of individual muscles in the lumbar part of the erector spinae muscle. *Electromyography*, **10**, 5–21.

KADEFORS, R. (1978) Application of electromyography in ergonomics: new vistas. *Scand. J. Rehabil. Med.*, **10**, 127–33.

KAMON, E. (1966) Electromyography of static and dynamic postures of the body supported on the arms. *J. Appl. Physiol.*, **21**, 1611–18.

KAMON, E. & GORMLEY, J. (1968) Muscular activity pattern for skilled performance and during learning of a horizontal bar exercise. *Ergonomics*, **11**, 345–57.

KUMAR, S. (1971) Studies of trunk mechanics during physical activity. PhD thesis, University of Surrey, UK.

KUMAR, S. (1979) Variations in stress and response due to lifting in different planes. In *Sciences in Weight Lifting* (Ed. J. TERAUDS), Human Kinetics Pubs, Champaign, Il. pp. 31–41.

KUMAR, S. (1980) Physiological response to weight lifting in different planes. *Ergonomics*, **23**(10), 987–93.

KUMAR, S. (1988) Preventative research – an effective therapy. In *Ergonomics in Rehabilitation* (Ed. A. MITAL & W. KARWOWSKI), pp. 183–97. Taylor & Francis, London, New York and Philadelphia.

KUMAR, S. (1993) The role of the intra-abdominal pressure in the mechanics of the human trunk. In *Advances in Idiopathic Low Back Pain* (Ed. E. ERNST, M.I.V. JAYSON, M.H. POPE and R.W. PORTER), pp. 86–90. Blackwell Scientific Publications, Vienna.

KUMAR, S. (1994) A computer desk for bifocal lens wearers with special emphasis on selected telecommunication tasks. *Ergonomics*, **37**, 1669–78.

KUMAR, S. (1995) Electromyography of spinal and abdominal muscles during garden raking with two rakes and rake handles. *Ergonomics*, **38**, 1793–1804.

KUMAR, S., NARAYAN, Y. & ZEDKA, M. (1995) An electromyographic study of unresisted trunk rotation with normal velocity among healthy subjects. *Spine* (in press).

KUMAR, S. & CHENG, R.C.K. (1991) Biomechanical analysis of raking and comparison of two rakes. *Int. J. Indust. Ergonom.*, **7**, 31–9.

KUMAR, S. & DAVIS, P.R. (1983) Spinal loading in static and dynamic postures: EMG and intra-abdominal pressure study. *Ergonomics*, **26**(9), 913–22.

KUMAR, S. & SCAIFE, W.G.S. (1975) Frequency amplitude relationship of signals in isometric surface electromyograms. EMG Clin. Neurophysiol., **15**, 539–44.

KUMAR, S. & SCAIFE, W.G.S. (1979) A precision task posture and strain. *J. Safety Res.*, **11**(1), 28–36.

LESTIENNE, F. (1979) Effects of inertial load and velocity on the braking process of voluntary limb movements. *Exp. Brain Res.*, **35**, 407–18.

LINDSTROM, L.R. (1970) *On the Frequency Spectrum of EMG Signals.* Technical Report-Research Laboratory of Medical Electronics, Chalmers University of Technology, Goteborg, Sweden.

LIPPOLD, O.C.J. (1952) The relation between integrated action potentials in a human muscle and its isometric tension. *J. Physiol. (London)*, **117**, 492–99.

LLOYD, A.J. & VOOR, J.H. (1973) The effect of training on performance efficiency during a competitive isometric exercise. *J. Motor Behav.*, **5**, 17–24.

LOEB, G.E. & GANS, C. (1986) Signal processing and display. In *Electromyograph for Experimentalists* (Eds G.E. LOEB & C. GANS. University of Chicago Press, Chicago, IL.

MANNING, D.P., MITCHELL, R.G. & BLANCHFIELD, L.P. (1984) Body movements and events contributing to accidental and nonaccidental back injuries. *Spine*, **9**,

734–49.

MATON, B., LEBOZEC, S. & CNOCKAERT, J.C. (1980) The synergy of elbow extensor muscles during dynamic work in man: II. Braking elbow flexion. *Eur. J. Appl. Physiol.*, **44**, 279–89.

METRAL, S. & CASSAR, G. (1981) Relationship between force and integrated EMG activity during voluntary isometric anisotonic contraction. *Eur. J. Appl.Physiol.*, **346**, 185–98.

MILNER-BROWN, H.S. & STEIN, R.B. (1975) The relation between the surface electromyogram and muscular force. *J. Physiol. (London)*, **246**, 549–69.

MILNER-BROWN, H.S., STEIN, R.B. & YEMM, R. (1973) Changes in firing rate of human motor units during linearly changing voluntary contractions. *J. Physiol.*, **230**, 371–90.

MIRKA, G.A. (1991) The quantification of EMG normalization error. *Ergonomics*, **34**(3), 343–52.

MORITANI, T. & DEVRIES, H.A. (1978) Reexamination of the relationship between the surface integrated electromyogram (IEMG) and force of isometric contraction. *Am. J. Phys. Med.*, **57**, 263–77.

ORTENGREN, R., ANDERSSON, G., BROMAN, H., MAGNUSSON, R. & PETERSEN, I. (1975) Vocational electromyography: studies of localized muscle fatigue at the assembly line. *Ergonomics*, **18**(2), 157–74.

PACHELLA, R.C. & PEW, R.W. (1968) Speed–accuracy trade-off in reaction time: effect of discrete criterion times. *J. Exp. Psychol.*, **76**, 19–24.

PERRY, J. & BEKEY, G.A. (1981) EMG–force relationships in skeletal muscle. *CRC Crit. Rev. Biomed. Eng.*, **7**, 1–22.

PERSON, R.S. (1958) Electromyographical study of coordination of the activity of human antagonist muscles in the process of developing motor habits (Russian text). *Jurn Vys'cei Nervn Dejat*, **8**, 17–27.

RADCLIFFE, C.W. (1962) The biomechanics of below-knee prostheses in normal, level, bipedal walking. *Artif. Limbs*, **6**, 16–24.

ROBERTSHAW, C., KUMAR, S. & QUINNEY, H.A. (1986) Electromyographic and cinematographic analysis of rising and lowering between sitting and standing. In *Trends in Ergonomics*. Human Factors III (Ed. W. KARWOWSKI, pp. 738–49. North-Holland, Amsterdam, New York.

SCHULTZ, A.B., ANDERSSON, G.B.J., HADERSPECK, K., ÖRTENGREN, R., NORDIN, M. & BJÖRK, R. (1982) Analysis and measurement of lumbar trunk loads in tasks involving bends and twists. *J. Biomech.*, **15**, 669–75.

STOKES, I.A.F., RUSH, S., MOFFROID, M., JOHNSON, G.B. & HAUGH, L.D. (1987) Trunk extensor EMG–torque relationship. *Spine*, **12**(8), 770–6.

SUURKULA, J. & HAGG, G.M. (1987) Relations between shoulder/neck disorders and EMG zero crossing shifts in female assembly workers using the test contraction method. *Ergonomics*, **30**, 1553–69.

VREDENBREGT, J. & RAU, G. (1973) Surface electromyography in relationship to force, muscle length and endurance. In *New Developments in Electromyography and Clinical Neurophysiology*, (Ed. J.E. DESMEDT)., Karger, Basel, Switzerland.

YANG, J.F. & WINTER, D.A. (1983) Electromyography reliability in maximal and submaximal contractions. *Arch. Phys. Med. Rehabil.*, **64**, 417–20.

CHAPTER TWO

Physiological basis and concepts of electromyography

ALWIN LUTTMANN

Institut für Arbeitsphysiologie an der Universität Dortmund, Dortmund

2.1 THE PHYSIOLOGICAL BASIS OF ELECTROMYOGRAPHY

2.1.1 General Functions of Muscles and Electromyography

Muscles can contract and produce force. In a living organism movements are performed by means of muscular activation. Through coordinated movement of its body parts the organism can change its position in space and exert mechanical forces on its environment. In addition to locomotion, activation of muscles is also required for transport processes within the body such as the convection of fluids in the cardiovascular and gastrointestinal systems or the transportation of gases in the respiratory system.

Muscle contraction and force production are due to a change in the relative positions of various thread-like molecules or filaments arranged in the interior of the muscle. This sliding of the filaments is triggered by an electrical phenomenon known as action potential. As is explained in greater detail in the following, an action potential results from a change in the membrane potential which exists between the interior of a muscle cell and the external space. The recording of the pattern of muscle action potentials is called electromyography – the record itself is called an electromyogram (EMG). Electromyography therefore records an electrical phenomenon which is causally related to muscle contraction. On account of the close functional relationship between the electrical action potential and the electromyogram, on the one hand and the mechanical muscle contraction, on the other, the EMG represents an adequate parameter for the analysis of muscle activities. Statements about the time course of a muscle contraction, the contraction force, and the coordination of the activities of several muscles can be derived from the amplitude and the frequency spectrum of the EMG as well as from the changes in these values over time. In ergonomics, conclusions about the level of muscular strain and the occurrence of muscular fatigue can be drawn from electromyograms. In addition, indications about work design can be derived for purposes of reducing strain and fatigue.

In order to understand the characteristic features of EMGs it is necessary to explain the electrical phenomena occurring at muscle cells, to describe the structure and function of muscles and their components as well as to discuss the relationship between electrical excitation and the production of mechanical force by muscles.

2.1.2 Basic Physiology of Excitable Membranes

All cells in the human body are surrounded by a cell membrane which separates the intracellular from the extracellular space. Its structure and function is fundamentally the same for all body cells. The membrane represents a diffusion barrier which prevents the various composite fluids in the intracellular and extracellular spaces from mixing. It also possesses structures which enable a specific exchange of both substances and information between both compartments.

2.1.2.1 Membrane Structure and Membrane Properties

The basic structure of a cell membrane is shown in Figure 2.1. The membrane comprises a double layer of phospholipids. Both surfaces of this layer are covered with proteins. In addition, proteins are embedded into the lipid bilayer, permeating it either fully or partially. The bilayer structure of the cell membrane and the properties of the lipid molecules are important for the

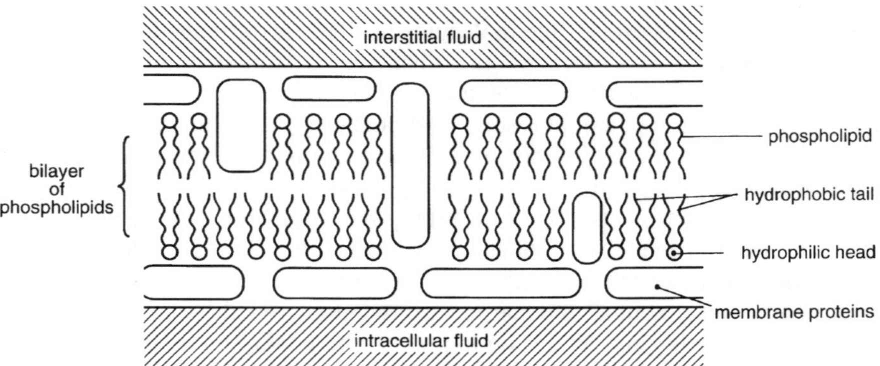

Figure 2.1 Schematic representation of a cell membrane, consisting of a bilayer of phospholipids, covered and penetrated by membrane proteins.

means by which an exchange between the intracellular and the interstitial compartments is restricted. Lipid molecules are elongated and unsymmetrical in structure. They possess a polar head and a non-polar tail (Eckert and Randall, 1983). The polar heads are hydrophilic, i.e. water soluble, whereas the non-polar tails are hydrophobic, i.e. water insoluble. Within the double layer the lipid molecules are positioned in such a way that the hydrophobic ends face each other in the middle of the membrane whereas the hydrophilic ends are immersed in the aqueous solutions present in the intracellular and extracellular spaces. The hydrophobic property of the tails in the inner lipid layer means that the membrane represents an almost insuperable barrier for water, water-soluble molecules, and ions.

The membrane proteins (cf. Figure 2.1) play a significant part in the exchange processes between the two compartments. At the functional level, a distinction can be made between two groups of proteins, transport proteins, and receptor proteins. Transport proteins allow substances to travel across the membrane from one compartment to the other. They are generally characterized by high specificity, i.e. each transport protein enables the transportation of only one or a small number of specific substances. Transport proteins are termed carrier molecules, membrane pumps, or membrane channels according to their particular characteristic. By contrast, receptor proteins combine specifically with certain molecules (e.g. hormones) and serve in the transfer of information across the membrane.

The fluids in the intracellular and extracellular spaces are composed differently, in particular with regard to the concentrations of ions in both compartments. The unequal distribution of ions results in an electrical potential difference known as a membrane potential. The order of magnitude of the membrane potential in most cells of the body is between −60 and −90 mV. The negative sign indicates that the interior of the cell is negative in comparison with the exterior space. Most organs comprise cells whose membrane potential remains largely constant over time. Slight variations may occur as a result of changes in the ional composition of the fluids surrounding the membrane. A completely different behaviour, however, is associated with the so-called excitable membranes found in nerve and muscle cells. The membrane potential of these cells, starting from a resting potential of approximately −60 to −90 mV between the intracellular and extracellular spaces, can change within fractions of a millisecond to approximately +20 to +50 mV. Such rapid change in the transmembrane potential is termed an action potential. The duration of an action potential amounts to a few milliseconds (with the exception of the heart muscle where it is several hundred milliseconds). Action potentials are responsible for fast information transfer. In muscles they also have the task of triggering muscular contractions.

The electrical signals derived in electromyography are directly related to the action potentials of a muscle. In order to improve understanding of the electrical events occurring during muscular activation, further explanations of the physiological processes which determine the membrane potential and the development of action potentials in particular are provided in the following.

2.1.2.2 Ion Concentrations in Muscles

Part of a muscle fiber is represented in the upper part of Figure 2.2. The fiber is surrounded by a cell membrane which divides the intracellular fluid from the interstitial fluid. A measurement system comprising an electrode and a voltage display indicates a transmembrane potential of approximately −75 mV. The symbols for the ions in the upper part of Figure 2.2 vary in size as a sign of the unequal distribution of the relevant ions between the intracellular and the interstitial phase. The quantitative information on typical ion concentrations is to be found in the table in the lower part of Figure 2.2. The composition of the intracellular fluid is characterized by a high concentration of potassium cations (K^+) and protein anions (A^-) whereas the interstitial fluid is rich in sodium cations (Na^+) and chloride anions (Cl^-). (The concentration of further ions such as calcium, magnesium, bicarbonate

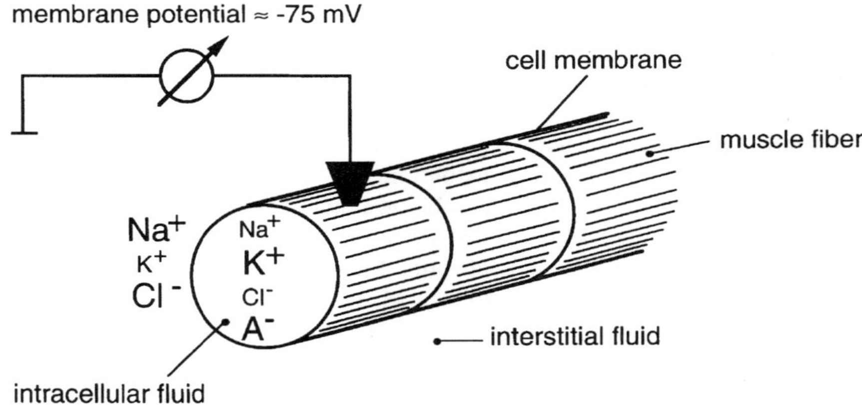

ions	intracellular concentration in mmol/l	interstitial concentration in mmol/l	ratio inside: outside
Na^+	12	145	1:12
K^+	155	4	40:1
Cl^-	4	120	1:30
A^-	155	—	—

Figure 2.2 Concentrations of the most relevant ions in the intracellular and interstitial fluids of a muscle (data according to Woodbury (1965)) and schematic representation of the measurement of transmembrane potential.

and phosphate is not provided here since these substances only have a slight influence on the levels of the membrane potentials and on the electrical phenomena during the development of the action potential at the membranes.) Approximate ratios of the concentrations (internal : external) are provided in the right-hand column of the table in Figure 2.2. They range between 1 : 30 for Cl^- and 40 : 1 for K^+. Negatively charged protein ions (A^-) are only found inside the cells. Their concentration in the extracellular space is negligible.

The unequal distribution of the cations potassium and sodium results from the fact that the membrane contains transport proteins which are capable of active transport. The term 'active' is applied to transport in which ions are transported across a membrane against its concentration gradient. In the present case Na^+ ions are transported from the intracellular to the extracellular space. By contrast, transportation of K^+ takes place in the inward direction. In both instances ions are transported from the space with the lower concentrations to the space with the higher concentrations. Metabolic energy is consumed during active transport. In accordance with the 'uphill' direction, such a mechanism is called an ion pump.

Whereas the uneven distribution of the cations results from the active transport of Na^+ and K^+, the uneven distribution of the anions between the two phases is due to the fact that part of the anions, namely the protein anions, cannot pass through the cell wall on account of their size and are therefore trapped in the interior of the cell. Consequently, for reasons of electroneutrality, other negative charge carriers which are able to permeate the membrane (in particular Cl^-) accumulate in the external phase. A further reason for the accumulation of the Cl^- ions in the external phase is the transmembrane potential (cf. Section 2.1.2.3). As a result of the potential gradient, Cl^- ions (as well as other membrane-permeating ions not considered here) are transported into the external space.

2.1.2.3 Membrane Potentials

The Resting Potential The concentration gradient for Na^+ and K^+ results in the creation of an electrical potential difference across the membrane. This is illustrated by Figure 2.3 which shows a schematic membrane separating intracellular and extracellular spaces. The development of an electrical voltage is to be explained firstly by reference to the uneven distribution of one type of ion only, namely potassium. As in reality, it is assumed that the potassium concentration is higher inside the cell than outside. Again in accordance with the real situation, ion channels, allowing the passage of K^+ ions, are present in the membrane. On account of the concentration gradient, a force, F_{con} is exerted on the ions in the outward direction. This results in an outflux of potassium. Since potassium is present in ional form, i.e. it is electrically charged, the net number of positive charges increases in the exterior space and

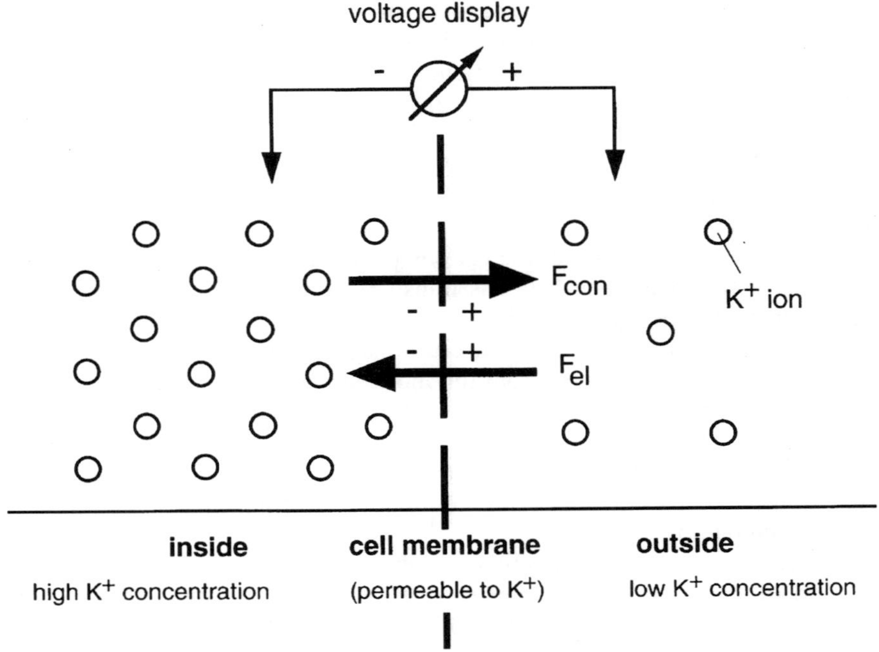

Figure 2.3 Formation of a transmembrane equilibrium potential: the force on the ions caused by the concentration gradient F_{con} is balanced by the electromotive force F_{el}.

decreases in the interior space. Accordingly the interior of the cell is negatively charged and an electrical potential difference is produced across the membrane. The potential difference exerts an inwardly directed force, F_{el}, on the ions. This force has the opposite effect to the force F_{con} which is directed outwards. Further ion flux towards the exterior is thereby restricted. The potential difference at which an equilibrium is established between the two forces is termed the equilibrium potential. Its level V_m can be calculated as follows using an equation derived by Nernst in the last century:

$$V_m = -\frac{R \cdot T}{z \cdot F} \cdot \ln \frac{c_i}{c_o} \qquad \text{(Nernst equation)}$$

where V_m is the membrane voltage, R is the gas constant, T is the absolute temperature, z is the valence of the ions under test, F is the Faraday constant, and c is the concentration of the ions under test (i = inside, o = outside).

The Nernst equation shows that the level of the membrane voltage is basically determined by the ratio of the ion concentrations in both phases.

In the presence of several types of ions – as is the case with biological cells – the membrane potential can be estimated using an extension of the Nernst equation developed by Goldman (1943). Considering the three most relevant ions Na^+, K^+, and Cl^-, the following equation accordingly results:

$$V_m = -\frac{R \cdot T}{F} \cdot \ln \frac{p_K \cdot [K^+]_i + p_{Na} \cdot [Na^+]_i + p_{Cl} \cdot [Cl^-]_o}{p_K \cdot [K^+]_o + p_{Na} \cdot [Na^+]_o + p_{Cl} \cdot [Cl^-]_i} \qquad \text{(Goldman equation)}$$

where p is the permeability (concentrations are indicated by [ ]). Comparison with the Nernst equation shows that in the Goldman equation all of the types of ions contribute towards the potential. However, their concentrations are weighted according to the permeability of the membrane for the particular type of ion. Therefore ions of low permeability, i.e. ions which can only pass the membrane with difficulty, have less influence on the potential difference. Measurements have revealed the following permeabilities of the membrane for the three ions considered before in the Goldman equation: $p_K : p_{Na} : p_{Cl} = 1 : 0.03 : 0.1$ (Kuffler *et al.*, 1984). This means that the membrane is most permeable for potassium and that the membrane potential is mainly determined by potassium ions. Using the concentrations provided in Figure 2.2 and the permeabilities given in the above, a membrane potential of approximately −75 mV is calculated on the basis of the Goldman equation.

The Action Potential Figure 2.4(a) illustrates a schematic set-up for the description of the processes involved in the triggering of action potentials: the right-hand section shows a voltage display unit for recording the potential difference across the membrane. A stimulus unit, with which a stimulus current can be passed through the membrane, is shown in the left-hand section. It is assumed that the current can pass the membrane in either the outward or the inward direction. An inward current (dashed arrow in Figure 2.4(a) and dashed lines in Figure 2.4(c)) causes the voltage across the membrane to increase (=hyperpolarization). The time curves for the membrane potential during stimulation by inward currents are indicated in Figure 2.4(b) by dashed lines. The membrane voltage reached during the stimulation period depends on the intensity of the current.

Reversal of the stimulus current produces an outward current (continuous arrow in Figure 2.4(a) and continuous lines in Figure 2.4(c)) and leads to a decrease in the voltage across the membrane (=depolarization). A small depolarizing current results in a small local response in the membrane voltage (Figure 2.4(b)). If the current is intensified, the depolarization increases. If depolarization becomes so strong that a certain value, the so-called threshold, is exceeded, membrane excitation is induced. An active response is elicited from the membrane. As a result, the transmembrane potential is reversed within fractions of a millisecond and a positive value is produced. This active response is known as an action potential.

The active response of the excitable membranes in the nerve and muscle fibers is produced by sodium and potassium channels opening in response to a stimulus. The permeability of these channels is dependent on the potential difference

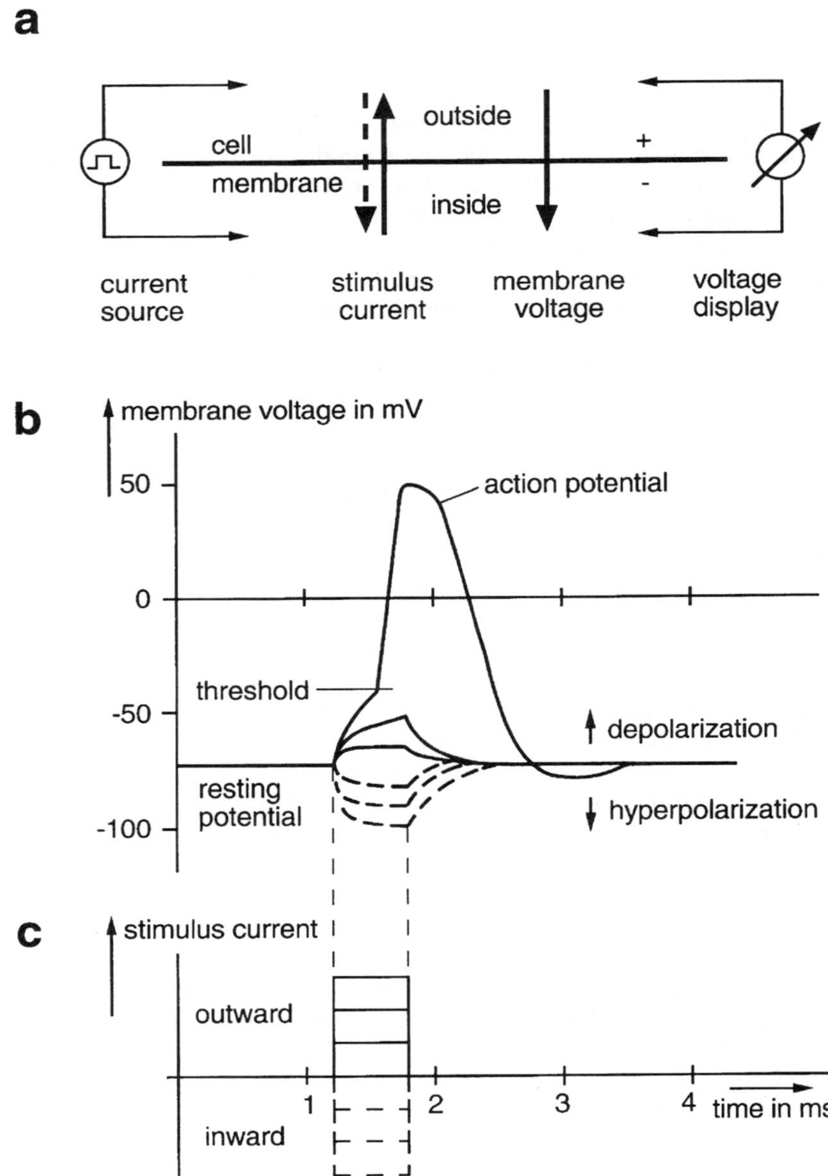

Figure 2.4 Release of action potentials by electrical stimulation.

across the membrane. If, as a result of external influences (e.g. stimulus current), the membrane voltage is reduced to the threshold voltage, sodium channels open at first. Subsequently, in accordance with the high concentration gradient for sodium, sodium ions flow inwards across the membrane. As a result, the potential difference between the inner and the outer surface of the membrane is reduced further and the permeability of the membrane for sodium is increased still further. Therefore, once the membrane voltage has reached the threshold potential, the subsequent process is regenerative or self-perpetuating. The membrane response operates according to an all-or-none principle.

The peak value of the action potential can be estimated by applying the Goldman equation. It is to be assumed here that the ratio of $p_K : p_{Na}$, which amounted to $1:0.03$ in the resting state, changes to approximately $1:20$. By including these numerical values in the Goldman equation and adopting the ion concentrations provided in Figure 2.2, it becomes clear that the membrane potential, which was determined in the resting state by the potassium ions, is dependent in the excited state on the sodium ions. The peak value is calculated at approximately +50 mV.

The sodium channels are only activated for approximately 1 ms. They subsequently close again. This results in a drop in the membrane potential after the peak has been reached (repolarization). The repolarization process is accelerated by the fact that potassium channels also open under the influence of the action potential. The resultant K^+ efflux accelerates the return of the membrane potential to the resting potential.

The significance of the ion channels for the development of the action potential has been investigated during the last 50 years using a wide variety of experimental procedures. The models developed as early as the 1940s and 1950s by Hodgkin and Huxley (1952) in order to show that the permeability of the membrane for particular ions is voltage dependent are still valid today. They have, however, been supplemented by essential insights gained from the use of the so-called patch-clamp technique (Neher and Sakmann, 1976). Since this technique permits recording of ion currents passing through single ion channels, it enables detailed analysis of the characteristics of the channels.

2.1.2.4 Release of Action Potentials

In the previous section it was shown on the basis of Figure 2.4 that an action potential is always triggered when the membrane is depolarized to the threshold level. External application of the depolarizing current via a stimulus unit was assumed. *In vivo* the action potentials are triggered in a similar manner by local depolarizing currents. Action potentials occur at muscle fibers as a result of two processes: action potential propagation along muscle fibers and neuromuscular transmission of excitation at the motor end-plates.

Action Potential Propagation Along Muscle Fibers The principle of action potential propagation along an excitable fiber is illustrated in Figure 2.5. It is assumed

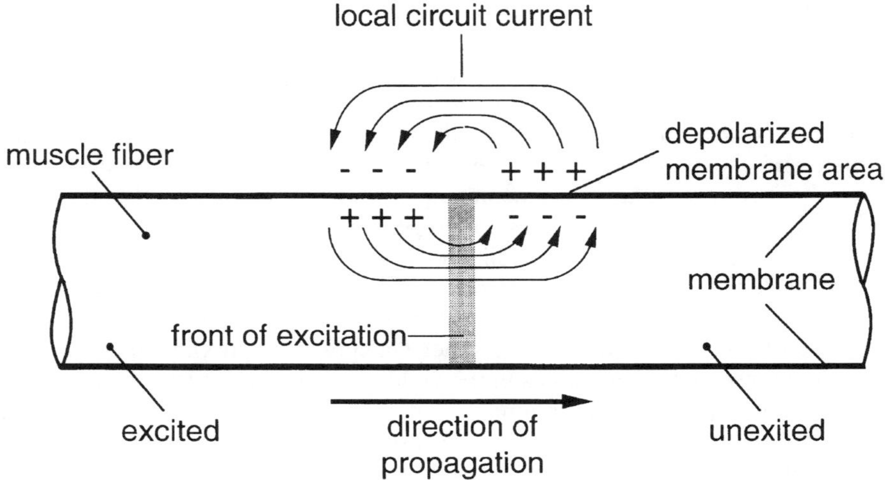

Figure 2.5 Mechanism of action potential propagation along a muscle fiber.

that the left side of the schematically represented muscle fiber is excited. In this state the inner surface of the muscle fiber membrane is positive in relation to the outer surface. In the non-excited area in the right-hand section of the diagram the polarity is reversed. As can be seen from Figure 2.5, a potential gradient exists between the excited and the non-excited areas on account of the difference in the charge distribution. The difference in the potential results in a local circuit current. As the arrows in Figure 2.5 indicate, in the unexcited area the current passes the membrane in an outward direction. An outward current leads – as was shown for the stimulus current in Figure 2.4 – to membrane depolarization in a narrowly defined area. The depolarization is sufficient for the threshold potential to be reached, thus inducing excitation in the previously non-excited area. As a result, an action potential occurs there, too. In the example shown in Figure 2.5 the front of excitation moves to the right. This newly initiated action potential may, in turn, excite a point further to the right along the membrane. In the manner described in the above, an excitation produced at one point on a muscle fiber can spread over the entire muscle fiber.

The velocity with which the action potential is reproduced along the length of the fibers (action potential propagation velocity) is dependent on the diameter of the fibers. It is higher for thicker fibers than for thinner ones. An additional influencing factor is the composition of the fluids adjacent to the muscle fiber membrane. The K^+ concentration in the extracellular fluid is of particular importance for the action potential propagation velocity (Kössler *et al.*, 1990).

Kössler *et al.* (1990) demonstrated that an increased K^+ concentration in the interstitial fluid reduces the propagation velocity. This relationship is significant in terms of explaining a frequently described finding, namely that the propagation velocity is reduced during muscle fatigue (for further explanations see Section 2.1.4.3).

Neural Control of Muscle Contraction and Neuromuscular Transmission The activity of the skeletal muscles is controlled via the motor nervous system. This is represented schematically in the upper part of Figure 2.6. The efferent nerve fibers of the motor system issue from the ventral horn of the spinal cord and terminate in the musculature. Only two nerve fibers are indicated in Figure 2.6. Activation of the musculature is effected via action potentials which occur at the cell bodies in the spinal cord. The action potentials are routed to the muscle via the nerve fibers and are then transmitted to the muscle fibers via special connecting elements known as neuromuscular synapses or motor end-plates. In each case one nerve fiber is connected to several muscle fibers via end-plates. All of the muscle fibers belonging to one nerve fiber are excited simultaneously and contract simultaneously as well. A nerve fiber, together with the associated group of muscle fibers, therefore forms the smallest functional unit of a muscle – the motor unit. Dependent on the function of the particular muscle, there is a large difference in the number of fibers per motor unit. In muscles which execute a

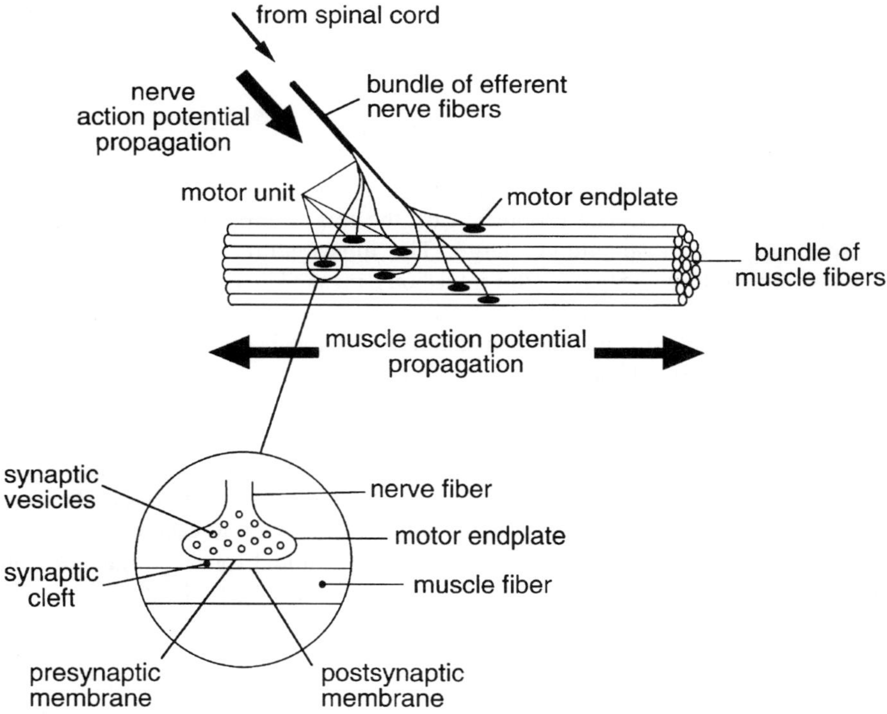

Figure 2.6 Neural control of muscular activity and principal structure of the neuromuscular synapse.

movement with great precision, the number of fibers per motor unit is very small (e.g. between five and ten in the external eye muscles) whereas the number is considerably greater for muscles which produce large forces with a low degree of precision (e.g. 750 in the m. biceps brachii; Rüegg 1987).

An enlarged view of a motor end-plate is presented in the lower part of Figure 2.6. The principle of synaptic transmission is to be described on the basis of this diagram. The presynaptic terminal of the nerve contains a large number of synaptic vesicles, each containing a transmitter substance. In motor end-plates the transmitter substance has been identified as acetylcholine. Excitation of the nerve fiber results in the content of the vesicle being released into the synaptic cleft. The transmitter substance diffuses through the cleft to the postsynaptic membrane where it combines with membrane receptors. As a result of the bond, the permeability of the membrane for the sodium and potassium cations alters. A local potential change, which is at first limited to a narrow area near the synapse, consequently occurs at the postsynaptic membrane. This change is called end-plate potential. The end-plate potential is poled in such a way that the postsynaptic membrane is depolarized. Under normal circumstances the degree of depolarization is such that the threshold is exceeded and an action potential is triggered.This action potential, which is at first limited to the vicinity of the motor end-plate, then spreads to both sides along the entire muscle fiber in the manner described in the previous section.

2.1.3 Fundamentals of Muscle Physiology

All movements which are performed by or in the body are the result of the activation of muscles. The proportion of the total body accounted for by the musculature is correspondingly large. The musculature amounts to between approximately 40 and 50% of the total body mass. The energy expenditure in the muscles differs very widely according to the particular activity. At rest approximately 25% of the oxygen taken up by the body is consumed in the muscles. A 10–20-fold increase in the oxygen consumed by the musculature may occur during exhausting muscular work.

Various types of musculature are found within the body. In spite of the structural differences, the mode of action is fundamentally the same for all muscles. Contraction and force generation by a muscle result from elongated contractile proteins or arrays of proteins shifting in relation to each other, thereby producing a tensile force. The mechanical energy required for this originates from the conversion of the chemical energy obtained from the ingested food.

2.1.3.1 Types of Muscles

Three different types of muscles can be identified within the human body on the basis of histological and functional differences: the skeletal muscles, the cardiac muscle, and the smooth muscles.

The skeletal muscles are connected to the bones of the skeletal system via tendons. Muscles link various parts of the body's osseous support structure. Voluntary activation of the muscles by the motor nervous system causes the bones to move in relation to each other. The skeletal muscles are responsible for the movement of body segments or the entire body. An arrangement of light and dark stripes running perpendicularly to the direction of the muscle fibers can be detected microscopically in skeletal muscle. On account of this striping, which is ascribable to a regular arrangement of structural elements to be described in greater detail in the following, the skeletal muscles are described as cross-striated muscles.

The cardiac muscle is responsible for the convection of the blood in the blood vessels. Like the skeletal muscle it is also cross-striated. In terms of its functional characteristics it differs from the skeletal muscle in the way the action potentials are released and transmitted.

Smooth muscles are found in the intestinal region, in the blood vessels, and in the ciliary muscle of the eye. Controlled by the autonomic nervous system, the activity of the smooth muscles is involuntary. In contrast to the skeletal and cardiac muscles no striation is apparent.

The further explanations relating to structure and characteristics focus on the skeletal muscle since ergonomic applications of electromyography are limited to this part of the musculature.

2.1.3.2 Fine Structure of Cross-striated Muscles

The structure of a muscle is represented in Figure 2.7. A skeletal muscle consists of elongated, mainly parallel muscle cells (muscle fibers) which extend over the length of the muscle and turn into a tendon at each end (Figure 2.7a). The muscle fibers are enveloped by an excitable cell membrane. The diameter of the fiber amounts to between 20 and 200 μm (Figure 2.7b). Within the fibers lie cylindrical subunits of the muscle fibers, so-called myofibrils. They are also arranged in parallel and are approximately 1 μm in diameter. A regular arrangement of light and dark transverse stripes can be seen on the fibrils (Figure 2.7c). The light-coloured stripes are known as the I band and the dark-coloured ones as the A band. In the middle of the dark A band there is a relatively narrow lighter area which is termed the H zone. A narrow dark structure, called the Z line, is located within the light I band.

The characteristic cross-striation of a muscle is due to a regular arrangement of the proteins from which the muscle is formed. The muscle structure represented in Figure 2.8 is based on longitudinal and cross-sections through a myofibril. The muscle consists of two subunits, known as filaments, which are each arranged in groups and alternate with each other in a regular sequence (Figure 2.8a). A distinction can be made between the thick and thin filaments. Named after their most important constituent, the thick filaments are also known as myosin filaments and the thin ones as actin filaments. The two groups of filaments slide past each other. This explains the differences in the visual

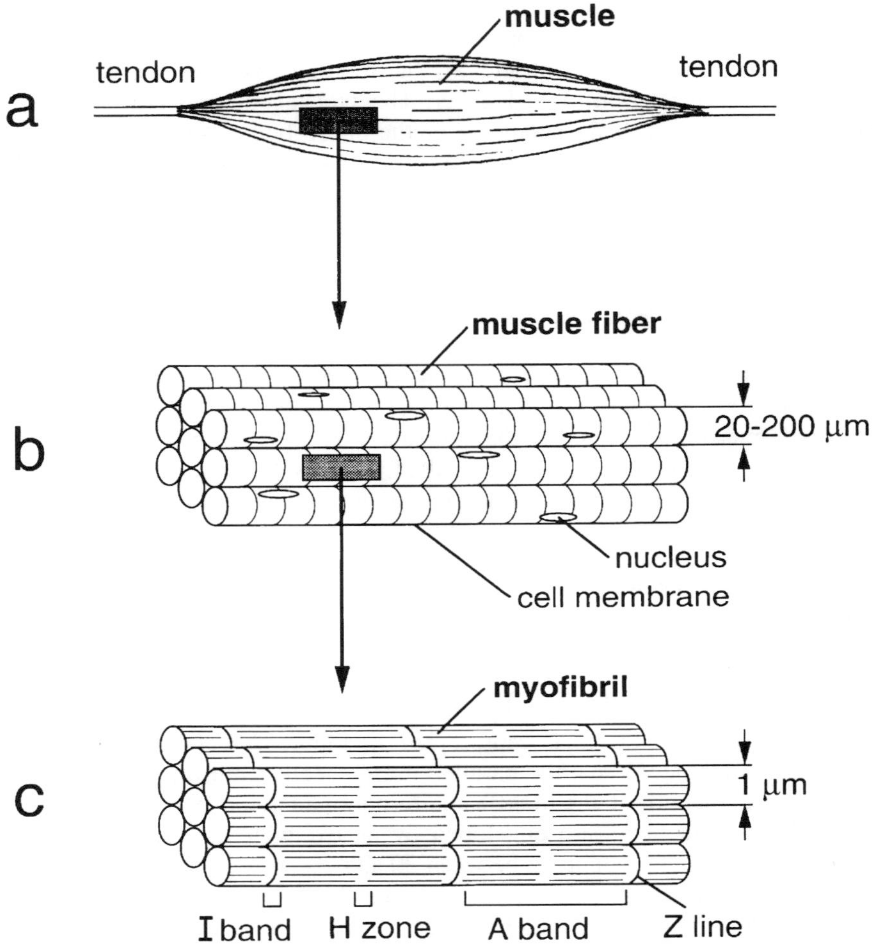

Figure 2.7 Fine structure of a cross-striated muscle consisting of muscle fibers and myofibrils.

appearance described in the above: microscopically, the areas within the A bands in particular, in which the myosin and the actin filaments overlap, appear dark. By contrast, the H zone within the A band, in which only the myosin filaments are located, appears lighter. The areas in which only actin filaments occur form the light I bands. The unit located between two Z lines is called the sarcomere.

A regular arrangement of the filaments is not only to be found in the direction of the muscle fibers but also in a transverse direction. This becomes clear from the filament arrangement in the cross-section shown in Figure 2.8(b): six thin actin filaments are grouped around each thick myosin filament

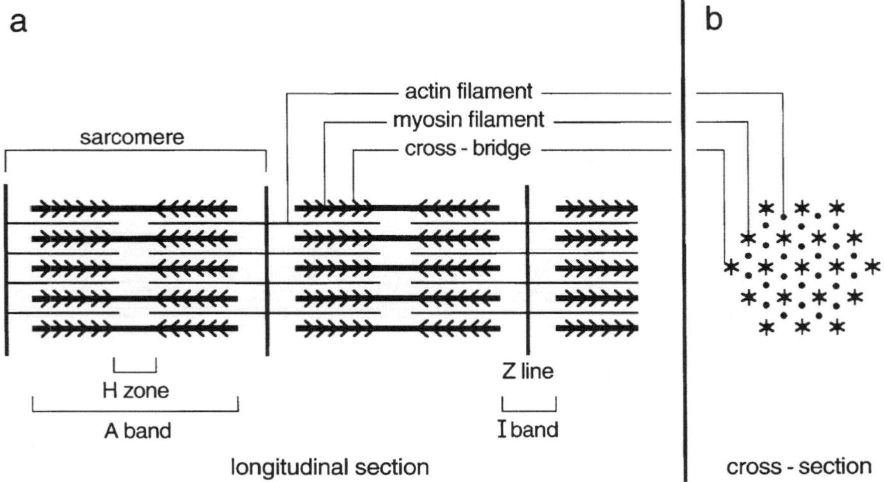

Figure 2.8 Longitudinal and cross-sections of a myofibril consisting of actin and myosin filaments.

in a regular hexagonal formation. The myosin filaments possess head-like elements which point in the direction of the thin filaments. During contraction of the muscle so-called cross-bridges are formed between the myosin filament and the neighbouring six actin filaments.

The regularly arranged and interdigitating sets of filaments represent the structural basis for muscle contraction. The decisive process in the production of muscular force and the shortening of a muscle is the formation of the cross-bridges. To facilitate understanding of the processes involved here, the following sections describe the molecular structure of the filaments, the structural changes accompanying muscle contraction, and the physiological conditions under which activation of cross-bridges occurs.

2.1.3.3 Molecular Structure of Myofibrils

The thick filaments are formed from proteins known as myosin and the thinner filaments mainly comprise the protein actin. The two proteins, actin and myosin, are essentially responsible for the ability of a muscle to contract. They are therefore also known as contractile proteins. As is shown in Figure 2.9, myosin and actin molecules combine in a characteristic manner to create the filaments. A single myosin molecule is represented schematically in Figure 2.9(a). It comprises thin, elongated parts, the 'tail' and the 'neck', and a globular 'head region'. If a number of myosin molecules are introduced into an aqueous solution, they aggregate to form a molecular bundle. During this process the tails of the individual molecules arrange themselves so that they lie parallel to each

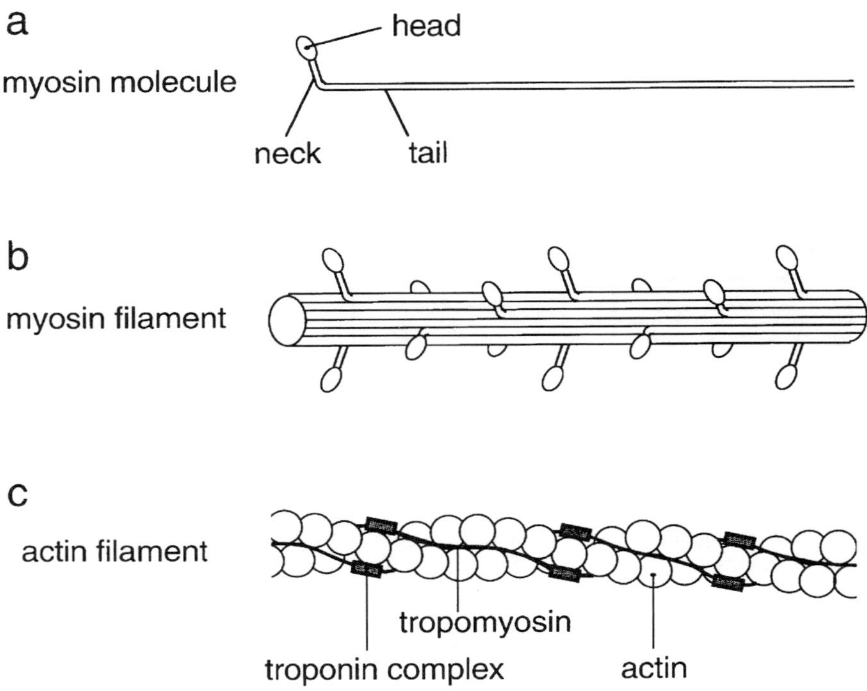

Figure 2.9 Molecular structure of myosin and actin filaments.

other. The necks and the heads then protrude out of the sides of the bundle. A very similar arrangement is to be found in the myosin filaments of the muscle. A section of a myosin filament is shown in Figure 2.9(b). The long tails of approximately 200–300 myosin molecules have arranged themselves in parallel to form a bundle. The heads protrude out of the side of the bundle. The diagram indicates the regular arrangement of the heads on the filament. In each case two heads face each other around the circumference of the filament. Each of the 'pairs of heads' is set at an angle of approximately 60° in relation to the preceding one. The heads of the myosin filaments form the cross-bridges to the actin filaments referred to in Figure 2.8.

The thin filaments mainly consist of approximately 400 globular actin molecules. As shown in Figure 2.9(c), actin molecules combine with elongated tropomyosin molecules to form a two-stranded helix. A third protein, the troponin complex, accumulates at certain intervals along this double helix.

2.1.3.4 Sliding Filament Theory

In the 1950s the evaluation of findings from light and electron-microscopic investigations led to an interpretation of the processes involved in muscle contraction which is still valid today, the sliding filament theory. It was observed that with the increasing contraction of a muscle the length of the sarcomeres, i.e. the distance between two Z lines, continued to shorten but that, at the same time, the length of the thick and thin filaments remained unchanged. From these observations it was concluded that the sets of filaments slide past each other during contraction. The principle involved in this sliding of filaments is presented in Figure 2.10. During the contraction of a muscle the thick myosin-filament group and the thin actin-filament group move towards each other. As a result, the distance between the Z lines shortens and the I band and the H zone narrow. By contrast, the A band, whose width is determined by the length of the thick filaments, remains unchanged. During the elongation of a muscle the opposite changes in the optical appearance occur: the length of the sarcomeres increases as does the width of the H zone and the I band whereas the width of the A band again remains constant.

2.1.3.5 Molecular Mechanisms of Contraction

In the sliding filament theory it is proposed that the sliding of the thin filaments towards the center of the thick filaments results from active forces developed between the actin and myosin filaments. The force arises from the repeated attaching and detaching of the myosin heads to binding sites on the actin filament. The principle involved here is to be explained with reference to Figure 2.11.

Figure 2.11 shows parts of two myosin filaments, each with a protruding myosin head and faced by an actin filament. When the muscle contracts a

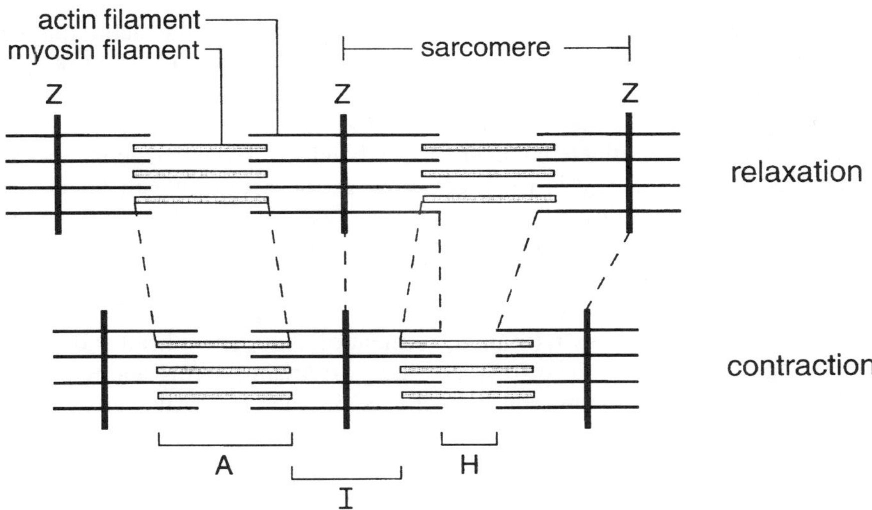

Figure 2.10 Principle of the shortening of sarcomeres due to the sliding of the filaments.

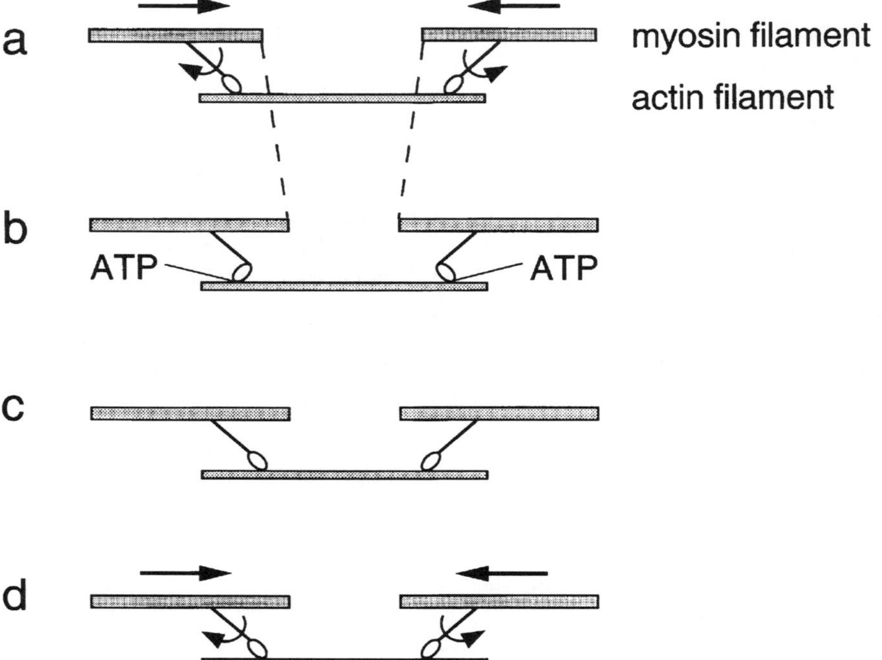

Figure 2.11 Sequence of events in the attachment of cross-bridges between the myosin and actin filaments during the sliding of the filaments.

chemical bond is created between the myosin heads and the actin filament (Figure 2.11a). This process involves tilting of the myosin heads (Figure 2.11a and b). The rotational movement of the myosin heads produces a force which enables the actin and myosin filaments to move towards each other. After the movement has been effected the chemical bond between the actin and the myosin is broken. Energy is required to break the actin–myosin bond. It is derived from the splitting of the energy-rich adenosine triphosphate (ATP). After the actin–myosin complex has been broken the myosin head swings back into its original position (Figure 2.11c) and the molecules are available for a further cross-bridge cycle (Figure 2.11d). The bonding of actin and myosin is once again accompanied by a tilting movement of the myosin head and further shortening of the muscle. The force and the shift arising during a single cross-bridge cycle is small. However, the effects of the successive cross-bridge cycles overlap each other within a sarcomere. In turn, the contractions of a large number of sarcomeres connected in parallel and in series add up within a muscle. The cumulative effect therefore enables large forces to be produced. The mechanical energy produced by the muscle during the contraction originates from the chemical energy released during the splitting of ATP.

The contraction state of a muscle is controlled by the concentration of Ca^{2+} ions in the area surrounding the actin and myosin filaments. At a low Ca^{2+} concentration ($<10^{-7}$ mol l^{-1}) the muscle is relaxed. Only when the Ca^{2+} concentration is increased by a factor of approximately 100 to more than 10^{-5} mol l^{-1} is it possible for the cross-bridge cycle described in the above to proceed. Reference to the schematic representation of the actin filament in Figure 2.9 is recommended to facilitate understanding of the dependence of the muscle's contractility on the Ca^{2+} concentration. The actin filaments are shown to consist of globular actin molecules which, together with thread-like tropomyosin molecules, form a double helix. Troponin complexes occur at regular intervals. An actin filament is shown in cross-section in Figure 2.12. In the left-hand section (Figure 2.12a) the configuration of the actin and troponin molecules is provided for an absence of Ca^{2+}. In this state the tropomyosin threads are wrapped around the actin molecules in such a way that the potential binding sites for the myosin heads are covered. Attachment of the myosin heads to the actin is then not possible. The muscle is in a relaxed state. If the Ca^{2+} concentration is increased in the area surrounding the actin filaments, Ca^{2+} is bound to the troponin. Due to this binding process the troponin complex undergoes a change in conformation and causes the tropomyosin thread to shift position (Figure 2.12b). As a result, the binding sites on the actin are exposed and the actin–myosin interaction, which leads to the contraction of the muscle, is enabled.

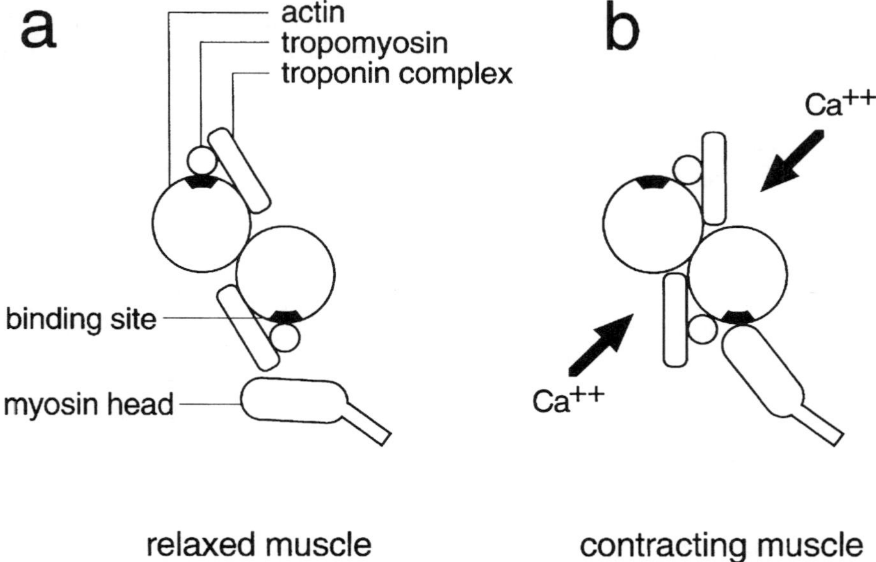

Figure 2.12 Cross-section of the actin filament and the influence of calcium on the actin–myosin interaction.

2.1.3.6 *Excitation–Contraction Coupling*

The previous sections focused on both the origin of the excitation of the muscle fiber membrane and the molecular mechanism of muscle contraction. The following section sets out to explain how the connection between the electrical phenomenon 'excitation' and the mechanical phenomenon 'contraction' is established. This connection is termed excitation–contraction coupling. It results from the release of Ca^{2+} ions from intracellular depots by the membrane action potentials.

The process is to be described with reference to the longitudinal section through a muscle fiber provided in Figure 2.13. The figure shows that invaginations, called transverse tubules (t-tubules), are located at regular intervals along the muscle fiber membrane close to the Z lines. Further spaces, forming a close network around each fibril, are each located between two transverse tubules in the longitudinal direction of the muscle. This network of longitudinal tubules is called the sarcoplasmic reticulum. The membrane of the sarcoplasmic reticulum is able to transport Ca^{2+} ions actively into the inner space of the tubules. The sequestering activity of the sarcoplasmic reticulum ensures that the intracellular concentration of Ca^{2+} in the resting muscle is kept below 10^{-7} mol l^{-1}. At this low calcium concentration no Ca^{2+} is bound to the troponin and an actin–myosin interaction is inhibited. In the relaxed state of the muscle Ca^{2+} is stored in the interior of the longitudinal tubules.

In order to trigger a contraction calcium must be released from the tubuli. This is effected by the action potential. The close contact between the terminal cisternae of the sarcoplasmic reticulum and the t-tubules is indicated in Figure 2.13. If an action potential spreads over the muscle fiber membrane it is carried

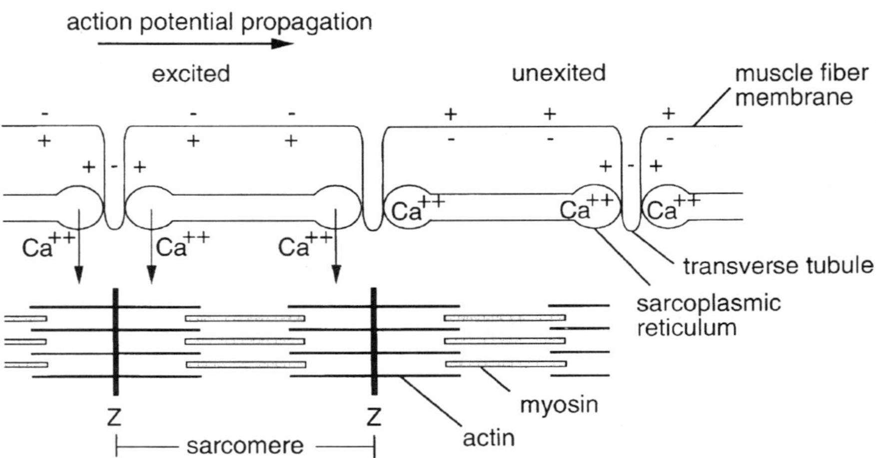

Figure 2.13 Electromechanical coupling: Ca^{2+} ions are released from the sarcoplasmic reticulum under the influence of the membrane excitation and diffuse to the sarcomere.

deep into the muscle fibers down the t-tubules. As a result of a mechanism which has not been fully explained yet, the change in the electrical field at the contact point between the transverse tubules and the longitudinal tubules leads to the release of calcium from the sarcoplasmic reticulum. Ca^{2+} diffuses along its concentration gradient to the binding sites on the troponin complex, thereby triggering the actin–myosin interaction. The calcium concentration in the intracellular space of the muscle is increased for as long as action potentials pass along the muscle fiber membrane. As soon as the excitation stops, the calcium concentration in the muscle is sufficiently lowered again as a result of the active transport of Ca^{2+} back into the sarcoplasmic reticulum. The muscle then relaxes back to the resting state.

2.1.3.7 Excitation and Contraction Summarized

The events controlling the contraction of the cross-striated muscle are as follows:

1. A nerve action potential arriving at the neuromuscular synapse releases acetylcholine and triggers a muscle action potential.
2. The muscle action potential spreads over the total muscle fiber membrane.
3. The action potential is carried deep into the muscle fiber along the transverse tubules.
4. Calcium is released from the sarcoplasmic reticulum and diffuses into the myofibrils.
5. The binding of calcium to troponin results in a conformational change in the troponin and causes a shift in the position of the tropomyosin molecules.
6. As a result, the binding sites on the actin filaments are exposed and cross-bridges are formed between the actin and the myosin filaments.
7. The formation of the cross-bridges causes a tilting of the myosin heads. A longitudinal pull force between the actin and myosin filaments is generated in the process.
8. The pull force causes the actin and myosin filaments to slide past each other. The muscle contracts.
9. The actin–myosin bond is broken with the splitting of ATP. The myosin head is free once more for a new attachment to the actin filament, again generating a pull force between the actin and myosin filaments.
10. The Ca^{2+} concentration in the myofibrils remains increased as long as action potentials are triggered on the muscle fiber. Cross-bridge cycles are thereby generated continuously. The muscle remains contracted.
11. If the muscle fiber membrane is no longer excited, the Ca^{2+} level in the myofibrils is lowered through active calcium uptake by the sarcoplasmic reticulum. The calcium bound to the troponin is removed, the conformational change in the troponin ceases and the tropomyosin returns to its original position. Consequently, further cross-bridge attachments are prevented and the muscle relaxes.

2.1.4 Fundamentals of Occupational Electromyography

The aim of electromyographical investigations is to study the electrical excitation processes in the muscles during real or simulated work. The preferred approach is to derive EMGs through the application of non-invasive methods. In most investigations the electrodes are attached to the surface of the skin. This technique – surface electromyography – has some inherent limitations: it is only possible to derive EMGs from the layer of muscle nearest the surface of the skin. If muscles overlap, a signal is either not derivable from the lower muscle at all or there is considerable cross-talk from the muscles closer to the electrodes. In such cases muscle activity can be recorded by inserting needle or wire electrodes into the deeper-lying muscles. Such invasive measurement methods are employed in particular in clinical medicine in the diagnosis of nervous and/or muscular diseases. However, this methodology is less appropriate for ergonomic studies.

2.1.4.1 Recording the Electromyograms

A typical action potential of an excitable membrane is shown in Figure 2.4. The time curve and the potential values provided in this figure are obtained if the potential difference between the interior and exterior of the membrane is recorded. It is only possible to measure the potential difference across the membrane if one of the electrodes is located in the interior of the muscle However, such intracellular measurement methods cannot be applied in humans on account of the muscle cell diameters. These are in the range 20–200 μm (cf. Figure 2.7). Consequently, the tip of electrodes suitable for intracellular derivation would have to be only a few micrometers in diameter. Such thin electrodes cannot be made from metal. They usually consist of a glass capillary filled with a conductive liquid. If such a glass electrode was introduced into a muscle it would break with the first small movements. The use of glass electrodes is therefore limited to isolated muscle preparations with severely restricted contractability. Humans and intact animals are definitely excluded.

Surface electrodes with a diameter in the order of up to 1 cm or needle or wire electrodes measuring up to a few millimeters are usually employed in the derivation of electromyograms from humans. This means that the electrodes are considerably larger than the diameter of the muscle cells. Electrodes of this size cannot therefore be introduced into the intracellular space. They can merely be used to record potential differences in the extracellular space of the muscle or on the surface of the skin.

The principle of EMG derivation using surface electrodes is shown in Figure 2.14. For purposes of simplification and clarification, the muscle is only represented by one muscle fiber (shown considerably enlarged). In the region of the front of excitation a current flows between the unexcited and the excited area of the muscle fiber. The current is indicated by arrows in Figure 2.14

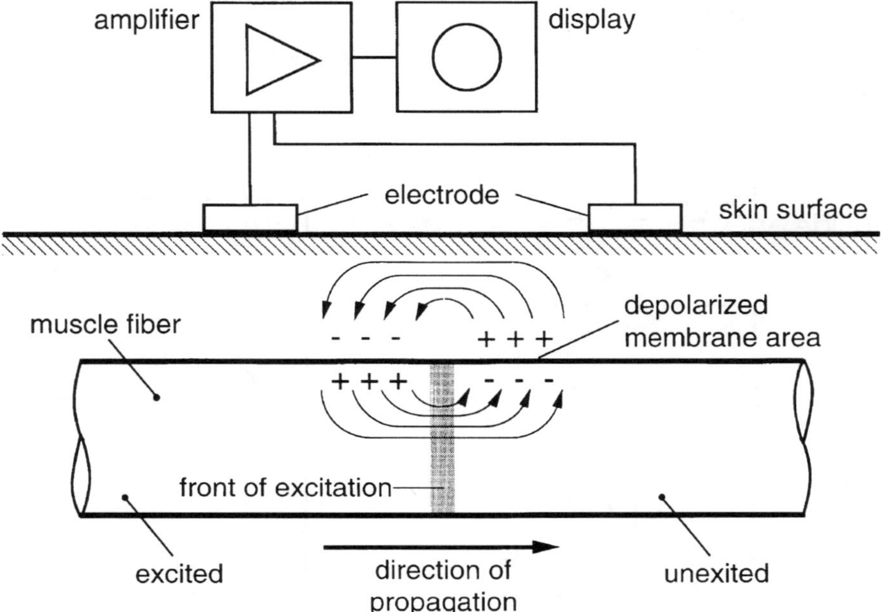

Figure 2.14 Principle of extracellular action potential recording using surface electrodes.

(local current circuit, cf. Figure 2.5). The currents induce local potential differences, which can be measured using two electrodes, within the extracellular space. The potential differences are causally related to the action potential of the muscle fiber. They can therefore be regarded as an image of the muscular excitation. Either invasive electrodes inserted into the extracellular space or – as shown in Figure 2.14 – surface electrodes applied to the surface of the skin can be employed in the measurement of the potential difference.

The time curves of the electrical potential derived using extracellular electrodes is considerably different from the curve of the intracellularly recorded transmembrane potential shown in Figure 2.4. The development of the signal curve typical for extracellular derivation is to be explained on the basis of Figure 2.15. A section of a muscle fiber is presented in all parts of the figure. Two electrodes are located near the muscle fiber membrane. The polarity distribution across the membrane is indicated for certain points in time during the propagation of an excitation along the muscle fiber. The time curve for the potential difference measured between the electrodes is shown in the right-hand section. In the unexcited state (Figure 2.15a) the potential difference between the electrodes is zero. As soon as the excitation reaches the left electrode (Figure 2.15b) a potential difference is measured. When the excitation reaches the

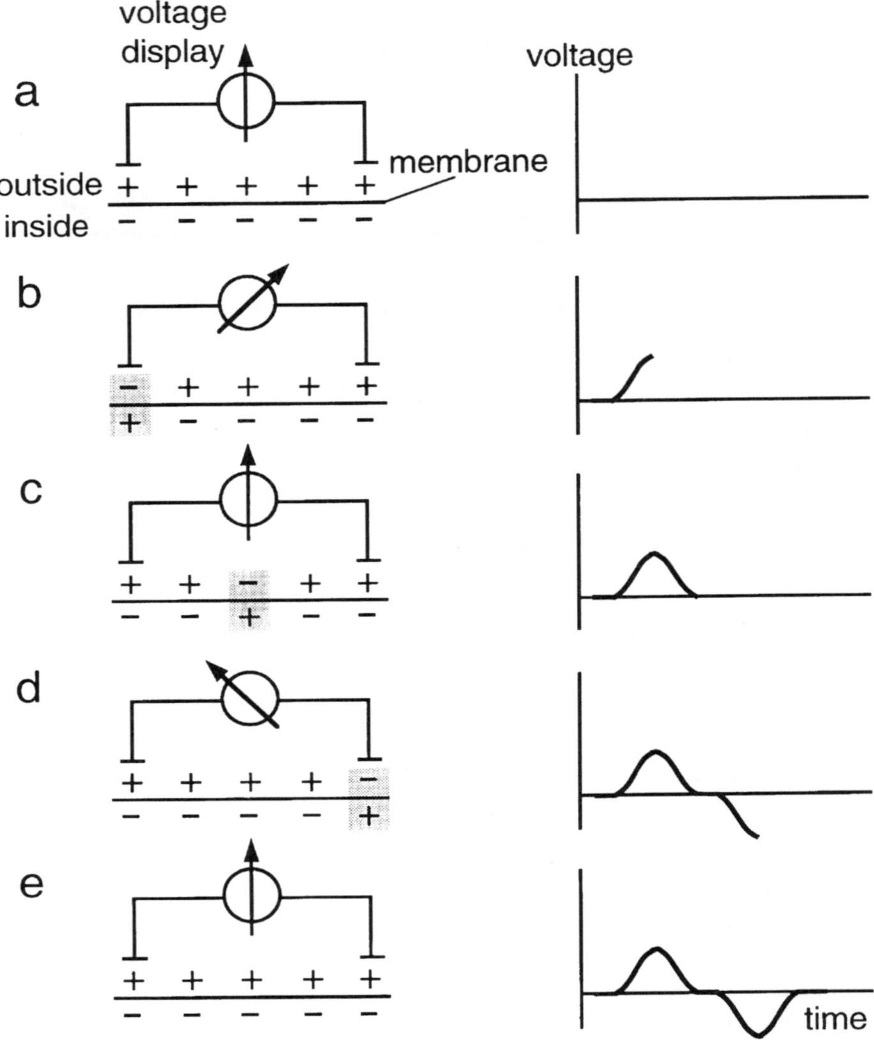

Figure 2.15 Sequence of events during the formation of a biphasic action potential. The potential measurement is performed using extracellular electrodes (adapted from Laurig, 1977).

membrane region in the middle of the electrodes (Figure 2.15c), the voltage display indicates zero once more. As the excitation progresses further the polarity of the voltage is reversed (Figure 2.15d) until the membrane is in a resting state again and the same potential exists at both of the electrodes (Figure 2.15e). The typical signal curve recorded using extracellular electrodes is therefore characterized by a reversal in the polarity. Such a time curve is described as a biphasic action potential, whereas the signal curve obtained using an intracellular electrode is called monophasic (cf. Figure 2.4).

Although the time curve of the voltage across the membrane during excitation is the same for all muscle fibers, the action potentials measured with extracellular electrodes can differ widely. The reason can be deduced from Figure 2.14: the potential difference between two points on the skin surface or in the extracellular depends on the distance between the electrodes, the distance of the electrodes from the muscle fiber, and on the direction of the fiber relative to the electrodes.

In real EMG recordings action potentials of many muscle fibers are superposed. Excitation occurs simultaneously in all of the muscle fibers

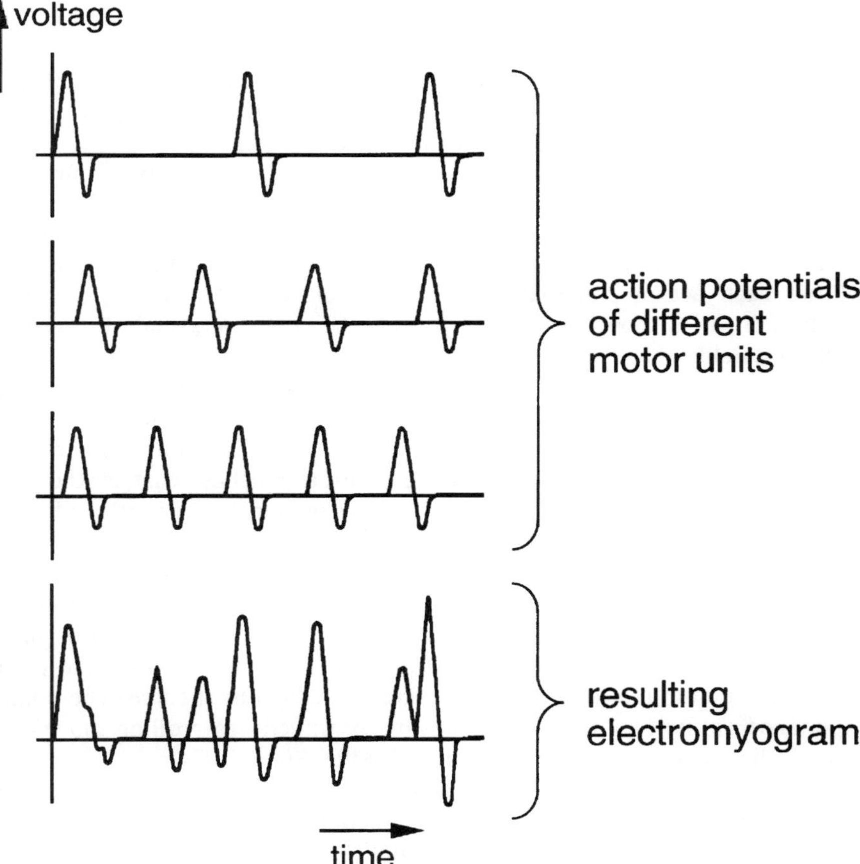

Figure 2.16 Formation of the electromyogram by superposition of the biphasic action potentials originating from different motor units (adapted from Laurig, 1977).

belonging to a single motor unit. According to the definition of a motor unit, these fibers are connected to the same nerve fiber. An action potential on this nerve fiber therefore triggers a muscle action potential in all of the muscle fibers simultaneously. The various motor units of a muscle are, however, excited at different times. The activity of three motor units is shown in the upper traces in Figure 2.16. The amplitude of the biphasic potentials and the 'firing rate' of the motor units are assumed to be different. The superposition of the three motor unit action potentials results in an unsteady signal which is called an interference pattern (Laurig, 1977). This unsteady signal represents the EMG of a muscle containing three active motor units. In the EMG the activity pattern of the individual motor units is generally no longer recognizable. Only in the case of very low muscular activation may it be possible to identify the effect of individual motor units.

2.1.4.2 Gradation of Muscular Force

The contraction force of a muscle can be varied within wide limits and thereby adapted to particular requirements. This adaptation is controlled by two different mechanisms, known as summation and recruitment.

The principle of summation is to be described on the basis of Figure 2.17: each excitation of a motor unit – represented by a schematic action potential – elicits a contraction of short duration. The temporal relationship between both of the processes is recognizable in Figure 2.17(a). With a time lag a twitch follows each action potential. The muscle relaxes between the individual twitches. If, as is shown in Figure 2.17(b), the action potentials follow each other so quickly that the previous contraction has not ended before the next action potential arrives, the twitches overlap. The resulting contraction force is then greater than during individual contractions. If the action potentials occur in even more rapid succession (Figure 2.17c), the individual contractions merge to an even greater extent and the muscle reaches what is known as tetanus. In this state the total contraction force can increase to 5–10-fold of the twitch force (Peiper, 1994).

In summation the muscular force is controlled by the firing rate, i.e. the frequency with which the motor units are activated (rate coding). In the second type of gradation of the muscular force – recruitment – the number of motor units which are excited varies according to the force requirement, the partial forces of different motor units overlapping to create a total force. The relative importance of the two neural mechanisms for directly controlling movements – rate coding and recruitment – was described by Stein (1974).

With regard to the EMG, both summation and recruitment have the same effect. In both cases a greater number of action potentials are produced in a muscle per unit of time. The EMG amplitude therefore increases with an increase in the force. Accordingly, the EMG amplitude is closely related to the level of force. In order to demonstrate this relationship, Figure 2.18 (upper trace) presents the result of a simple laboratory experiment in which the EMG

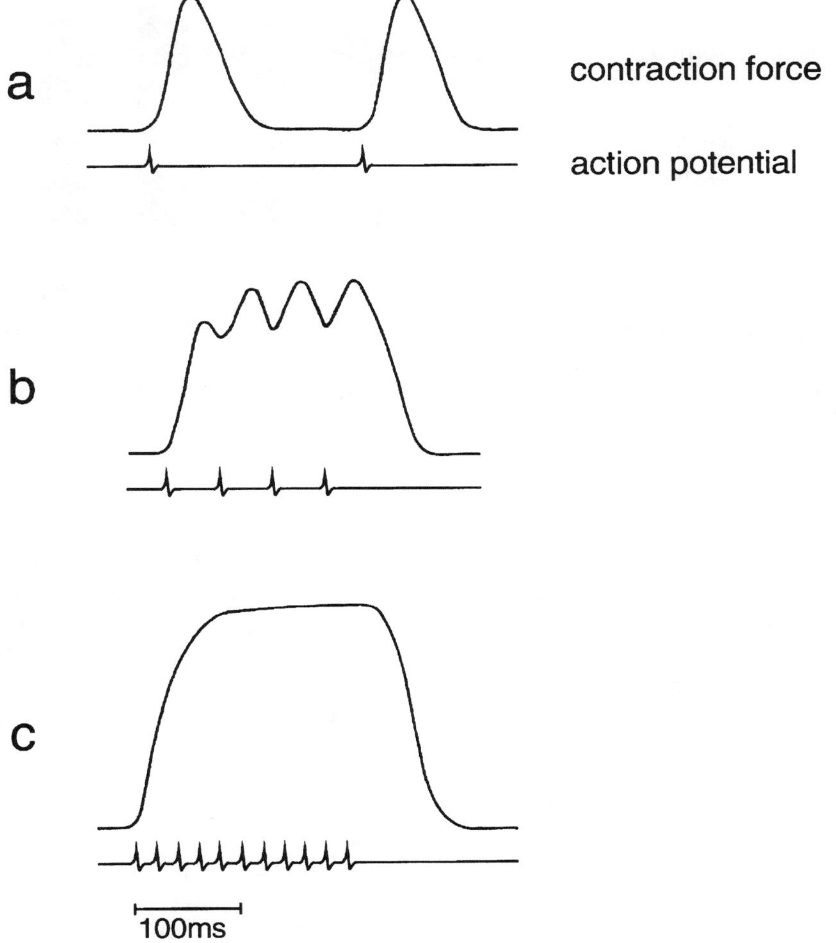

Figure 2.17 Summation of muscle fiber twitches for different firing rates of the muscle fiber.

is provided for the right m. biceps brachii during the holding of various weights with an elevated forearm. The EMG amplitude increases with increasing weights.

A frequently used procedure for processing the EMG further in order to achieve data reduction is shown in the lower traces. The EMG is rectified (middle trace) and then integrated via a time window of adjustable length (here 400 ms). The resulting Electrical Activity (EA) essentially represents the envelope curve of the EMG (lower trace). Another method frequently employed in the reduction of EMG data is the calculation of the root mean square (RMS) within predetermined time sections.

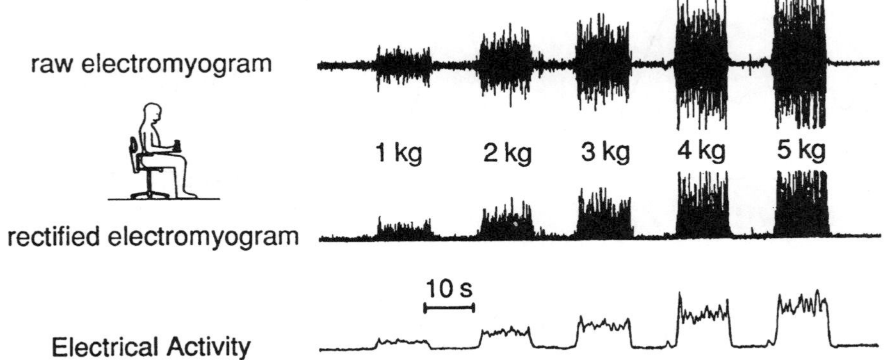

Figure 2.18 Electromyogram and Electrical Activity of the m. biceps brachii during the holding of loads of different weights (adapted from Luttmann *et al.*, 1992a).

2.1.4.3 Fatigue-induced Changes in EMG

A muscle's ability to produce muscular force decreases with sustained or frequently repeated contractions. This process is termed fatigue. Only two typical fatigue-induced EMG changes are to be shown here and a discussion of the possible physiological interpretations takes place against the background of the fundamental muscle functions provided in the initial sections. A further description and quantification of the changes occurring in EMG during fatigue are provided in Chapter 6 of this volume by Hägg and Kadefors.

EMG Changes in the Time Domain An EMG derived from the m. biceps brachii during static loading is shown in Figure 2.19. The test person holds a 5 kg weight in the manner shown in the diagram for as long as possible. The recordings reveal that the EMG and the EA of the biceps brachii increase during this activity even though the muscular force required to hold the load remains constant. Such an increase in the electromyogram during contraction was described as early as 1923 by Cobb and Forbes. Since then it has been determined and evaluated for many muscles under widely differing conditions, both in the laboratory (for a review see Basmajian and De Luca 1985) and in occupational field studies at a variety of real work-places (e.g. Christensen, 1986; Luttmann *et al.*, 1988, 1989, 1996a, 1996b; Luttmann and Jäger, 1992).

When interpreting the EMG increase in the course of the fatiguing process, discussion centers essentially on two hypotheses.

1 For reasons still to be discussed in a later section, the increase in the EMG is due to the fact that there is an increase in the low-frequency components and a decrease in the high-frequency components in the motor unit action potentials. Since the muscle tissue acts like a spatial low-pass filter, low-frequency components are attenuated less than high-frequency components. The spectral shift in the motor unit action potentials is expressed in the

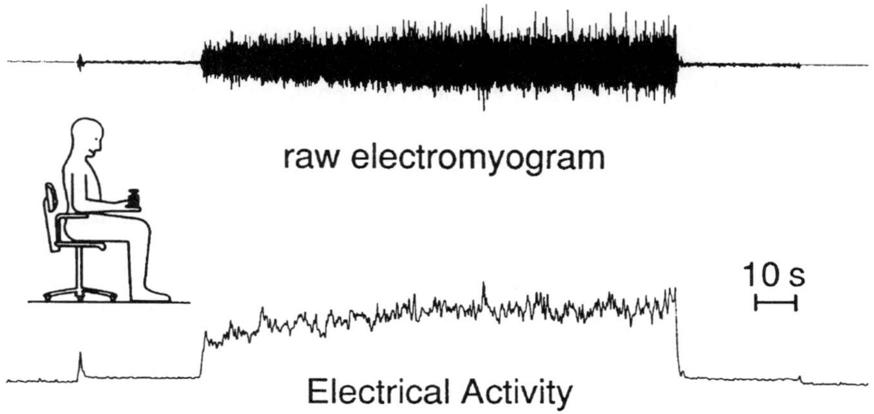

Figure 2.19 Electromyogram and Electrical Activity of the m. biceps brachii during the holding of a load of 5 kg for as long as possible (adapted from Luttmann *et al.*, 1992a).

EMG as an increase in amplitude. (De Luca, 1979; Basmajian and De Luca, 1985).

2 The increase in the EMG amplitude indicates a rise in the number of action potentials produced per time unit in the course of a sustained contraction. The additional activation of the muscle fibers is necessary because, as a result of fatigue, either the force produced per action potential drops or many action potentials remain ineffective and do not induce a twitch in the muscle fiber. In order to maintain a constant force either the action potential firing rate is increased or additional motor units are recruited (Edwards and Lippold, 1956; Eason, 1960; Vredenbregt and Rau, 1973; Laurig and Sauernheimer, 1975; Maton 1981).

The biochemical changes which may be responsible for a reduction in the force-generation capacity during muscular fatigue have been discussed for many years in the literature (cf. review by Vøllestad and Sejersted, 1988). Possible causes of fatigue cited in the literature include the depletion of the muscle's energy supplies and the accumulation of metabolic products such as lactate and hydrogen ions. However, although they may be of significance for the development of fatigue when large forces are generated, measurements show that no major depletion of ATP, the mediator of energy conversion, was established even at the end of a fatiguing activity (cf. Sjøgaard, 1990). In addition, an increase in the EMG for constant loading of the muscle can also be determined for forces which are far below the maximum force and for which the energy supply of the muscle should suffice (Jørgensen *et al.*, 1988). Two possible reasons for the force failure are discussed: a disturbance in the spread of the action potential across the muscle fiber membrane and a reduction in the electromechanical coupling.

With regard to the first reason, each action potential is associated with an outflow of K^+ ions and an inflow of Na^+ ions and, consequently, with a change in the ion concentrations in the intra- and extracellular space. The changes in the ion concentrations effected by a single action potential are very small. In addition, at a low firing rate the restoration of the ionic gradients across the membrane is ensured at all times by the active transport of Na^+ and K^+. However, a long sequence of action potentials occurring in quick succession may lead to a clear change in the intra- and extracellular cation concentrations on account of the insufficient capacity of the ion pumps (for a review of the literature see Vøllestad and Sejersted, 1988). As regards to potassium, it is to be assumed that the increase in the extracellular K^+ concentration is particularly high within the t-tubules as the latter only contain a small amount of fluid on account of their narrow lumen (cf. Figure 2.13) and the exchange rate through the openings is restricted (Almers, 1980). In addition, the membrane of the t-tubules only contains a small number of Na^+/K^+-ion pumps (Venosa and Horowicz, 1981).

The reduction in the ional gradients across the muscle fiber membrane leads to a local depolarization and, possibly, to a conduction block. Direct intracellularly recorded membrane potentials from isolated frog muscles have confirmed that the resting potential is reduced from approximately −90 mV to approximately −40 mV after a fatiguing electrical stimulation (Lännergren and Westerblad, 1986; Westerblad and Lännergren, 1986). An impairment of the action potential propagation, particularly in the area of the transverse tubules, has been suggested as a possible cause of fatigue by several authors (Bigland-Ritchie *et al.*, 1979; Jones, 1981). The consequence of complete or partial blocking of the action potentials at the transverse tubules is that some of the action potentials do not result in the release of Ca^{2+} from the sarcoplasmic reticulum. These action potentials do not therefore induce a muscle twitch. If, however, a certain force is required, as for example during the holding of a weight, additional muscle fibers must be activated. The required force can only be produced by means of an increased number of action potentials. An increase in the amplitude of the EMG signal is then detected (see Figure 2.19).

With regard to the second reason, the changes in the ional composition of the intra- and extracellular spaces during sustained contractions result in a reduction in both the resting potential and in the peak action potential and, consequently, in a reduction in the membrane voltage pulses occurring during excitation (Lännergren and Westerblad, 1986; Westerblad and Lännergren, 1986). Studies by Vergara *et al.* (1978) have shown that the amount of Ca^{2+} released from the sarcoplasmic reticulum is highly dependent on the pulse amplitude. Since the muscle contraction is controlled via the release of Ca^{2+} ions, a reduction in the action potential amplitude may lead to a reduction in electromechanical coupling and a diminishment of the force response. This is confirmed by several studies: after fatiguing tetanic stimulations in single frog muscle cell preparations a regular sliding of the filaments was only observed in the peripheral layers of myofibrils, a largely irregular myofibrillar pattern occurring in the central part (Edman and Lou, 1992). The application of fluorescent microscopy in the direct measurement of the intracellular calcium distribution in single frog muscle cells revealed a non-uniform calcium distribution, with greater concentration levels near the outer parts than near the center of a fiber, after fatiguing contractions (Westerblad *et al.*, 1990). A failure in the inward spread of the action potentials along the t-tubules, leading to a diminished electromechanical coupling and, consequently, to a force failure was concluded from both findings. With progressive fatigue more action potentials are required to produce a constant force, this resulting in an increase in the EMG amplitude (see Figure 2.19).

The occurrence of a muscle tremor (Sälzer, 1973) and a possible increase in temperature as a result of the biochemical processes of energy conversion (Petrofsky and Lind, 1980; Schneider *et al.*, 1988) are further possible reasons for an increase in the EMG amplitude during fatiguing contraction.

EMG Changes in the Frequency Domain A typical EMG change in the frequency domain is presented in Figure 2.20. It shows the frequency spectrum of an EMG of the m. erector spinae at the beginning and the end of holding a load in a bent-forward position. After the fatiguing contraction the power spectral density is increased in the low-frequency range and decreased at higher frequencies.

An increase in low-frequency components in the EMG during fatiguing voluntary contractions was described as early as the beginning of the century by Piper (1909, 1912). The first quantitative evaluations of the fatigue-related spectral shift to the left were undertaken by Kogi and Hakamada (1962). Since

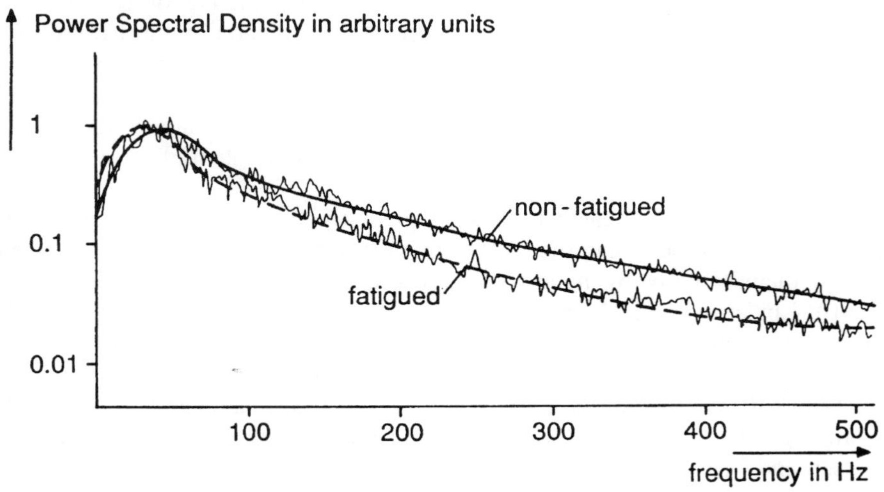

Figure 2.20 EMG spectra of the m. erector spinae at the beginning and the end of a sustained contraction in a bent-forward position (adapted from Jäger *et al.*, 1984).

then the fatigue response in the frequency domain have been investigated in a large number of studies (for a review see Basmajian and De Luca, 1985). Apart from use in basic research investigations into muscle physiology, the spectral shift has been included in ergonomic studies, in particular as evidence of muscle fatigue (Örtengren *et al.*, 1975; Hägg *et al.*, 1987; Hägg, 1991; Jørgensen *et al.*, 1991; Öberg *et al.*, 1992).

For many years there has been intense discussion about the physiological processes responsible for the shift to the left in the EMG spectrum during fatigue. Two main interpretations have emerged: synchronization or grouping of motor unit firing and reduction in the action potential propagation velocity along the muscle fibers.

With regard to the first reason, in the interpretation of the spectral shift in terms of synchronization or grouping it is assumed that the sequence of the motor unit discharges is more irregular in the non-fatigued state and that the tendency to discharge at the same or almost the same time increases during the fatiguing process. This results in a decrease in the high-frequency components and an increase in the power at low frequencies (Kogi and Hakamada, 1962; Chaffin, 1973; Sakamoto *et al.*, 1982). The increase in a muscle tremor during fatigue is also associated with muscle fiber activity which is to a large extent synchronous and associated with an increase in low-frequency components (Laurig, 1969).

With regard to the second reason, the influence of the conduction velocity on the time course of the biphasic action potential and on the power spectrum is to be explained with reference to Figure 2.21. In the case of high conduction velocity (upper part of Figure 2.21) the front of excitation covers the distance between the two extracellular electrodes within a short space of time. At a low conduction velocity (lower part of Figure 2.21) more time is required to cover the same distance. Accordingly, the biphasic action potential is short for a high conduction velocity and long for low velocity. In the spectrum the high-frequency components are emphasized at high velocity and at low velocity the low-frequency components are pronounced. The changes shown here for a single muscle fiber apply equally to a mixture of action potentials from various fibers.

The relationship, indicated here only qualitatively, between a reduction in the conduction velocity and the shift to the left in the power spectrum has been studied quantitatively by means of model calculations by Lindström *et al.* (1970), Lindström and Magnusson (1977), and Stulen and De Luca (1981). It has also been demonstrated in a large number of experimental studies that, in the course of sustained contractions, the conduction velocity decreases and the spectrum shifts to the left (e.g. Naeije and Zorn, 1982; Sadoyama *et al.*, 1983; Broman *et al.*, 1985; Mucke and Zöllner, 1986; Zwarts *et al.*, 1987; Arendt-Nielsen and Mills, 1988). However, the question of whether the reduction in the conduction velocity can be regarded as the only reason for the left-shift in the spectrum is still unresolved (Krogh-Lund, 1993; Krogh-Lund and Jørgensen 1993).

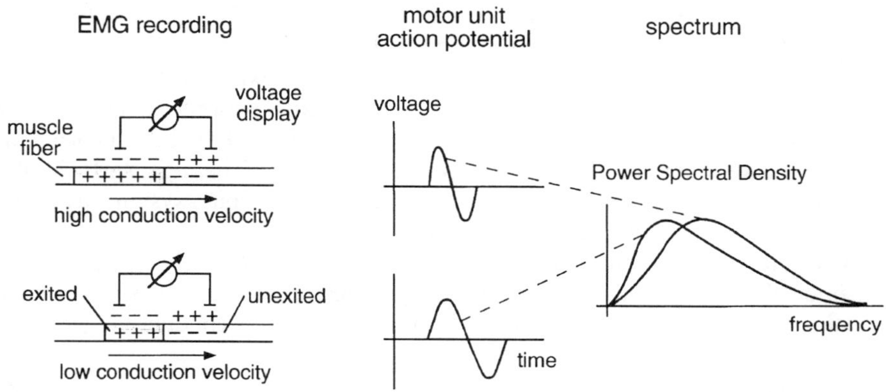

Figure 2.21 Influence of the action potential conduction velocity on the time curve and the spectrum of the biphasic action potential. Typical changes have been shown larger than in reality for reasons of clarity.

Initially, the accumulation of lactic acid in the muscles was considered to be essentially responsible for the reduction in the propagation velocity during fatigue (Mortimer *et al.*, 1970). However, in more recent studies (for a review see Vøllestad and Sejersted, 1988; Sjøgaard, 1990) preference is given to the hypothesis that the slowing of the conduction velocity is caused by the changes in the concentrations of the cations, in particular K^+ accumulation in the extracellular space. This hypothesis is confirmed by *in vitro* studies on isolated rat muscles exposed to different potassium concentrations (Kössler *et al.*, 1990).

The causes of and the evidence for a K^+ accumulation in the interstitial space during fatigue have already been explained in the preceding section. It is to be examined in the following how the extracellular K^+ accumulation can affect the propagation velocity: considering the mechanism of the propagation of excitation (cf. Figure 2.5), propagation of the action potential essentially depends on adjacent points along the membrane being depolarized to the threshold level by the local circuit current and new action potentials developing there. A certain delay is involved in this process since the circuit current first has to reverse the charge of the membrane capacity. The 'depolarizing time' which passes at each membrane element until the threshold is reached determines the conduction velocity. The depolarizing time is directly dependent on the intensity of the local circuit current which is, in turn, determined by the amplitude of the action potential and the steepness of the rising phase. As explained in the above, in the fatigued state both of these quantities are altered as a result of the change in the ion gradients. Therefore they cause lengthening of the depolarizing time and a reduction in the conduction velocity during fatigue.

The reduction in the conduction velocity is currently the most widely acknowledged interpretation of the fatigue-induced spectral change. However, in a series of studies Krogh-Lund and Jørgensen (1991, 1992, 1993) and Krogh-Lund (1993) have analyzed critically the validity of the synchronization and the conduction velocity hypotheses under various experimental conditions. The authors state that interpreting the spectral shift entirely in terms of a change in the conduction velocity is invalid or only partially valid, in particular for low- and moderate-level endurance contractions. The left shift in the spectrum was in part assumed to reflect central nervous system-mediated synchronization of motor unit firing as well.

When interpreting the spectral changes it is additionally necessary to consider that a rise in the muscle temperature possibly occurring during sustained contractions results in an increase in the conduction velocity

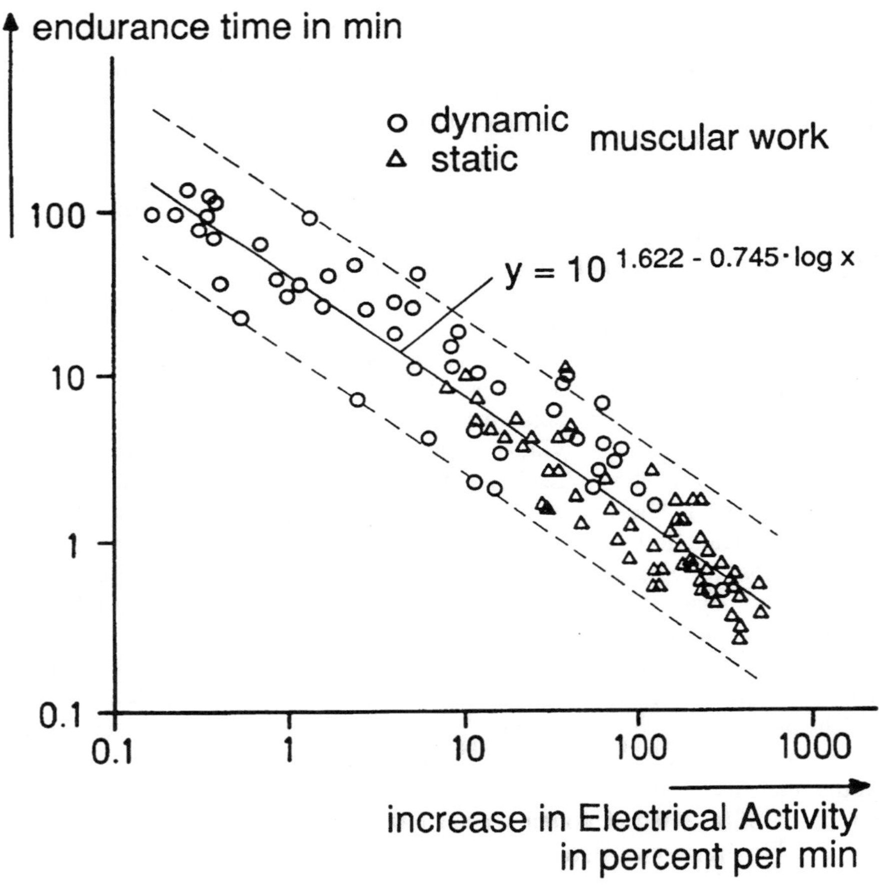

Figure 2.22 Relationship between the increase in Electrical Activity and the endurance time (extended version of a figure from Laurig (1974, 1975), including experimental findings by Knowlton *et al.* (1951), Scherrer and Bourguignon (1959), Rau and Vredenbregt (1970), Laurig (1974), Rhein *et al.* (1974) and Ballé (1976).

(Schneider *et al.*, 1988). Furthermore, the changes in the conduction velocity described in the above may be overlaid by the consecutive activation of different types of muscle fibers (fast twitch and slow twitch fibers) with differing conduction velocities.

Muscle Fatigue and Endurance Time An activity in which the working musculature is fatigued can only be performed for a limited time. After this time has elapsed, the musculature is exhausted and further contraction is not possible. The greater the generated force, the more quickly the state of exhaustion is reached (Rohmert, 1962). As shown in Figure 2.19, the development of fatigue is accompanied by an increase in the EMG and the EA. A close relationship exists between the steepness of the rise and the possible duration of a fatiguing activity: a flat increase in EA is connected with a long duration of activity whereas a steep rise is associated with a rapid fatiguing process and a short duration of activity. The relationship between the fatigue-induced increase in EA and the endurance time has been investigated quantitatively by Laurig (1974, 1975). Measurement results for a fairly large number of experiments are presented in Figure 2.22. The data are derived from various muscles for both static holding activities and dynamic load. The increase in EA is expressed as a percentage of the EA at the beginning of the fatiguing activity. In a double-logarithmic plot the measured values can be approximated by a straight line. The resultant regression function and the confidence interval are given in Figure 2.22.

2.2 CONCEPTS OF EMG FIELD STUDIES IN ERGONOMICS

2.2.1 Application of Electromyography in Field Studies

Electromyography is the adequate method for measuring the muscular activities of individual muscles or muscle groups in the work-place. In field studies in occupational medicine and ergonomics the non-invasive method of surface electromyography is almost exclusively employed since this measurement method allows the subjects to perform their activity almost unhindered.

The electromyograms are either recorded in a storage unit attached to the subject's body or they are relayed to a receiving station via a portable telemetric transmitter and stored there. The myoelectrical signals can be recorded in the form of the raw EMG or after data reduction, for example by means of averaging or integration. Raw EMGs are indispensable in field studies where the subjects move freely and the working conditions are not controlled by the investigator. The recording of averaged or integrated EMGs has the disadvantage that artefacts and disturbances produced by movement of the electrodes, power-line hum, or other electrically induced noise cannot be distinguished from the information signal. Recording the raw electromyogram does, however, require high storage capacity. Consequently, the use of telemetric transmitters and the recording of the data on stationary magnetic tape recorders are to be preferred to other methods when recording throughout whole work shifts.

Some applications of electromyography in ergonomic field studies are described in the following. The applications can be basically assigned to two areas: the determination of the level of muscle strain for various activities within a complex work sequence and the examination of whether an activity is associated with the occurrence of muscle fatigue. In both areas the aim is ultimately to reduce high muscular strain or fatigue through work design.

2.2.2 Determination of Muscular Strain

Quantitative comparison of the electromyographical recordings of various activities requires precise designation of activity sections. A method which has proven itself in practice involves producing an 'electronic protocol' which is recorded together with the EMGs on the same tape and included in computer-assisted evaluation of the data (Ballé *et al.*, 1982; Laurig *et al.*, 1987; Luttmann *et al.*, 1988, 1996a). In this method the subject is accompanied by an observer who produces an electrical code signal by means of a numerical keyboard. The activity code which is employed must always be adapted in advance to the conditions prevailing at the work-place as well as to the particular activity under investigation. The method allows important influencing factors such as the shape, size, or mass of objects moved by the person as well as the person's posture and movements at a given moment in time to be recorded and related to the physiological reactions.

An application of electromyography taken from a field study into surgeons in urology (Luttmann *et al.*, 1992b, 1996a) is presented in the following. It involved the examination of the muscular strain occurring during the performance of endoscopic operations using different operating methods. The measurements were taken during so-called transurethral resections. In this type of operation, tissue from the prostate or bladder is removed from the body using an endoscope. In classical 'direct endoscopy' a look-through endoscope is used. It is a rod-shaped optical instrument which allows the surgeon to look directly at the operating area. This requires that the eye is in direct contact with the aperture of the instrument. Steep inclination of the upper body and the head over long periods is frequently necessary during positioning and handling of the instrument. As a result, a high level of strain in the musculature of the shoulder–arm region is to be expected. The disadvantageous posture which the surgeon is forced to adopt can be avoided using a new method in which a video camera is mounted on the endoscope ('monitor endoscopy').

Short sections from an original recording of the electromyograms of four muscles during the application of both direct and monitor endoscopy are shown in Figure 2.23. The comparison reveals that the EMG amplitude is

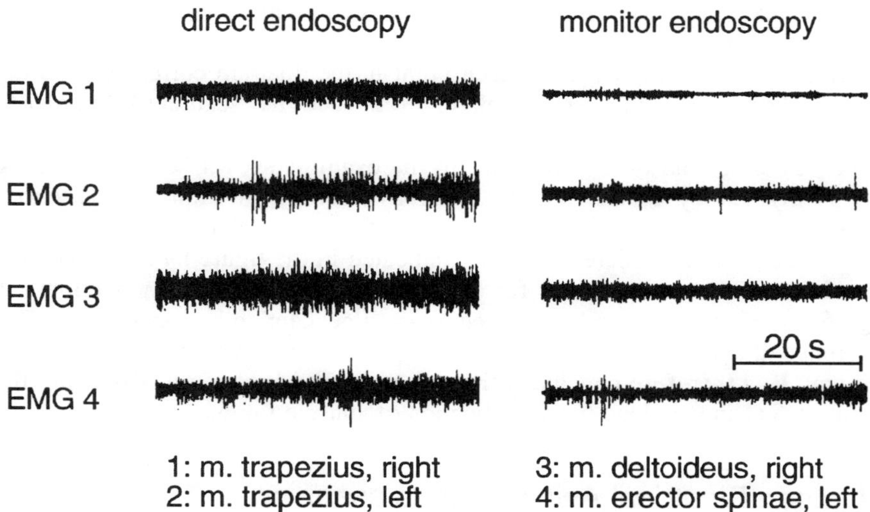

Figure 2.23 Sections of an original recording of electromyograms from different muscles of surgeons during the performance of urological operations using direct and monitor endoscopy (adapted from Luttmann *et al.*, 1992b).

clearly smaller for monitor endoscopy than for the direct method, particularly with regard to the shoulder muscles.

A quantitative evaluation of the Electrical Activity derived from four surgeons during 15 operations shows that during monitor endoscopy the average activity of the shoulder muscles (right and left trapezius, right deltoideus) is reduced to between one- and two-thirds of the EA occurring during direct endoscopy (Luttmann *et al.*, 1996a). From this finding it can be concluded that the muscular strain is significantly reduced when the monitor method is employed.

2.2.3 Determination of Muscular Fatigue

A second example illustrates the use of electromyography in field studies in order to detect muscular fatigue. The example is taken from an investigation into cashiers working at scanner check-outs (Luttmann *et al.*, 1989, 1991). In this investigation the electromyograms of four muscles of the left arm and shoulder (trapezius, deltoideus, biceps brachii and extensor carpi ulnaris) were recorded on all six working days of one week. As previously described, an activity code was additionally recorded. It allowed the following activities to be differentiated: price recording using the scanner, price recording via the keyboard, cash registering, moving heavy articles, waiting and subsidiary activities.

As an example, the time curve for the Electrical Activity during price recording using the scanner ('scanning') is indicated for both an arm and a shoulder muscle in Figure 2.24. Each point in the diagram corresponds to the average activity in a time section relating to the activity 'scanning'. In order to quantify the work-load the number of scanned articles per minute is included in the lower section of the diagram. In the example shown here the working day is divided by breaks into five activity sections. Within these sections regression lines are indicated for both the EA and the number of scanned articles. A significant increase in EA (indicated in the figure by an asterisk) is found in at least one muscle for four out of the five activity sections. A significant change in working speed over time cannot be established. The inference drawn from this is that the muscle load does not change throughout the day. It is therefore concluded that the measured increase in EA is not the result of a change in load but an indication of muscle fatigue.

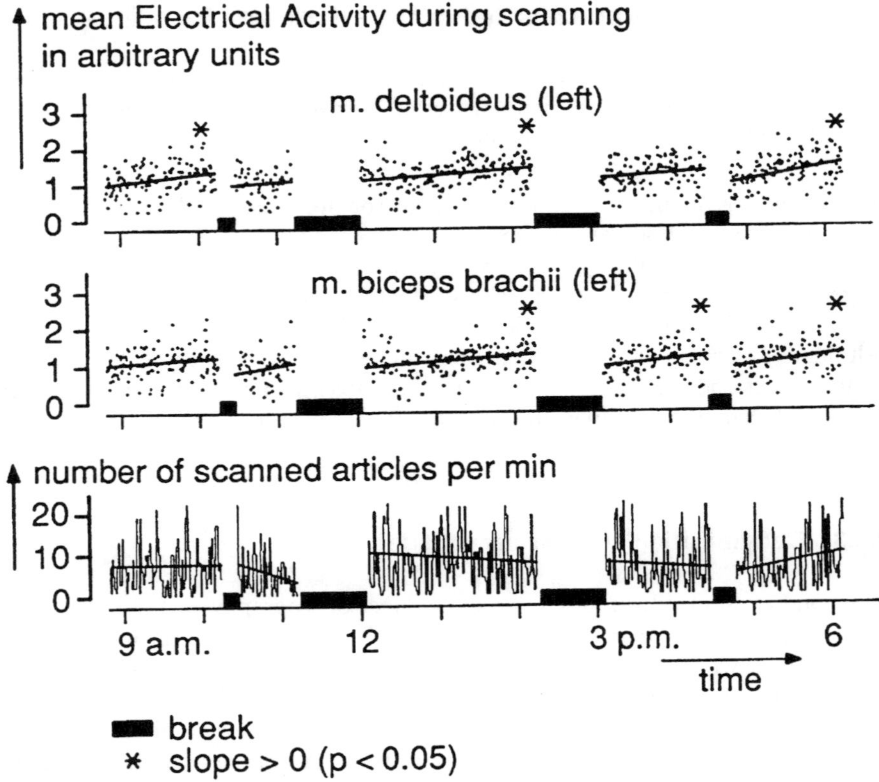

Figure 2.24 Mean Electrical Activity during scanning at a supermarket check-out and the number of articles scanned during cash-desk work over a whole working day (adapted from Luttmann *et al.*, 1992a).

A fatiguing activity can only be performed for a limited time. It should therefore be interrupted by regular breaks. The task of ergonomic planning of the work sequence is to develop a physiologically justified work-rest regimen. This can be achieved by referring to the relationship between EA increase and endurance time as shown in Figure 2.22. The regression function and the confidence interval provided there are also included in Figure 2.25. The range for the measured slopes of the EA increase during the cash-desk activity is indicated on the abscissa by shading. The respective endurance values can be read from the ordinate. They amount to between 50 and 110 min. Uninterrupted cash-desk work should not exceed this period of time.

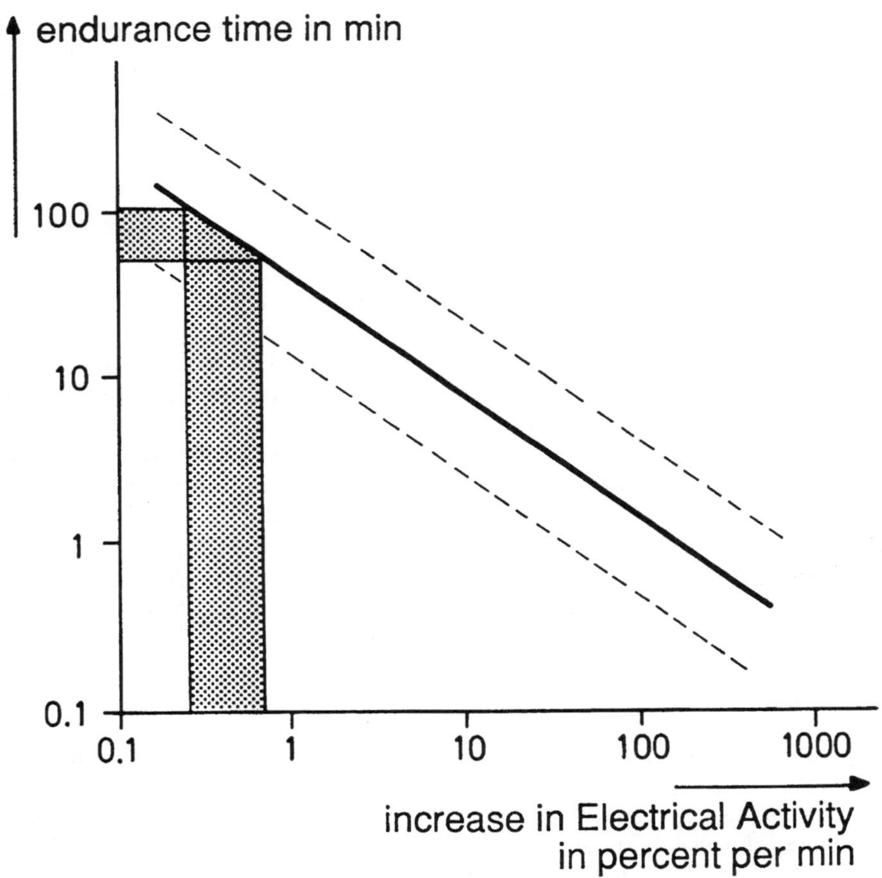

Figure 2.25 Relationship between the increase in Electrical Activity and the endurance time (regression function and confidence interval taken from Figure 2.22). Shaded area: range for the slopes of the EA time curves measured during cash-desk activities and the resulting endurance time (adapted from Luttmann *et al.* 1992a).

An average duration of the activity sections amounting to 98 min for a range extending from 13 to 256 min was established in this field study (Luttmann *et al.*, 1989). The average duration of the activity is therefore within the region of the endurance time derived on the basis of the muscle-physiological findings. However, on account of the wide time range, considerably larger values are also to be found. Consequently, in the redesign of the work sequence it was recommended that a system of rest breaks should be introduced in order to avoid the long activity sections.

In the time domain the development of muscle fatigue was inferred from the change in the EMG amplitude. In the frequency domain it is equally possible to conclude the existence of muscle fatigue from the spectral shift which occurs during occupational work (e.g. Örtengren *et al.*, 1975; Hägg *et al.*, 1987; Jørgensen *et al.*, 1991). Studies on the relationship between the fatigue-induced spectral shift and the endurance time have been performed by Hagberg (1981). These data can also be used to estimate the possible duration of activity and to derive indications for a physiologically justified work-rest regimen.

2.2.4 Conclusion

The applications of surface electromyography indicated here clearly demonstrate this method's suitability for use, not only in laboratory experiments, but also in ergonomic field studies. The method enables the documentation of the strain of individual muscles or muscle groups during complex activities as well as in application of a suitable method for protocolling the activity sequence, the identification of those activities which result in a high level of muscular strain. Furthermore, electromyography is suitable for detecting the development of muscle fatigue during an activity. Comparison with earlier investigations enables the determination of the possible duration of the particular activity, this, in turn, providing the basis for the establishment of a physiologically justified work-rest regimen.

References

ALMERS, W. (1980) Potassium concentrations changes in the transverse tubules of vertebrate skeletal muscle. *Fed. Proc.*, **39**, 1527–32.

ARENDT-NIELSEN, L. & MILLS, K.R. (1988) Muscle fibre conduction velocity, mean power frequency, mean voltage and force during submaximal fatiguing contractions of human quadriceps. *Eur. J. Appl. Physiol.*, **58**, 20–5.

BALLÉ, W. (1976) Laboruntersuchungen über erholungswirksame Pausen bei einseitig dynamischer Armarbeit. Diplomarbeit Technische Hochschule Darmstadt.

BALLÉ, W., SMOLKA, R. & LUTTMANN, A. (1982) Aufbau und Erprobung eines Meßfahrzeugs zur Datenerfassung in arbeitsphysiologischen Felduntersuchungen. Zbl. Arbeitsmed., **32**, 214–18.

BASMAJIAN, J.V. & DE LUCA, C. (1985) Muscles Alive – Their functions revealed by electromyography (5th edn). Williams and Wilkins, Baltimore.

BIGLAND-RITCHIE, B., JONES, D.A. & WOODS, J.J. (1979) Excitation frequency and muscle fatigue: electrical responses during human voluntary and stimulated contractions. *Exp. Neurol.*, **64**, 414–27.

BROMAN, H., BILOTTO, G. & DE LUCA, C.J. (1985) Myoelectric signal conduction velocity and spectral parameters: influence of force and time. *J. Appl. Physiol.*, **58**, 1428–37.

CHAFFIN, D.B. (1973) Localized muscle fatigue – definition and measurement. *J. Occupat. Med.*, **15**, 346–54.

CHRISTENSEN, H. (1986) Muscle activity and fatigue in the shoulder muscles during repetitive work. *Eur. J. Appl. Physiol.*, **54**, 596–601.

COBB, S. & FORBES, A. (1923) Electromyographic studies of muscular fatigue in man. *Am. J. Physiol.*, **65**, 234–51.

DE LUCA, C.J. (1979) Physiology and mathematics of myoelectric signals. *IEEE Trans. Biomed Engng.*, **26**, 313–29.

EASON, R.G. (1960) Electromyographic study of local and generalized muscular impairment. *J. Appl. Physiol.*, **15**, 479–82.

ECKERT, R. & RANDALL, D. (1983) *Animal Physiology*. W.H. Freeman, New York.

EDMAN, K.A.P. & LOU, F. (1992) Myofibrillar fatigue versus failure of activation during repetitive stimulation of frog muscle fibres. *J. Physiol.*, **457**, 655–73.

EDWARDS, R.G. & LIPPOLD, O.C.J. (1956) The relation between force and integrated electrical activity in fatigued muscle. *J. Physiol.*, **132**, 677–81.

GOLDMAN, D.E. (1943) Potential, impedance, and rectification in membranes. *J. Gen. Physiol.*, **27**, 37–60.

HAGBERG, M. (1981) Muscular endurance and surface electromyogram in isometric and dynamic exercise. *J. Appl. Physiol.*, **51**, 1–7.

HÄGG, G.M. (1991) Comparison of different estimators of electromyographic spectral shifts during work when applied on short test contractions. *Med. Biol. Engng. Comput.*, **29**, 511–16.

HÄGG, G.M., SUURKÜLA, J. & LIEW, M. (1987) A worksite method for shoulder muscle fatigue measurements using EMG, test contractions and zero crossing technique. *Ergonomics*, **30**, 1541–51.

HODGKIN, A.L. & HUXLEY, A.F. (1952) A quantitative description of membrane currents and its application to conduction and excitation in nerve. *J. Physiol.*, **117**, 500–44.

JÄGER, M., LUTTMANN, A. & LAURIG, W. (1984) Räumliche und zeitliche Aktivitätsverteilung der Rückenmuskulatur im Lendenwirbelsäulenbereich. *Biomed. Techn. Ergänzungsband*, **29**, 87–8.

JONES, D.A. (1981) Muscle fatigue due to changes beyond the neuromuscular junction. In *Human Muscle Fatigue: Physiological Mechanism* (Ed. R. PORTER & J. WHELAN), pp. 178–96. Pitman Medical, London.

JØRGENSEN, K., FALLENTIN, N., KROGH-LUND, C. & JENSEN, B. (1988) Electromyography and fatigue during prolonged, low-level static contractions. *Eur. J. Appl. Physiol.*, **57**, 316–21.

JØRGENSEN, K., JENSEN, B.R. & KATO, M. (1991) Fatigue development in the lumbar paravertebral muscles of bricklayers during the working day. *Int. J. Indust. Ergonom.*, **8**, 237–45.

KNOWLTON, G.L., BENNET, R.L. & MCCLURE, R. (1951) Electromyography of fatigue. *Arch. Phys. Med.*, **32**, 648–51.

KOGI, K. & HAKAMADA, T. (1962) Frequency analysis of the surface electromyogram in muscle fatigue. *J. Sci. Labour (Tokyo)*, **38**, 519–28.

KÖSSLER, F., CAFFIER, G. & LANGE, F. (1990) Probleme der Muskelermüdung – Beziehung zur Erregungsleitungsgeschwindigkeit und K^+-Konzentration. *Z. Ges. Hyg.*, **36**, 354–56.

KROGH-LUND, C. (1993) Myo-electric fatique and force failure from submaximal static elbow flexion sustained to exhaustion. *Eur. J. Appl. Physiol.*, **67**, 389–401.

KROGH-LUND, C. & JØRGENSEN, K. (1991) Changes in conduction velocity, median frequency, and root mean square-amplitude of the electromyogram during 25% maximal voluntary contraction of the triceps brachii muscle, to limit of endurance. *Eur. J. Appl. Physiol.*, **63**, 60–9.

KROGH-LUND, C. & JØRGENSEN, K. (1992) Modification of myo-electric power spectrum in fatigue from 15% maximal voluntary contraction of human elbow flexors, to limit of endurance: reflexion of conduction velocity variation and/or centrally mediated mechanism? *Eur. J. Appl. Physiol.*, **64**, 359–70.

KROGH-LUND, C. & JØRGENSEN, K. (1993) Myo-electric fatigue manifestations revisited: power spectrum, conduction velocity, and amplitude of human elbow flexor muscles during isolated and repetitve endurance contractions at 30% maximal voluntary contraction. *Eur. J. Appl. Physiol.*, **66**, 161–73.

KUFFLER, ST. W., NICHOLLS, J.G. & MARTIN, A.R. (1984) *From Neuron to Brain* (2nd edn). Sinauer Associates, Sunderland.

LÄNNERGREN, J. & WESTERBLAD, H. (1986) Force and membrane potential during and after fatiguing, continuous high-frequency stimulation of single *Xenopus* muscle fibres. *Acta Physiol. Scand.*, **128**, 359–68.

LAURIG, W. (1969) Analyse des Unterarmtremors bei statischer Arbeit durch Korrelationsfunktion und Leistungsspektrum. *Sportarzt und Sportmedizin*, **11**, 434–44.

LAURIG, W. (1974) *Beurteilung einseitig dynamischer Muskelarbeit.* Beuth, Berlin.

LAURIG, W. (1975) Methodological and physiological aspects of electromyographic investigations. In *Biomechanics* (Ed. P.V. KOMI), pp. 219–30. V-A. University Park, Baltimore.

LAURIG, W. (1977) Elektromyographie. In *Grundkurs Datenerhebung 1* (Ed. K. WILLIMCZIK), pp. 67–94. Limpert, Bad Homburg.

LAURIG, W. & SAUERNHEIMER, W. (1975) Simulation eines antagonostischen Muskelsystems für den Sonderfall ermüdender isometrischer Muskelarbeit. *Biol. Cybern.*, **220**, 17–26.

LAURIG, W., LUTTMANN, A. & JÄGER, M. (1987) Evaluation of strain in shop-floor situations by means of electromyographic investigations. In *Trends in Ergonomics/Human Factors IV* (Ed. S.S. ASFOUR), pp. 685–92. Elsevier Science Publishers, Amsterdam.

LINDSTRÖM, L.H. & MAGNUSSON, R.I. (1977) Interpretation of myoelectric power spectra: a model and its applications. *Proc. IEEE*, **65**, 653–62.

LINDSTRÖM, L., MAGNUSSON, R. & PETERSÉN, I. (1970) Muscular fatigue and action potential conduction velocity changes studied with frequency analysis of EMG signals. *Electromyography*, **10**, 341–56.

LUTTMANN, A. & JÄGER, M. (1992) Reduction of muscular strain by work design: electromyographical field studies in a weaving mill. In *Advances in Industrial Ergonomics and Safety IV* (Ed. S. KUMAR), pp. 553–60. Taylor and Francis, London, Washington.

LUTTMANN, A., JÄGER, M. & LAURIG, W. (1988) Surface electromyography in work-physiological field studies for the analysis of muscular strain and fatigue. In *Electrophysiological Kinesiology* (Ed. W. WALLINA, H.B.K. BOOM, & J. DE VRIES), pp. 301–4. Elsevier, Amsterdam.

LUTTMANN, A., JÄGER, M. & LAURIG, W. (1989) Elektromyographische Untersuchungen an Kassenarbeitsplätzen mit Scannern. *Z. Arbeitswiss.*, **43**, 234–40.

LUTTMANN, A., JÄGER, M. & LAURIG, W. (1991) Electromyographical studies on check-out work. In *Electromyographical Kinesiology* (Ed. P.A. ANDERSON, O.J. HOBERT & J.V. DANOFF), pp. 145–8. Elsevier, Amsterdam.

LUTTMANN, A., JÄGER, M. & LAURIG, W. (1992a) Elektromyographie als Werkzeug zur ergonomischen Arbeitsgestaltung. *Med. Orth. Techn.*, **112**, 318–22.

LUTTMANN, A., LAURIG, W. & SÖKELAND, J. (1992b) Electromyography on surgeons during urological operations. In *Proceedings of the 14th Annual International Conference of the IEEE Engineering in Medicine and Biology Society.* (Ed. J.P. MORUCCI, R. PLONSEY, J.L. COATRIEUX, & S. LAXMINARAYAN), pp. 1430–2. Paris.

LUTTMANN, A., SÖKELAND, J. & LAURIG, W. (1996a) Electromyographical study on surgeons in urology, part I: influence of the operating technique on muscular strain. *Ergonomics*, in press.

LUTTMANN, A., JÄGER, M., SÖKELAND, J. & LAURIG, W. (1996b) Electromyographical study on surgeons in urology, part II: determination of muscular fatigue. *Ergonomics*, in press.

MATON, B. (1981) Human motor unit activity during the onset of muscle fatigue in submaximal isometric isotonic contractions. *Eur. J. Appl. Physiol.*, **46**, 271–81.

MORTIMER, J.T., MAGNUSSON, R. & PETERSÉN, I. (1976) Conduction velocity in ischemic muscle: effect on EMG spectrum. *Am. J. Physiol.*, **219**, 1324–29.

MUCKE, R. & ZÖLLNER, I. (1986) Muscle fibre conduction velocity during fatiguing and nonfatiguing isometric arm contractions. *Biomed. Biochim. Acta*, **45**, 77–80.

NAEIJE, M. & ZORN, H. (1982) Relation between EMG power spectrum shifts and muscle fibre action potential conduction velocity changes during local muscular fatigue. *Eur. J. Appl. Physiol.*, **50**, 23–33.

NEHER, E. & SAKMANN, B. (1976) Single-channel currents recorded from membranes of denervated frog muscle fibers. *Nature*, **260**, 799–802.

ÖBERG, T., SANDSJÖ, L., KADEFORS, R. & LARSSON, S.-E. (1992) Electromyographic changes in work-related myalgia of the trapezius muscle. *Eur. J. Appl. Physiol.*, **65**, 251–7.

ÖRTENGREN, R., ANDERSSON, G., BROMAN, H., MAGNUSSON, R. & PETERSÉN, I. (1975) Vocational electromyography: studies of localized muscle fatigue at the assembly line. *Ergonomics*, **18**, 157–74.

PEIPER, U. (1974) Muskulatur. In *Lehrbuch der Physiologie* (Ed. R. KLINKE & St. SILBERNAGL), pp. 73–98. Thieme, Stuttgart.

PETROFSKY, J.S. & LIND, A.R. (1980) The influence of temperature on the amplitude and frequency components of the EMG during brief and sustained isometric contractions. *Eur. J. Appl. Physiol.*, **44**, 189–200.

PIPER, H. (1909) Über die Ermüdung bei willkürlichen Muskelkontraktionen. *Archiv für Anatomie und Physiologie*, Physiologische Abteilung, 491–98.

PIPER, H. (1912) *Elektrophysiologie Menschlicher Muskeln*. Springer, Berlin.

RAU, G. & VREDENBREGT, J. (1970) *Electromyographic Activity During Voluntary Static Muscle Contractions*. Instituut voor Perceptie Onderzoek, Report No. 192, Eindhoven.

RHEIN, A., BUCHHOLZ, C., KRAMER, H. & KÜCHLER, G. (1974) Veränderungen der biolelektrischen Muskelaktivität während langdauernder Willkürkontraktionen bei vorgegebener Haltekraft. *Acta Biol. Med. Germ.*, **33**, 231–9.

ROHMERT, W. (1962) *Untersuchung über Muskelermüdung und Arbeitsgestaltung*. Beuth, Berlin.

RÜEGG, J.C. (1987) Muskel. In *Physiologie des Menschen* (23rd edn) (Ed. R.F SCHMIDT & G. THEWS), pp. 66–86. Springer, Berlin.

SADOYAMA, T., MASUDA, T. & MIYANO, H. (1983) Relationships between muscle fibre conduction velocity and frequency parameters of surface EMG during sustained contraction. *Eur. J. Appl. Physiol.*, **51**, 247–56.

SAKAMOTO, K., USUI, T., HAYAMI, A. & OHKOSHI, K. (1982) The wave analysis with the fast Fourier transform on surface electromyogram and tremor during an acute and an accumulative fatigue. *Electromyogr. Clin. Neurophysiol.*, **22**, 207–228.

SÄLZER, M. (1973) *Tremoruntersuchungen als Methode in der Arbeitswissenschaft*. Beuth, Berlin.

SCHERRER, J. & BOURGUIGNON, A. (1959) Changes in the electromyogram produced by fatigue in man. *Am. J. Phys. Med.*, **38**, 148–58.

SCHNEIDER, J., SILNY, J. & RAU, G. (1988) Noninvasive measurement of conduction velocity in motor units influenced by temperature and excitation pattern. In *Electrophysiological Kinesiology*. (Ed. W. WALLINGA, H.B.K. BOOM & J. DE VRIES), pp. 251–4. Elsevier, Amsterdam.

SJØGAARD, G. (1990) Exercise-induced muscle fatigue: the significance of potassium. *Acta Physiol. Scand.*, **140**, (Suppl. 53).

STEIN, R.B. (1974) Peripheral control of movement. *Physiol. Rev.*, **54**, 215–43.

STULEN, F.B. & DE LUCA, C.J. (1981) Frequency parameters of the myoelectric signal as a measure of muscle conduction velocity. *IEEE Trans. Biomed. Engng.*, **28**, 515–23.

VENOSA, R.A. & HOROWICZ, P. (1981) Density and apparent location of the sodium pump in frog sartorius muscle. *J. Membr. Biol.*, **59**, 225–32.

VERGARA, J., BEZANILLA, F. & SALZBERG, B.N. (1978) Nile blue fluorescence signals from cut single muscle fibers under voltage or current clamp conditions. *J. Gen. Physiol.*, **72**, 775–800.

VØLLESTAD, N.K. & SEJERSTED, O.M. (1988) Biochemical correlates of fatigue. *Eur. J. Appl. Physiol.*, **57**, 336–47.

VREDENBREGT, J. & RAU, G. (1973) Surface electromyography in relation to force, muscle length, and endurance. In *New Developments in Electromyography and Clinical Neurophysiology*, Vol. 1 (Ed. J.E. DESMEDT), pp. 607–22. Karger, Basel.

WESTERBLAD, H. & LÄNNERGREN, J. (1986) Force and membrane potential during and after fatiguing, intermittent tetanic stimulation of single *Xenopus* muscle fibres. *Acta Physiol. Scand.*, **128**, 369–78.

WESTERBLAD, H., LEE, J.A., LAMB, A.G., BOLSOVER, S.R. & ALLEN, D.G. (1990) Spatial gradients of intracellular calcium in skeletal muscle during fatigue. *Pflügers Arch.*, **415**, 734–40.

WOODBURY, J.W. (1965) The cell membrane: ionic and potential gradients and active transport. In *Physiology and Biophysics* (Ed. TH.C. RUCH & H.D. PATTON), pp. 1–25. W.B. Saunders Company, Philadelphia.

ZWARTS, M.J., VAN WEERDEN, T.W. & HAENEN, H.T.M. (1987) Relationship between average muscle fibre conduction velocity and EMG power spectra during isometric contraction, recovery and applied ischemia. *Eur. J. Appl. Physiol.*, **56**, 212–16.

CHAPTER THREE

Noise and artefacts

ROLAND ÖRTENGREN

Chalmers University of Technology, Göteborg

3.1 INTRODUCTION

All measurements are to some extent exposed to disturbances and this is particularly the case in electrophysiological measurements because the signal amplitude is low compared to the disturbances. The disturbances have different origins; some are generated in the environment, some arise as a result of the experimental conditions, and others such as noise are inherently generated in the measurement system. To reduce the disturbances to acceptable levels can be difficult and requires well-designed equipment and a qualified measurement engineer to plan the measurement system and also to carry out or survey the measurements.

For standardized measurements of, for example, cardioelectric signals and signals from the brain, nerves, and muscles, which are made in special laboratories with dedicated equipment, the measurement conditions are well controlled and the problems few. But when myoelectric signals are recorded for work-load assessment for studies in the ergonomics laboratory and in the field, useful results can only be obtained if the disturbances can be controlled. The purpose of this contribution is to describe different disturbances, their origin, and ways to reduce their influence. The examples mentioned are meant to illustrate principles and not as a basis for design since this requires a more thorough treatment. Neither are aspects on patient safety explicitly covered here, despite the requirement that such aspects must be considered in parallel when designing for disturbance elimination. Safety may, for example, require that isolation amplifiers are used, but this is no drawback since they have excellent disturbance elimination properties and simplify connecting the components of the measurement system together.

Measurement systems for recording myolectric and other signals designed according to the principles given were used in laboratory studies by Andersson and Örtengren (1974) and Andersson *et al.* (1976) and in field studies at work sites by Kadefors *et al.* (1978), Malmqvist *et al.* (1981), and Andersson and Örtengren (1984). For further treatment of the subject from a more clinical point of view, the reader is referred to Basmajian and De Luca (1985). An excellent treatment of the principles for grounding and shielding in measurement intrumentation is given in the classical book by Morrison (1977). The book is old but the principles are still valid.

3.2 DISTURBANCE SOURCES

In measurement theory, it is customary to consider all unwanted signals that are inevitably recorded together with the wanted signals as noise. In this case artefacts and interference from power lines are denoted deterministic noise as they – at least in principle – have controllable causes. The third component, random noise, has several origins but it can be said to be generated internally in the electronic circuitry and the measurement object. The random noise is dominated by the thermal noise caused by the irregular motion of thermally agitated electrons or ions moving in conducting materials and electrolytes. The spectral composition of this noise is independent of the frequency; therefore it is often referred to as white noise.

Since the different types of disturbances have different origins, it is somewhat misleading to use the term noise for all. In electromyography the disturbances can be divided into noise, artefacts, and interference from electromagnetic phenomena in the environment, including disturbances from power lines. Different methods and techniques are used for their control and elimination.

3.3 NOISE AND NOISE REDUCTION

In electromyography (EMG) contributions to the total noise come from the tissue, the tissue–electrode junction, and the electronic circuitry used, of which the amplifier is the most important. Due to the stochastic nature or randomness of the noise, noise signals from several sources are added on a power basis and not linearly. (An illustration of this principle is given below.) If the EMG amplifier is connected to an oscilloscope and the sensitivity increased, the noise can be seen as a thicker than normal baseline. The amplitude variations are normally distributed with a variance equal to the noise signal power. The square root of the variance, the standard deviation, thus equals the root mean square (RMS) voltage. Of course the amplitude of the myoelectric action potentials must be several times larger than the noise level for meaningful observations.

In a resistive material the minimum thermal noise voltage v_e is given by

$$(v_e)^2 = 4kTR\,\Delta f \tag{1}$$

where k is Boltzmann's constant (1.381×10^{23} J K^{-1}), T is the absolute temperature (K), R the resistance (Ω), and Δf is the bandwidth (Hz). For a resistance of 1000 Ω and a bandwidth of 1000 Hz at room temperature the noise voltage is 1.3 μV. As the resistance used in this calculation is about as low as the resistive component of the source impedance can be in recordings of myoelectric signals, the only way to reduce the noise component is to reduce the bandwidth.

An amplifier is characterized by its noise voltage, which is measured at the output with the input short circuited and the gain set to maximum. The measured voltage divided by the gain is the amplifier input noise voltage which is added to all signals connected to the input.

The noise of the amplifier is determined by the type and quality of the transistors used and the design of the input stage of the amplifier. With an input stage equipped with modern field effect transistors, noise levels below 0.5 μV over a 500 Hz bandwidth have been accomplished. This noise voltage level determines the sensitivity of the amplifier and in this case it is possible to record myoelectric signals with amplitudes down to a few microvolts. Since the myoelectric signal with contributions from many motor units also has noise properties, the signals are added, as noted before, on a power basis. This means that when a noise signal of 0.5 μV and a myoelectric signal of 1.5 μV are added together the amplitude becomes 1.58 μV, that is, an error of approximately 5%. As 0.5 μV is at the same level as the tissue noise, it is perhaps not necessary to reduce the amplifier's noise much further.

Another important amplifier characteristic is the signal-to-noise ratio or, more properly, the dynamic range, which is also referred to the input. It is usually defined as the ratio between the maximum RMS voltage of an input sine wave signal that does not saturate the amplifier and the RMS value of the input noise voltage and expressed in decibels (dB), i.e. 20 times the logarithm of the voltage ratio.

In practice it is not possible to distinguish the tissue noise from the noise of the tissue–electrode junction. To reduce the noise as much as possible, the impedance of the tissue–electrode junction should be kept as low as possible. In the case of surface electrodes, the impedance is determined by the size of the electrode area and the preparation of the skin. To prepare the skin well is very important in order to reduce the noise as well as the sensitivity to other disturbances. The impedance should always be checked before any recordings are made and is measured by using an alternating current impedance meter with a frequency in the range 100–200 Hz. Impedance values above a few kilo-ohms should not be accepted. In the case of intramuscular electrodes the contact area is very small and the electrode material different. Therefore much higher impedances must be accepted.

There is one additional source of disturbance that can severely degrade the quality of the recordings and it is the quantification error of the analog to digital conversion process which must take place before the signals can be digitally analyzed in a computer. This error is sometimes referred to as quantification noise. The error has to do with the fact that the converter has a fixed number of levels. If the number of binary digits of the A/D converter is, for example, 12 bits, the number of amplitude levels that can be distinguished is 4096, or rather when allowing for a symmetrical swing ±2048. If the voltage range is set to, for example, 2 V peak-to-peak, the maximum resolution becomes 0.5 mV. If the conservative engineer sets the gain to a low value in order not to overdrive the amplifiers so that the voltage swing of the signal only utilizes, for example, 0.2 V, the effective resolution will be 1/400. Then, if a safe margin against low signal is set to 5, the dynamic range will be limited to 1/80, which might be too little in applications where the signal amplitude varies a lot.

Noise of the thermal and other types do not cause serious problems in amplifier chains for direct recording under direct supervision. The investigator can normally select the bandwidth and adjust the gain so that signal utilizes the available dynamic range of say 80 dB of the recording system adequately. This is because under normal recording conditions, the dynamic range of the signal is not very large. This means that the gain can be set so that an output amplitude is obtained safely under saturation of the amplifier or clipping and distortions of the signal. If the safety margin is selected to be 10 dB and a minimum signal-to-noise ratio of 30 dB is required, an 80 dB dynamic range system still provides 40 dB for signal amplitude variations, i.e. a variation ratio of 1 : 100. A dynamic range of 80 dB is equal to an amplitude variation ratio of 1 : 10000.

A 16 bit A/D converter can (excluding zero) express numbers from ±1 to ±32 768. Numerically this corresponds to a dynamic range of 90 dB, but for a varying signal such as a sine wave or random noise at least 12 dB should be set off for the signal variations. With a minimum signal-to-noise ratio of 30 dB, this results in a net dynamic range of 48 dB which should be enough even for vocational electromyography, i.e. recording during work.

An A/D converter with 16 bit resolution gives a much slower sampling rate than, for example, a 12 bit converter. However, the net dynamic range of the 12 bit converter is 24 dB lower or maximally 24 dB. If this range is too small, it is possible to remedy the situation by using amplifiers with automatic gain control and together with the digitized signal also record the instantaneous gain setting. The gain values are used during analysis to calculate comparable amplitude values.

It is worth noting that an ordinary analog tape recorder with an FM recording system has a maximum dynamic range of 48 dB. The dynamic range is determined as the ratio in decibels between the maximum sine wave output amplitude, usually 1 V and the RMS value of the output noise voltage over the total bandwidth that the tape recorder allows, for example 10 kHz at 30 inches per second of tape speed. The noise level then is 4.0 mV and the spectral density's 40 $\mu V \sqrt{Hz}^{-1}$. Surface EMG recorded by means of bipolar electrodes with an interelectrode distance of 25–30 mm has a bandwidth of, at most, approximately 500 Hz. By recording the EMG at high speed and reducing the tape recorder bandwidth by filtering with a cut-off frequency of 500 Hz, the noise level will be 0.9 mV resulting in a dynamic range of 61 dB which is indeed much better than 48 dB. The cost is that a lot more tape is needed, namely 16 times more than at the speed that has the minimum sufficient 625 Hz bandwidth.

3.4 ELECTROMAGNETIC INTERFERENCE

Indoors and maybe particularly in a laboratory the environment is full of electric and magnetic fields from power lines, incandescent lamps, and particularly, fluorescent lamps with their inductors, other electric equipment and machinery, ovens, radio transmitters, etc. A sensitive amplifier for electrophysiological signals is very likely to amplify signals induced by these fields and saturate if nothing is done. The coupling can be either capacitive responding to electric fields or inductive responding to magnetic fields. Radio signals can usually be disregarded because their frequencies are much above the physiological range. Electric fields are capacitively coupled to the human body (see Figure 3.1). Because the surface resistance is high, the voltage of the surface can become quite high. There will be a weak surface current causing a voltage gradient. An ordinary amplifier, having a gain high enough for EMG, connected between an electrode on the body surface and a ground electrode attached to the wrist or other reference position, would immediately saturate. To enable recordings of the weak signals from a source having a different ground reference, the differential amplifier was invented. The differential amplifier has three input connections, two high impedance signal input connections and one signal ground connection. The amplifier amplifies the voltage difference between the signal input connections and attenuates the voltage between these two inputs and the ground reference connection. These voltages are referred to as the differential mode voltage and the common mode voltage, respectively. In addition to the differential gain A, the common mode rejection ratio (CMRR) of the differential amplifier is an important parameter. The CMRR is usually expressed in decibels, sometimes denoted as CMR (common mode rejection). With notations according to Figure 3.1, the output voltage v_o can be expressed as

$$v_o = A\ (v_i + \text{CMRR}\ v_c) \tag{2}$$

A modern differential amplifier has a CMRR of at least 80 dB and the value can be as high as 120 dB. If the desired signal has an amplitude of 50 μV and the common mode voltage disturbance is 0.5 V, the amplifier must have a CMRR of 100 dB, for the desired signal to be ten times higher than the common mode disturbance voltage at the output.

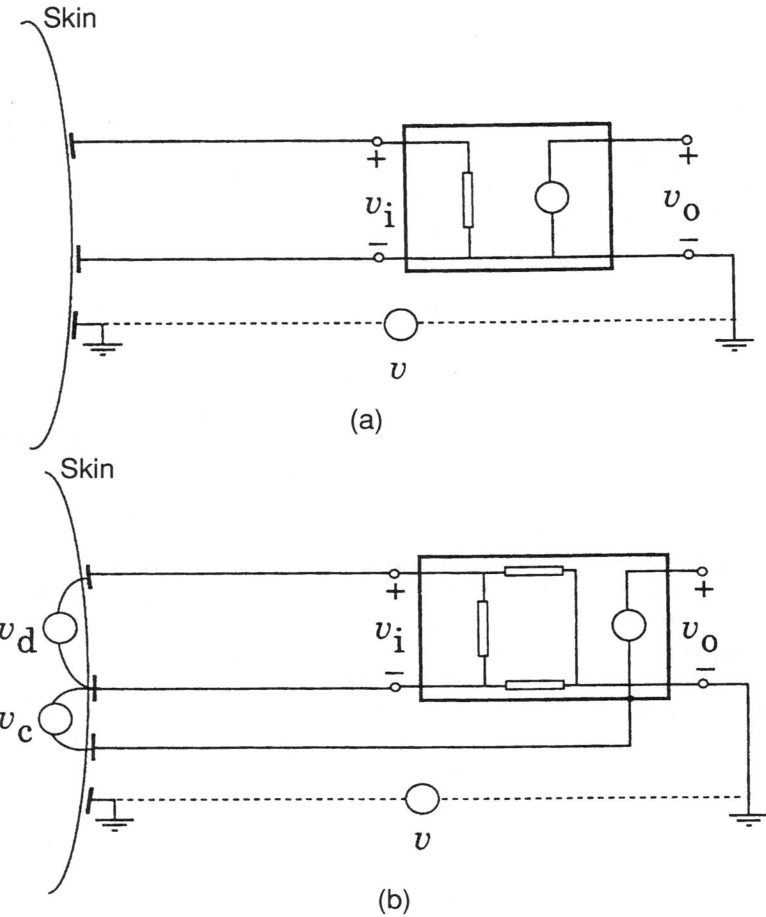

Figure 3.1 (a) Single input amplifier. The potential difference *v* between the subject's body surface and the amplifier ground results in a current that causes a voltage drop in the ground input lead which is added to the desired signal and amplified as well. (b) Differential input amplifier. In this case the voltage drop in the ground lead becomes a common mode voltage which is attenuated by the amplifier. The amplifier input leads have the same and very high impedance to the ground.

The proper way to handle electric fields is to carry out the recordings in a shielded cage, a Faraday cage, which creates an equipotential barrier and effectively closes out all electric fields. If the cage is connected to the systems ground, very little interference is caused. In many laboratories the investigation rooms are made as Faraday cages, but it is expensive and with today's amplifiers not really necessary. In field studies when problems arise, other measures must be undertaken. One measure is to shield all wiring as far as possible on the high-sensitivity side of the amplifiers. The purpose of the shield is to create an equipotential protective surface and, therefore, the shield should not carry any current, which means that the shield should be open at one end. To obtain as low a potential difference between the signal source and the shield as possible, the shield should be connected to the signal ground at the low side of the signal source, that is the side that is closest to ground. This means that the shield should be open on the amplifier side. The point to which the shield is connected should also be connected to the amplifier's signal ground to provide a reference for the signal source ground. An example of proper shielding is shown in Figure 3.2.

In electromyography this principle is difficult to follow when recordings from several electrode sites are being made since only one electrode wire shield can be properly connected. Any one electrode on the body can be used as a reference ground, but it is common practice to use a special ground electrode which is connected to the amplifier ground. There will always be a potential difference – but normally small – between the different electrodes. Therefore all electrode wire shields are connected to the reference ground and to simplify the wiring, this is done at the amplifier end.

A complement to shielding is grounding. The purpose of grounding is to connect all metal objects in the environment to the common ground so that they will obtain the same potential. In this way there will be no voltage difference between the subject and the various objects and, thus, no electric field that can cause interference, nor for that matter, any currents through the subject's body. Potential differences can be caused by, for example, stray currents in conducting pipes, ground currents in unsymmetrically loaded three-phase power systems, and currents induced in ground loops. In a measurement system consisting of pre-amplifiers, connecting cables, amplifiers, filters, RMS-DC converters, and analog-to-digital converters, etc., it is important to control the ground currents both from the disturbance and the safety point of view. It is

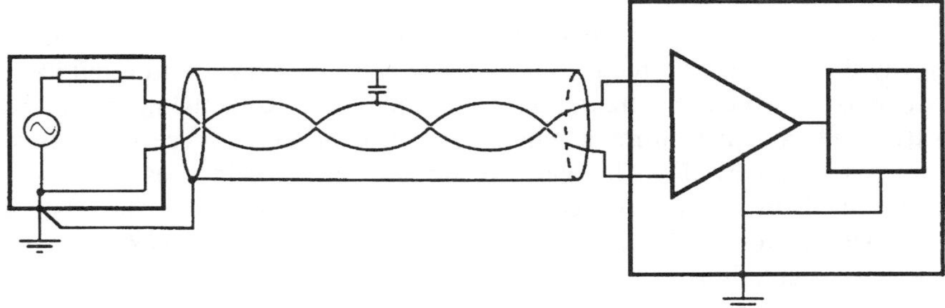

Figure 3.2 Example of correct connection of a shield to a differential amplifier in a measurement situation. The inner wires of the measurement cable should be twisted to reduce the effect of magnetic field induction. The signal source ground and amplifier ground should be connected by a separate wire to reduce the common mode voltage. For application in electromyography, see the text.

difficult to separate a signal ground, a supply ground, and a protective ground. Since a supply ground may carry leakage currents from transformers, power line wires, and cables, it is advisable to use differential amplifier inputs in all units receiving input signals of low amplitude. Ground loops must be avoided, but to do that may require that protective ground is disconnected from all units but one and oversized ground wires between the units are used instead. If this solution cannot be permitted for safety reasons, depending of how the system is built together, it will be necessary to use isolation amplifiers between the subject and the equipment.

All electric fields, either continuously present or the result of transient events, cause currents to flow on the body surface to the ground. Depending on the impedance on the body surface and from there to the ground, the potential difference between the body and the ground as well as between two points on the body surface can be quite high. By attaching a ground reference electrode to the body the potential difference between that point and the ground is reduced to zero. The potential difference between two points on the body depends on the size of the current and the impedance, and increases with increasing distance. Therefore, a larger common mode voltage between a bipolar surface electrode and the ground electrode on the body can arise, the effect of which is reduced by the amplifier's common mode rejection. Since the two electrode surfaces of the bipolar electrode will be close together, the differential mode voltage caused by the surface current is likely to be small. To reduce this voltage, the interelectrode impedance should be reduced as far as possible by proper skin treatment.

Electric fields have a long range, which means that increasing the distance to the source will not decrease the influence. Magnetic fields, on the other hand, decline considerably with distance from the source. At 0.5–0.7 m from a current-conducting wire the magnetic field has very little effect. However, contrary to the electric fields, the magnetic fields cannot be shielded out by conducting metals. The magnetic field induces current in wire loops, such as electrode wires. The larger the area of the loop, the larger the induced current. One common way to reduce this area as far as possible is to twist the wires. In this way the area will be reduced, but it will also be divided into a number of small areas, in which, depending on the orientation towards the magnetic field, positive and negative currents will be induced; however these currents will cancel each other. For bipolar electrodes the twisting can easily be accomplished by using twisted and shielded microphone cable, but for monopolar recordings the proper placement of the electrode wires may need some experimentation.

3.5 ARTEFACTS

Artefacts are transient disturbances in the signal to be recorded, which have a specific origin. Artefacts normally have a low frequency content and are quite easy to see in the raw myoelectric signal. On an oscilloscope screen or in a paper recording they appear as relatively short monopolar or bipolar deviations from the baseline. After detection, i.e. RMS conversion or full-wave rectification and low-pass filtering, it becomes very difficult to see irregularly occurring artefacts. However, periodically occurring artefacts, e.g. ECG disturbances, can be identified by means of the regular appearance of peaks in the recording. If the time constant of the low-pass filter is short enough so that the trace can return to the level of the EMG amplitude between every heartbeat, the influence of the ECG disturbances can be disregarded when reading the EMG amplitude. This principle was used by Andersson and Örtengren (1974) in their studies of myoelectric activity in paraspinal back muscles in sitting postures.

Movement artefacts are caused by changes of the skin or tissue to electrode connection. Typically a movement artefact is caused by a change in electrode–skin distance due to movement of the electrodes, for example in a gait analysis situation. The effect can easily be demonstrated by patting on a surface electrode or moving a needle electrode in the tissue. In the case of a surface electrode, the tissue–electrode junction functions as a capacitor. When the electrode moves, the skin–electrode distance is likely to change causing a change in capacitance C. Since the charge, Q, of the capacitor cannot change instantaneously, the voltage V over the junction must change according to

$$Q = C\,V \tag{3}$$

$$\Delta Q = C\,\Delta\,V + V\,\Delta\,C = 0 \tag{4a}$$

$$\Delta\,V = -V/C\,\Delta\,C \tag{4b}$$

The voltage change is added in series with the intended signal and amplified as well.

Other artefacts are caused by electromagnetic impulses from the environment, e.g. closing or opening a relay circuit causing an impulse that can be picked up on the amplifier's high-sensitivity side. The starting or stopping of a heavy electric machine may cause a voltage drop in the power line. Ideally the stabilized voltage of the amplifier power supply should eliminate the effect of this voltage drop, but unfortunately in many cases the impulse is led through the power supply to the amplifier, where an output artefact is caused.

The cardioelectric signal has a much higher amplitude than myoelectric signals normally have and, in particular at electrode locations close to the heart, the EMG electrodes are likely to pick up ECG disturbances. Surface electrodes, in particular, with their large interelectrode distance, can pick up high-amplitude ECG disturbances. One way to avoid this pick up, which can be used when the EMG electrode position with respect to the muscle fibers is not critical, is to orient the bipolar electrode perpendicular to the cardioelectric field. Then the ECG voltage between the two electrode surfaces of the bipolar electrode will be zero. Another possibility, which requires special electronic circuits, is to pick up the ECG signal, change the polarity, and, after suitable amplification, feed a current back to the body surface to cancel the currents caused by the cardioelectric field. Good elimination with this technique is in principle obtained only at one location.

The myoelectric and cardioelectric signals have different power spectra, the cardioelectric signal having a lower frequency content, In principle it should be possible to filter out the cardioelectric signal but the spectra overlap too much, so it is difficult to obtain a clean recording without too much myoelectric signal amplitude reduction. A fourth possibility utilizes the fact that the duration of the cardioelectric wave is approximately constant and much shorter than the period time, at least at lower levels of physical effort. The highest peak of the ECG wave is easy to identify, either in the EMG recording or by using a separate pick-up electrode. To eliminate the ECG disturbance a suitable portion of the recording is simply cut out. In EMG recordings for work-load investigations, this loss of recording time does not cause any serious loss of information. Most types of analysis, such as amplitude estimation, amplitude distribution analysis, and spectral analysis, can be carried out without problems. For the analyses, however, digital signal processing methods must be applied.

3.6 AUTOMATIC DISTURBANCE REMOVAL

Myoelectric signals recorded during work often suffer from disturbances caused by electrode movements under the clothes or by electrical equipment used by the worker or in the vicinity. The recordings can be very long and made in several channels. Of course an observer must always be present to check on signal quality and take action if something goes wrong. Even then it is not possible to obtain disturbance-free recordings. In particular when the recordings are long, it becomes necessary to utilize some procedure for automatic artefact detection and removal. For the extensive studies of muscle fatigue in building work that were carried out in Göteborg in 1976–1979 (see Malmqvist *et al.*, 1981), an automatic method for disturbance removal based on discriminant analysis was developed. The method has been described by Kadefors *et al.* (1978) and Arvidsson *et al.* (1981). In short the method works as follows.

The recorded signal is divided into 0.5 s segments and for each segment a number of parameters are calculated both in the time domain and the frequency domain. In this application two parameters from the time domain and two parameters from the frequency domain were selected. A classification rule is defined in terms of the relation between these parameters of the segment and an undisturbed signal segment. The degree of disturbance is assessed by calculating the deviation in vector form between the parameters of the segment under test and the parameter values of undisturbed segments. When the deviation is less than a threshold value, the segment is classified as undisturbed and otherwise as disturbed. Thus, a segment is classified as disturbed because its properties resemble those of other disturbed segments and also because its properties deviate from those of undisturbed segments. The segments classified as undisturbed are stored together with the proper time coordinate for subsequent analysis.

To find the parameter values of undisturbed segments, two experienced signal analysts carefully examined a large number of signal segments and classified them as either undisturbed or disturbed. The performance of the method is given by the rate of classifying disturbed segments as undisturbed, which was estimated to be 18% and the rate of classifying undisturbed segments as disturbed, which was estimated to be 13%.

Thus, it is possible to develop automatic routines for removal of artefacts and other disturbances from recordings of myoelectric signals. Even if all possible measures have been undertaken there is no guarantee of artefact-free recordings and some environments can be quite difficult. The availability of a disturbance elimination routine is a prerequisite for an automatic analysis which is necessary when the work spells are long and the number of subjects large. Yet, it is better to design the equipment and the experimental conditions in such a way that disturbances are not likely to occur.

References

ANDERSSON, G.B.J. & ÖRTENGREN, R. (1974) Myoelectric back muscle activity during sitting. *Scand. J. Rehab. Med.*, Suppl. 3, 73–90.

ANDERSSON, G.B.J. & ÖRTENGREN, R. (1984) Assessment of back load in assembly line work using electromyography. *Ergonomics*, **27**, 1157–68.

ANDERSSON, G.B.J., ÖRTENGREN, R., NACHEMSON, A. & ELFSTRÖM, G. (1974) Lumbar disc pressure and myolectric back muscle activity during sitting. I Studies on an experimental chair. *Scand. J. Rehab. Med.*, **3**, 115–21.

ANDERSSON, G.B.J., ÖRTENGREN, R. & NACHEMSON, A. (1976) Quantitative studies of back loads in lifting. *Spine*, **1**, 178–85.

ARVIDSSON, A., LINDSTRÖM, L. & ÖRTENGREN, R. (1981) Automatic detection of signal disturbances. Application to myoelectric signals from a work-place environment. In *Proceedings of the 5th Nordic Meeting on Medical and Biological Engineering* (Eds P. Å ÖBERG *et al.*), pp. 34–6. Department of Biomedical Engineering, Linköping University, Linköping, Sweden.

BASMAJIAN, J.V. & DE LUCA, C.J. (1985) *Muscles Alive, Their Functions Revealed by Electromyography* (5th edn) Williams & Wilkins, Baltimore, MD.

KADEFORS, R., LINDSTRÖM, L., PETERSÉN, I. & ÖRTENGREN, R. (1978) EMG in objective evaluation of localized muscle fatigue. *Scand. J. Rehab. Med.*, Suppl. 6, 75–93.

MALMQVIST, R., EKHOLM, I., LINDSTRÖM, L., PETERSÉN, I. & ÖRTENGREN, R. (1981) Measurement of localized muscle fatigue in building work. *Ergonomics*, **24**, 695–709.

MORRISON, R. (1977) *Grounding and Shielding Techniques in Instrumentation* (2nd edn.) Wiley, New York.

CHAPTER FOUR

EMG interpretation

DAVID WINTER
University of Waterloo, Ontario

4.1 INTRODUCTION

Electromyography has been used as a tool in research and in applied areas to tell us many things about the activity of muscles during any movement of humans or animals. Neurologists use electromyogram (EMG) recordings to determine the integrity of neural pathways, neurophysiologists use EMGs to test control and reflex mechanisms and identify synergistic patterns, clinicians use EMGs in a wide range of pathologies to identify atypical motor patterns, biomechanists and bioengineers use EMGs in a wide variety of muscle models and to actuate myoelectric prostheses, and ergonomists use EMGs as indicators of stress and fatigue. There is considerable overlap between their techniques and interpretations. However, unfortunately there is no common and accepted terminology and standards, which has led to misleading and erroneous terms and units. The purpose of this chapter is to provide the background that is common to most of these users and focus on the understanding that ergonomists require for their interpretations. The chapter will cover the biophysical basis of electromyography, important considerations regarding problems of recording, various processing techniques, and interpretations related to fatigue and muscle tension.

4.2 DEFINITIONS

Although most readers may be familiar with the following terms, it remains important to clarify the specialized terms and definitions used in electromyography prior to a discussion of the detailed signal-processing techniques.

1 An alpha (α) motor neuron is the neuron whose cell body is located in the anterior horn of the spinal cord; via its axon and end-plate, the α motor neuron innervates a group of muscle fibers.

2 A motor unit (MU) is the term used to describe the smallest controllable muscular unit. It consists of a single α motor neuron, its neuromuscular junction, and the muscle fibers it innervates.
3 A muscle-fiber action potential or motor action potential (MAP) is the term that describes the detected waveform resulting from the propagation of a depolarizing/repolarizing wave along a single muscle fiber. Unless an individual muscle fiber is separated from the adjacent fibers, it is not possible to isolate an individual MAP.
4 A motor unit action potential (MUAP) is the term given to the detected waveform resulting from the propagation of a depolarizing/repolarizing wave along all fibers associated with a given MU. The shape and amplitude of the MUAP are not only a function of the MU (its size, spatial distribution of the fibers, and propagating velocity) but also the electrode type (contact area, material, interwire spacing) and its location (distance from MU and conductivity of intervening tissue).
5 A myoelectric signal is the name given to the total signal detected at an electrode or differentially between electrodes. It is the algebraic sum of all MUAPs from all active MUs within the pick-up zone of the electrodes.

4.3 RECORDING THE RAW EMG

Before any form of processing of the EMG can occur a 'clean' EMG signal must be secured. This means that the recording is not attenuated and is free of movement artefacts and interference such as hum. Full details of the major considerations are presented by Winter (1990) and a resummarized here. The bioamplifier considerations are the gain and dynamic range, input impedance, frequency response, and common mode rejection ratio (CMRR).

The amplifier gain in conjunction with its dynamic range must take into account the maximum peak-to-peak amplitude of the myoelectric signal seen at the electrodes. For surface EMG the maximum is 5 mV. For indwelling electrodes the maximum is 10 mV. Thus if the dynamic range of the amplifier output was 10 V, the maximum gain should be 2000 for surface recording and 1000 for indwelling. If too high a gain is used, the peaks of the EMG will be clipped off. Independent of what gain is used, the voltage of the EMG as reported in tables and computer plots must be that seen at the input to the bioamplifier from the electrodes, in millivolts or microvolts.

The input impedance (resistance) of the amplifier must be sufficiently high so as not to attenuate the EMG signal at the input terminals. The electrode–skin impedance must be low in comparison with the input impedance. As a rule of thumb, the electrode–skin impedance must be less than 1% of the amplifier input impedance. If the input impedance were 10 mΩ, then the electrode–skin impedance would have to be less than 100 kΩ. High-impedance indwelling electrodes require very high impedance amplifiers. The skin impedance for surface electrodes is reduced by rubbing with rubbing alcohol until the skin is slightly reddened. Properly prepared skin also reduces artefacts due to movement of the electrodes on the skin surface.

The frequency response of the amplifiers takes into account the spectrum of the myoelectric signal relative to that of movement artefacts, which result from movement of the electrodes themselves and sudden movements of the electrode cables. Low-frequency artefacts lie in the range of 0–15 Hz; thus the low-frequency cut-off of the bioamplifier should be set to 15 Hz or slightly higher or lower, depending on the quality of the cables, the vigor of the movement, and the quality of the skin preparation. The high-frequency cut-off is usually set to 500 Hz for a surface EMG and to 1000 to 2000 Hz for an indwelling EMG.

Probably the most important (and least understood) characteristic of bioamplifiers is the CMRR. The body acts as an antenna and picks up electromagnetic radiation from power cords, fluorescent lighting, and electrical machinery. Bioamplifiers are differential amplifiers and a common hum signal is often seen on both electrodes. The ability of the differential amplifier to subtract this common signal is called the CMRR. A perfect subtraction will result in zero hum at the output terminals, but this would require a CMRR of infinity. A CMRR of 10 000 means that the subtraction is accurate to 1 part in 10 000. Thus 2 V hum on the input terminals will result in $1/10^4$ or 200 μV hum at the output. A recommended minimum CMRR is 10 000 (or, as often specified, 80 dB).

4.4 BIOPHYSICAL BASIS

The EMG signal is what we see on the pen recorder, the oscilloscope, or the computer screen. As such, it is an amplified and filtered version of what is detected at our electrodes. Therefore we first must look at the myoelectric signal, the term used to describe the signal 'seen' at the electrodes. We assume that surface electrodes are being employed; however, some discussion is made later about the advantages and disadvantages of surface versus indwelling electrodes.

The myoelectric signal is the signal detected between the two active electrodes that are inputs to our bioamplifier and that signal is the mathematical difference between the signal detected at each electrode site. Both electrodes are usually placed over the muscle and each electrode will detect signals from the active motor units underlying each electrode; this is called a bipolar recording. In some situations it is desirable that one of the electrodes is not placed over the muscle, but rather on an inactive site overlying bone. This signal is referred to as a monopolar recording, but it is still a subtraction, except, in this case, one of the signals is zero.

The myoelectric signal detected at any electrode pair has been shown to be a function of many variables (Kadefors, 1973; McGill and Huynh, 1988; Fuglevand, *et al.*, 1992). Those under the control of the researcher are: electrode surface area and shape, spacing between electrodes (bipolar spacing), orientation of electrode pair over the muscle (along the muscle fibers or across the muscle fibers), and where the pair are located over the muscle (over the motor point or towards either tendon). The neuromuscular variables that influence the detected signal are thickness of the skin and fat layer overlaying the muscle, the pennation angle of the fibers, the bulk of the muscle, the proximity of other muscles (cross-talk), the recruitment profile of the active motor units, the size of the motor units, and the propogating velocity of the depolarizing wave. As will be shown, the propagating velocity is a function of fiber type and fatigue and also the length of the muscle fibers. In isometric calibrations the researcher has some control over the length of the muscle fibers because they can specify the joint angles for the calibration protocol.

Each motor unit within the pick-up range will result in a MUAP being detected. As more motor units are recruited and as the firing rate of already-recruited units increases the myoelectric signal increases. Essentially, the myoelectric signal is an algebraic summation (superposition) of all detected MUAPS. At low levels of contraction when only a few motor units are detected it is possible to visually recognize those motor units (De Luca *et al.*, 1982a, McGill and Dorfman, 1985; McGill *et al.*, 1985). However, it is important to recognize that all motor units that are active are not detected because some are too far from the electrodes and that the active motor units close to the surface will dominate over deeper motor units. Fuglevand *et al.* (1992) presented biophysical predictions which showed the following: (1) a drastic decrease in peak-to-peak amplitude of the MUAP as the electrode–fiber distance increased from 0.5 to 10 mm, (2) a significant increase in the time between the two detected peaks of the biphasic action potentials as this electrode–fiber distance increased from 0.5 to 10 mm, (3) smaller electrodes had higher amplitudes and shorter duration than larger surface area electrodes and (4) the depth of detected MUAPs increased with bipolar spacing. Kadefors (1973) also reported that the spectrum of the EMG shifts upwards as the bipolar spacing is reduced from 20 to 1 mm. In summary, all these variables influence the depth of pick-up of a surface electrode and the frequency content of those MUAPs that are detected. For a common electrode size (49 mm^2) and bipolar spacing (2 cm) the depth of pick-up was predicted to be 1.8 cm (Winter *et al.*, 1994). Thus each electrode can be considered to detect signals from all active fibers over a hemispheric volume underlying the electrode. Larger surface area electrodes increase the breadth of this volume but introduce some cancellation due to minor phase differences of the signal detected at each point of the electrode surface. An increase in the fat layer underlying the electrode will drastically reduce the amplitude of myoelectric signal, smaller surface area electrodes will increase the amplitude slightly but decrease the pick-up range slightly, and bipolar spacing will decrease the pick-up range. All these factors are important when considering potential cross-talk between adjacent muscles and the muscle you think you are recording from; more will be said later on the issue of cross-talk.

Finally, these biophysical models make predictions that are contrary to interpretations regarding the 'filtering' characteristics of the muscle as a volume conductor. These so-called filtering characteristics are important to understand, in particular in ergonomics when fatigue-related changes are measured. A number of researchers claim that the volume conductor acts as a low-pass filter (Kadefors *et al.*, 1968; Lindström *et al.*, 1970). They maintain that the myoelectric spectrum lies in a cut-off region of a low-pass tissue filter and document that when fatigue occurs the propagation velocity decreases and the spectrum shifts downwards. If this were true the spectrum would shift towards the band-pass region and the amplitude would increase and this would explain what was observed. However, the biophysical theory would predict that lower propogating velocities would produce the same shape MUAPs but with greater duration and the same amplitude. Thus, when the EMG amplitude were calculated the full-wave rectified signal would have longer duration MUAPs and thus the area under the curve would increase. Such a phenomena does not need the volume conductor to have low-pass filter characteristics; simple resistive volume conductor characteristics are sufficient to explain what we see experimentally. A recent presentation by Merletti *et al.* (1992) confirmed from M waves recorded during rest and fatigue that their amplitudes did not increase during fatigue, only the duration increased. Thus, the area under the rectified M wave would be predicted to increase but not because of any filtering effect.

4.5 PROBLEMS OF CROSS-TALK

In all EMG recordings, whether in ergonomic studies or elsewhere, the investigator must be concerned with potential cross-talk problems. Several researchers claim that surface electrodes are quite prone to cross-talk compared with indwelling records (Perry *et al.*, 1981; De Luca and Merletti, 1988). Just because two muscles lie close to each other or even are separated by a thin layer of fascia does not mean that cross-talk occurs between any given pair of indwelling or surface recording sites. Perry *et al.* (1981) claimed without any theoretical or empirical evidence that no cross-talk existed between indwelling electrodes inserted into the gastrocnemii and soleus muscles. Based on experimental results of the pick-up zone of indwelling electrodes to be a radius of approximately 0.5 cm (Buchthal *et al.*, 1959) one might estimate that the cross-talk may not exist, but it has not been documented. A surface electrode on the gastrocnemii is considerably further away from the soleus muscle than an indwelling electrode yet Perry *et al.* (1981) assumed cross-talk in the surface records only. They introduced an equation which quantified any common activation between the surface and indwelling to be a measure of cross-talk. As both of these muscles are plantar flexors, common neural drive (co-activation) was automatically declared to be cross-talk on the surface records only. De Luca and Merletti (1988) using stimulation techniques claimed cross-talk between the tibialis anterior and lateral gastrocnemius. However, a simple functional test (such as gait) or a simple manual resistance test can be used to demonstrate no cross-talk between these muscles (Winter *et al.*, 1994). In fact, these researchers show that judicial placement of surface electrodes demonstrates negligible cross-talk between even more closely spaced muscles (tibialis anterior versus peroneus longus and peroneus longus versus lateral gastrocnemius).

If a simple manual resistance test cannot be done and there is suspicion of cross-talk between any two recording sites researchers can use cross-correlation techniques to quantify any common signal and can also take steps to reduce cross-talk (Winter *et al.*, 1994). Reduction in cross-talk has been documented using three techniques: reduction in the surface area of the electrodes, decrease in the bipolar spacing, and mathematical time differentiation of the recorded signal. All three techniques bias the myoelectric signal in favour of the motor units close to the surface electrode and penalize the MUAPs from more distant fibres. Thus, the pick-up zone under the surface electrode is effectively reduced by these three techniques.

4.6 PROCESSING OF THE RAW EMG IN THE TIME DOMAIN

Some analog processing can be done, while other processing can only be done in the digital domain. The different types of analog processing have been summarized (Winter, 1990) and are schematically compared in Figure 4.1. They include (1) half- or full-wave rectification (the latter is also called absolute value), (2) linear envelope detector (rectifier followed by a low-pass filter), and (3) integration of the full-wave-rectified signal for a fixed time, reset to zero, then integration of the full-wave rectified signal to a pre-set level, reset to zero, and then the integration cycle repeated.

In Figure 4.1 the upper trace is the raw EMG and the second trace is the full-wave-rectified EMG, which is now the input to the remaining four processing schemes. If filtered through a low-pass filter, the output signal is smoothed and delayed. If a filter of the correct type and cut-off frequency is chosen, the linear envelope (LE) profile follows closely that of the muscle tension (Inman *et al.*, 1952; Calvert and Chapman, 1977; Olney and Winter, 1985; Winter and Yack, 1987). Note that the output of the first three records is in millivolts. A straightforward integration over the time of the record is shown in the fourth trace and its units (mV·s) represent the area under the full-wave-rectified signal. The fifth trace shows an integration over 50 ms followed by a short duration (<1 ms) reset. The resultant waveform is in μV·s and it closely resembles the full-wave and linear envelope waveforms. The choice of integration period depends on the goal of the researcher. If one follows the rise of bursts of EMG activity as demonstrated in Figure 4.1, the integration period should be short and not exceed 50 ms. However, if long duration efforts are being analyzed or isometric calibrations are being done, a longer integration

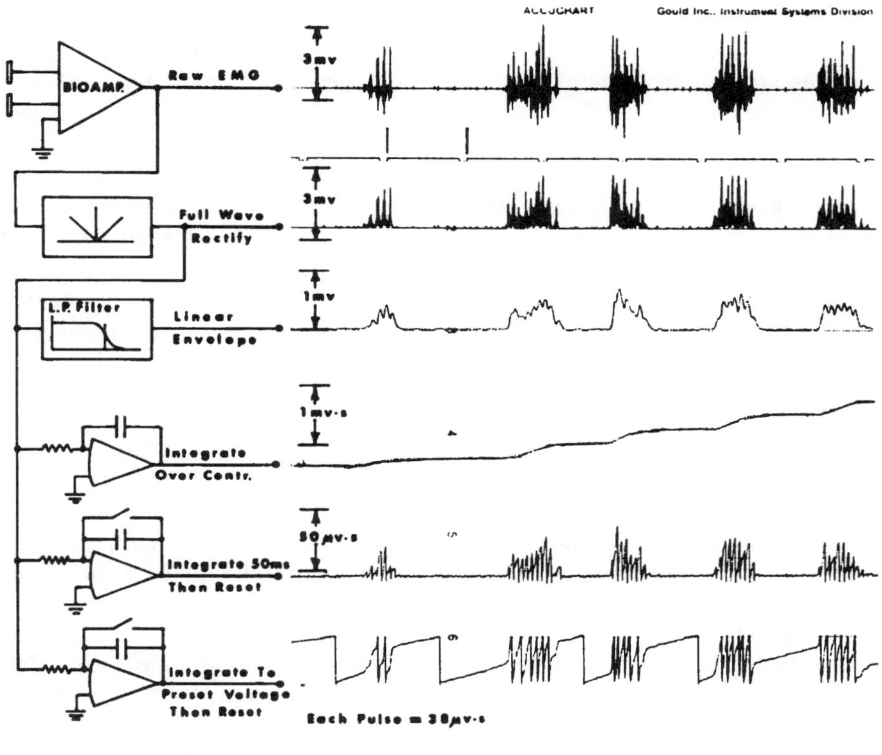

Figure 4.1 Schematic diagram of several common processing systems and the results of simultaneous processing of the same EMG through these systems. Reprinted with permission from Winter, D.A. (1990) *Biomechanics and Motor Control of Human Movement*. Wiley, New York.

time is desirable. Finally, the last trace integrates to a pre-set level (38 μV·s in this example), then resets. Thus, the area under a long direction record can be accurately determined simply by multiplying 38 μV·s by the number of resets. The choice of amplitude of reset should be high for high-level efforts and low for low-level efforts.

Digital techniques are becoming more common and require analog-to-digital conversion. Theoretically, according to the sampling theorem, the raw EMG must be sampled at least twice the highest frequency in the EMG. However, in practice it should be a minimum of four times the highest frequency. Thus, the surface EMG should be sampled at 2000 Hz and the indwelling EMG at 4000 Hz. All analog-processing schemes can be replaced by a digital equivalent. For example, digital filters can replace all analog integration techniques. However, there are some processing techniques that can only be accomplished in digital form, and these are now discussed.

An average EMG (AEMG) is regularly used and is simply the time integral of the full-wave-rectified signal divided by the integration period. Such single measures are simple ways of reporting the general level of activity of any given muscle over a predetermined time. The choice of integration time is left to the researcher. For example, in gait it may be desired to compare AEMG over the full stride period; thus the integration period would begin at heel contact and continue until next heel contact. Or in a lifting task the AEMG levels over the short period of lifting activity may be desired; thus the researcher would have to intervene to define the onset and end of activity from a visual record of each muscle's EMG. Trends over time of subsequent AEMGs have been used, for example, to quantify the onset of fatigue (De Luca, 1979).

4.7 AMPLITUDE PROCESSING OF THE EMG

Amplitude probability density functions quantify the relative probability that an EMG has a given amplitude. Since a raw EMG is a true alternating current signal, it should not have a direct current bias and should have the same amplitude distribution for positive and negative components. Thus amplitude probability density functions are usually calculated after full-wave rectification of the EMG (Dunfield and Shwedyk, 1978) or as amplitude probability distribution functions (Hagberg, 1979; Harba and Ibraheem, 1986; Louhevaara *et al.*, 1990). Amplitude probability distribution function (APDF) curves are inherently normalized to be 1.0 for the maximum amplitude of the EMG over the period of the contraction. Such normalization has been used to describe the time course of the contractions. Over a repetitive event it is important to describe statistically the shape of the contractions. Were they short-duration, high-amplitude bursts or were they longer duration, low-amplitude contraction, or were they a mixture of many types of contractions? Such information is demonstrated from the following time and amplitude plots. Two different types of efforts are represented in Figure 4.2(c). EMG 1 represents low-level activity fluctuating slowly about a mean; signal 2 represents EMG activity of the same AEMG but with short bursts of higher level activity. The amplitude density

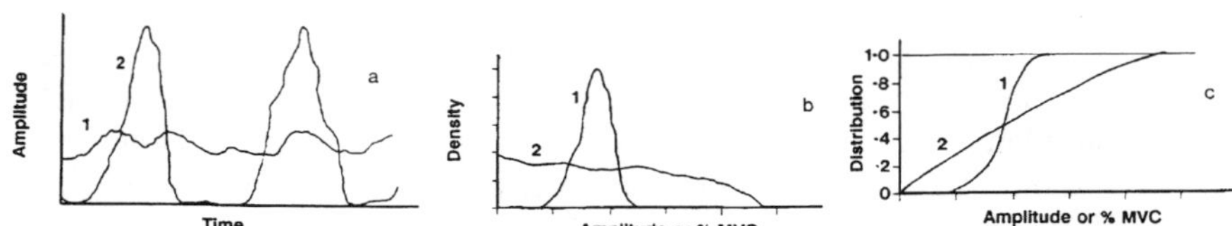

Figure 4.2 Use of amplitude, density and amplitude distribution functions to quantify the type of activity over extended periods of time. See text for details. Reprinted with permission from *Journal of Human Muscle Performance* (1991) **1(2)** 5–15, Aspen Publishers, Gaithersburg, MD.

plots for these two extremes are presented in Figure 4.2(b); here the probalility density (in observations/amplitude or probability/amplitude) is plotted against the amplitude. The area under each of these probability density curves is 1.0 Thus the integral of this curve yields the APDF curve, as shown in Figure 4.2(c). Thus, it is readily possible to determine what percent of the time the muscle is at a certain amplitude or less. Changes in these curves over time have also been used to quantify fatigue or adaptations as a result of fatigue (Hagberg, 1979). In Figure 4.2(c), for example, a shift of the curves to the right statistically quantifies a general increase in EMG activity.

4.8 MAXIMUM VOLUNTARY CONTRACTIONS

Maximum voluntary contractions (MVCs) are useful if we know what percent of the maximum possible activation is represented in an EMG or in the scaling of the linear envelope amplitude. MVCs have been used as a normalization protocol (Hagberg, 1979; Yang and Winter, 1984; Arsenault *et al.*, 1986; Harba and Ibraheem, 1986; Louhevaara *et al.*, 1990). There are considerable problems in actually measuring the MVC and even more problems in interpreting what it means. First, does it mean maximum force or maximum activation? Second, because of the inhibitory feedback from the Golgi tendon organs it is virtually impossible to recruit all motor units at their maximum firing rates. Thus, the MVC must be interpreted as that maximum value that a subject can voluntary generate during an isometric contraction with that inhibitory feedback present. Thus, from a group of successive MVCs the largest one should be defined to be the maximum. Such maximums can increase with training. The protocol for eliciting the MVC varies; some researchers used larger duration (5 s) maximums (Ericson *et al.*, 1986), while others used selected 1 s segments from a 3 s record (Arsenault *et al.*, 1986). It is therefore important to report the exact period selected and it is preferable that a shorter period is chosen to enable an isolated maximum to occur. In rapid, voluntary, dynamic movements, it is quite common to record EMGs that momentarily exceed the EMG associated with the isometric maximum. Such momentary maximums are usually elicited as a result of a rapid muscle lengthening superimposed on high-level eccentric activity; the excitatory stretch reflex of the spindles causes the EMG to momentarily exceed 100%.

Other problems associated with MVCs are whether such efforts are safe (in particular in the back muscles). In addition, it is impossible to elicit an MVC without some resultant co-activation that will reduce the joint moment and therefore result in an underestimate of the muscle force in the agonist muscles.

In spite of the inherent problems in achieving a valid and reliable MVC, the AEMG of such efforts is used as a standardized measure against which to compare subsequent efforts during the movement of interest. However, it has been documented (Enoka and Fuglevand, 1991) that the voluntary induced maximum is less than the maximum force capacity of the muscle.

4.9 ENSEMBLE AVERAGING

In cyclical movements such as walking, running, cycling, or any repetitive work situation there has been considerable effort to achieve an average EMG profile. Such cycle-to-cycle averages are called ensemble averages, in which the time base for each cycle is normalized to 100% and the profile for each successive cycle is averaged. A raw EMG cannot be averaged, but full-wave-rectified and LE EMGs are readily averaged. The degree of low-pass filtering depends on the goals of the research. If a measure of average latency was desired, there should be negligible filtering of the full-wave signal (a 100 Hz cut-off for a critically damped filter will result in only 2 ms delay in the LE). However, there are strong arguments for low-pass filters that model the twitch characteristic of the muscle. A 3 Hz critically damped low-pass filter, for example, has a transfer function that mimics muscles with twitches whose time to peak tension, T, is 53 ms. More of this processing will be presented in the later section on EMG versus muscle tension. A number of researchers have used the LE as a predictor of muscle tension (Inman *et al.*, 1952; Calvert and Chapman, 1977; Olney and Winter, 1985) and for processing the EMG prior to ensemble averaging (Yang and Winter, 1984; Kadaba *et al.*, 1985; Shiavi, 1985; Arsenault *et al.*, 1986; Ericson *et al.*, 1986).

The temporal precision of averages within the stride period in gait analyses has varied from 2% of the full cycle in some reports (Arsenault *et al.*, 1986) to 5% in others (Yang and Winter, 1984; Ericson *et al.*, 1986) and to coarse profiles averaged at as few as eight points over the cycle (Peat *et al.*, 1976). The precision of ensemble averaging is a function of how the EMG was processed prior to averaging that will reflect the goals of the research. As indicated above, averaged latency measures would require precision within a few milliseconds; thus the stride period would have to be averaged using approximately 400 samples over the stride. However, if the EMG linear envelope used a 3 Hz low-pass filter, it would be appropriate to average it at 5 or 2% intervals.

Ensemble averaging of myoelectric signals in gait has been reported for almost two decades (Peat *et al.*, 1976) and has been used to establish profiles of normal people (Shiavi, 1985; Winter and Yack, 1987) and as a database for comparison of pathological gait profiles (Winter, 1984). Such profiles have also been used in biomechanical models to decompose the force contribution of individual muscles in walking (Olney and Winter, 1985) and in running (Scott and Winter, 1990).

Normalization of the EMG for intersubject averages is an inherently desirable goal when all the factors which can cause the EMG to vary across subjects are considered: difference in muscle bulk and overlying fat tissue, intersubject differences in neural control strategies, bilateral dominance (Ounpuu and Winter, 1989), and minor differences in electrode location. Arsenault *et al.* (1986) rationalized that the underlying factor in common across subjects is % MVC; such acalibration should eliminate most of the biological differences that were known to exist due to differences in the muscle bulk and overlying fat tissue and the electrode location. Yang and Winter (1984) compared four different normalization techniques to see which would reduce the intersubject variability: 50% MVC, EMG amplitude per joint moment (μV Nm^{-1}), average value of LE over the stride period, and peak LE amplitude over the stride period. The reasons for choosing 50% MVC rather than MVC were 2-fold. First, it is considerably easier for a subject to maintain a 50% contraction at a steady value during calibration. Second, the linearity of EMG versus muscle tension is very good below 50% and most muscle activity in gait (except that of the plantar flexors) is well below 50%. The results compared with no normalization (μV) were surprising. For the six muscles tested, the % MVC and moment normalization either made no difference or actually increased the intersubject variability. Both the stride average and peak LE averaging techniques drastically decreased the variability in the intersubject ensemble average. In an extensive study of 16 muscles across a large number of subjects (Winter and Yack, 1987), the variability of intersubject averages was reduced by approximately 50%. Such a drastic reduction indicates that the major difference across subjects is amplitude alone and most of this can be attributed to differences in muscle bulk and thickness of the overlying fat tissue.

The variability of ensemble averages has been quantified to give a statistical measure of the stride-to-stride variability or intersubject variability (for databases). Two formulae have been proposed. Yang and Winter (1984) introduced a coefficient of variation (CV) score that is the root mean of the variance over the stride divided by the mean of the ensemble averaged signal. In effect, it is a linear variability-to-signal ratio. For stride-to-stride averages, CV scores can vary from 31 to 62%, while intersubject averages vary from 48 to 118% for unnormalized EMGs and 28 to 61% for normalized EMGs (Winter and Yack, 1987). Kadaba *et al.* (1985) have proposed an alternate measure called a variance ratio that is non-linear and is bounded between 0 and 1.

4.10 EMG CHANGES IN FATIGUE

One of the primary uses of electromyography in ergonomics is as an index of fatigue in muscles that are undergoing continuous loading. Over 25 years ago, Kadefors *et al.* (1968) reported shifts in the myoelectric spectrum from both surface and indwelling electrodes over the period of build-up of fatigue. Using a bank of analog filters they reported a decrease in the higher frequency spectrum (1000 Hz, 250 Hz) and an increase in the lower frequency components (63 Hz, 16 Hz). Thus, a shift in the myoelectric spectrum towards the low end was evident. They attributed these shifts to an increased synchronization of the motor units (which enhanced the lower frequency content), plus a replacement of some of the fast twitch units with lower frequency fatigue-resistant units. Such a drop-out of the larger fast twitch units and a replacement with smaller slow twitch units (or increased firing rate of already recruited small units) could explain the decrease in MVC tension. Two years later, Mortimer *et al.* (1970) attributed the lowering of the spectrum to a progressive decrease in the velocity of propagation of the depolarizing wave, thus increasing the duration of the detected MUAPs. The major cause of the decrease in the velocity was an accumulation of metabolic by-products in the muscle fibers. Lindström *et al.* (1970) in a 'maximum' load of the biceps brachii for 30 s reported a significant downward shift in the spectrum at the same time as the action potential conduction velocity decreased from 3.3 to 2.0 $m\,s^{-1}$. These conduction velocities are somewhat below the generally accepted value around 4 $m\,s^{-1}$ (Buchthal *et al.*, 1955; Sollie *et al.*, 1985) but the reduction in velocity certainly agrees with changes reported by others (Lynn, 1979). Petrofsky (1979) demonstrated another mechanism that can confound the velocity shifts. He independently controlled the temperature of the muscle and reported that the mean frequency of the EMG increased with temperature, which can be interpreted as an increase in the propogating velocity of the depolarizing wave. In spite of the fact that three mechanisms are involved, the net result is a downward shift in the frequency spectrum of the myoelectric signal.

The desire to report a single index of fatigue has resulted in several scores being proposed. Lindström *et al.* (1977) reported a regression coefficient of the exponential decay in conduction velocity to be a fatigue index. Another measure that integrates both the changes in conduction velocity and recruitment is the mean power frequency (MPF).

$$\mathrm{MPF} = \frac{\int_{f_1}^{f_2} fG(f)\mathrm{d}f}{\int_{f_1}^{f_2} G(f)\mathrm{d}f} \tag{1}$$

where $G(f)$ is the power spectral density function of the myoelectric signal and f_1 and f_2 are the minimum and maximum frequencies of the myoelectric signal. The percentage decrease in MPF over the fatigue period is considered an index of fatigue (Viitasalo and Komi, 1977; Lynn, 1979; Petrofsky and Lind, 1980; Hagberg, 1981). There are two major precautions that must be observed when pre- and post-fatigue MPFs are being recorded. Firstly, it is critical that the amplitude of the EMG used for this index calculation be the same (Komi and Viitasalo, 1976). Keeping the activation level constant ensures that the recruitment profile is maintained fairly constant. Second, the muscle length must remain the same because the detected myoelectric signal is very sensitive to fiber length (Lynn, 1979; Inbar *et al.*, 1987). In fact, changes in the MPF due to muscle length are of the same magnitude as the fatigue-induced changes. The conduction velocity decreases as the length of muscle fibers increases. Thus, if the fatigue record were taken with the muscle longer than the rest contraction, the downward shift of the spectrum would be exaggerated; if the fatigue record were recorded with the muscle shorter, the fatigue-induced MPF shift could be completely cancelled.

4.11 RELATIONSHIP BETWEEN EMG AND MUSCLE TENSION

During isometric contractions there are well-defined relationships between the amplitude of the EMG and the muscle tension (or the moment it produces at a given joint). If constant tension records are taken the AEMG can be plotted against tension. Both linear and non-linear relationships have been reported. Early studies by Lippold (1952) on the calf muscles of humans resulted in a linear relationship; Vredenbregt and Rau (1973), on the other hand, found a quite non-linear curve between EMG amplitude and elbow moment from the biceps brachii. The relationship was also very sensitive to muscle length as was seen from the plots at three different angles of elbow flexion.

During dynamic changes of isometric tension there is a phasic delay in the tension relative to the myoelectric signal. Inman *et al.* (1952) first demonstrated a lag of the tension behind the visually observed rises and falls in the myoelectric signal.They also showed that the LE signal rose and fell with the tension waveform. Gottlieb and Agarwal (1971) mathematically modeled

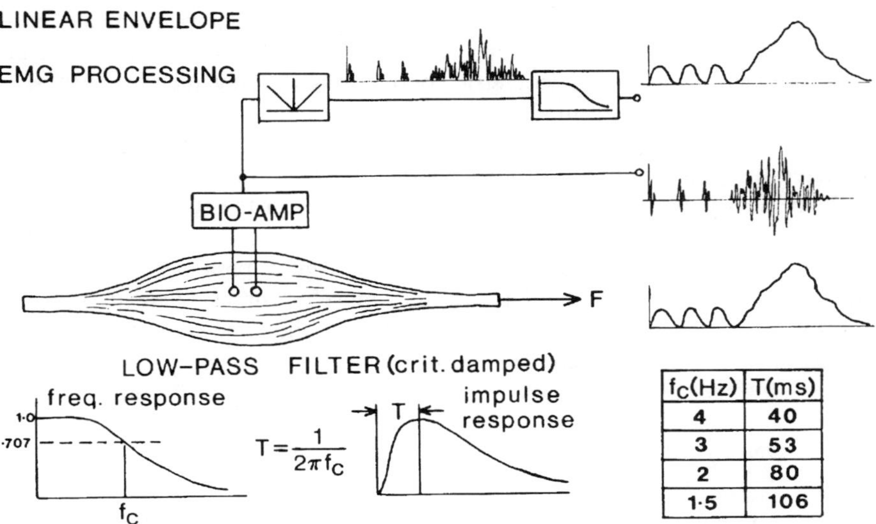

Figure 4.3 Linear envelope processing of the EMG using a second-order, critically damped, low-pass filter. This filter mimics the twitch response of the muscle such that the cut-off frequency of the filter, f_c, can be set to match the twitch time, T, of the muscle being recorded.

this relationship as a second-order low-pass system. Figure 4.3 is presented to demonstrate such processing which models the muscle's mechanical characteristics as a second-order, critically damped, low-pass filter. Such a relationship is not surprising since the frequency characteristics of muscle twitches has been shown to be a second-order, critically damped, low-pass system (Milner-Brown *et al.*, 1973). Each twitch can be considered to be the impulse response of that motor unit and associated with each twitch is the MUAP impulse. Thus, the full-wave-rectified EMG can be considered a summation of MUAPs which look like impulses of varying amplitude while the muscle tension is a graded summation of the twitches from all active motor units. Thus, if we chose a critically damped, low-pass filter of the correct frequency we could approximate the tension waveform. Because muscle twitches are close to being critically damped (or slightly overdamped) we have a simple relationship between the twitch time, T and the cut-off frequency of the filter, f_c:

$$f_c = \frac{1}{2\pi T} \tag{2}$$

There are close correlations between muscle force and the LE signal during isometric anisotonic contractions (Calvert and Chapman, 1977; Crosby 1978; Winter, 1987). The cut-off frequency of the filters was in the range of 2.3–8.5 Hz which is compatible with the 20–90 ms range of contraction times reported for twitch waveforms (Buchthal and Schmalbruch, 1970; Milner-Brown *et al.*, 1973). A comprehensive biophysical/biomechanical model predicted EMG–force relationships assuming four different recruitment and rate coding strategies during isometric contractions (Fuglevand *et al.*, 1993).

There have been a few models which have attempted to predict the muscle force under non-isometric, non-isotonic conditions. Olney and Winter (1985) predicted the joint moments during walking and the optimal cut-off frequency of the critically damped filters ranged from 1.0 Hz for the soleus to 2.8 Hz for the rectus femoris. In ergonomics tasks such as weight lifting, the trunk muscles were similarly modeled and muscle forces predicted (McGill and Norman, 1986). The prediction of the muscle forces then permits the researcher to use link segment modeling to predict the bone-on-bone forces and estimate whether they exceed desired limits.

4.12 CONCLUSIONS

Electromyography is one of the few techniques that allows an investigator to 'penetrate' the musculoskeletal system and see what is going on at the individual muscle level. Surface recordings are particularly easy to use as long as the proper recording techniques are observed in order to obtain a clean EMG free of hum, movement artefact, and cross-talk. Suitable processing of the

EMG enables the researcher to identify the phasic activation of each muscle and to model the muscle to predict its tension or its contribution to the joint moment-of-force. Also, frequency analysis techniques can readily identify and quantify localized muscle fatigue.

References

ARSENAULT, A.B., WINTER, D.A. & MARTENIUK, R.G. (1986) Is there a normal profile of EMG activity in gait? *Med. Biol. Engng. Comput.*, **24**, 337–43.

BUCHTHAL, F. & SCHMALBRUCH, H. (1970) Contraction times and fibre types in intact human muscle. *Acta Physiol. Scand.*, **79**, 435–52.

BUCHTHAL, F., GULD, C. & ROSENFALCK, P. (1955) Propagation velocity in electrically activated fibres in man. *Acta Physiol. Scand.*, **34**, 75–89.

BUCHTHAL, F., ERMINIO, P.D.L. & ROSENFALCK, P. (1959) Motor unit territory in different human muscles. *Acta Physiol. Scand.*, **45**, 72–87.

CALVERT, T.W. & CHAPMAN, A.E. (1977) Relationship between surface EMG and force transients in muscle: Simulation and experimental results. *Proc. IEEE*, **65**, 682–89.

CROSBY, P.A. (1978) Use of surface electromyography as a measure of dynamic force in human limb muscles. *Med. Biol. Engng Compat.*, **16**, 519–24.

DE LUCA, C.J. (1979) Physiology and mathematics of myoelectric signals. *IEEE Trans. Biomed. Engng*, **26**, 313–25.

DE LUCA, C.J. & MERLETTI, R. (1988) Surface myoelectric signal cross-talk among muscles of the leg. *EEG Clin. Neurophysiol.*, **69**, 568–75.

DE LUCA, C.J., LEFEVER, R.S., MCCUE, M.P. & Xenakis, A.P. (1982a) Behaviour of human motor units in different muscles during linearly varying contractions. *J. Physiol.*, **329**, 113–28.

DE LUCA, C.J., LEFEVER, R.S., MCCUE, M.P. & XENAKIS, A.P. (1982b) Control scheme governing concurrently active human motor units during voluntary contractions. *J. Physiol.*, **329**, 129–42.

DUNFIELD, V. & SHWEDYK, E. (1978) Digital EMG processor. *Med. Biol. Engng Comput.*, **16**, 745–51.

ENOKA, R.M. & FUGLEVAND, A.J. (1991) Neuromuscular basis of the maximum voluntary force capacity of muscle. In *Current Issues in Biomechanics: A Prospectus* (Ed. M.D. GRABINER), pp. 215–35. Human Kinetics, Champaign, IL.

ERICSON, M.O., NISELL, R. & ELKOLM, J. (1986) Quantified electromyography of lower-limb muscles during level walking. *Scand. J. Rehabil. Med.*, **18**, 159–63.

FUGLEVAND, A.J., WINTER, D.A., PATLA, A.E. & STASHUK, D. (1992) Detection of motor unit action potentials with surface electrodes: influence of electrode size and spacing. *Biol. Cybernetics* **67**, 143–53.

FUGLEVAND, A.J., WINTER, D.A. & PATLA, A.E. (1993) Models of recruitment and rate coding organization in motor unit pools. *J. Neurophysiol.*, **70**, 2470–88.

GOTTLIEB, G.L. & AGARWAL, G.C. (1971) Dynamic relationship between isometric muscle tension and the electromyogram in man. *J. Appl. Physiol.*, **30**, 345–51.

HAGBERG, M. (1979) The amplitude distribution of surface EMG in static and intermittent static muscular contractions. *Eur. J. Appl. Physiol.*, **40**, 265–72.

HAGBERG, M. (1981) Electromyographic signs of shoulder muscular fatigue in two elevated arm positions. *Am. J. Phys. Med.*, **60**, 111–21.

HARBA, M.I.A. & IBRAHEEM, A.A. (1986) EMG processor based on the amplitude probability distribution. *J. Biomed. Eng.*, **8**, 105–14.

INBAR, G.F., ALLIN, J. & KRANZ, H. (1987) Surface EMG spectral changes with muscle length. *Med. Biol. Engng Comput.*, **25**, 683–89.

IINMAN, V.T., RALSTON, H.J., SAUNDERS, J.B, FEINSTEIN, B. & WRIGHT, E.W., Jr (1952) Relation of human electromyogram to muscle tension. *EEG Clin. Neurophysiol.*, **4**, 187–94.

KADABA, M.P., WOOTEN, M.E., GAINEY, J. & COCHRAN, G.V.B. (1985) Repeatability of phasic muscle activity: performance of surface and intramuscular wire electrodes in gait analysis. *J. Orthop. Res.*, **3**, 350–59.

KADEFORS, R. (1973) Myo-electric signal processing as an estimation problem. In *New Developments in Electromyography and Clinical Neurophysiology*, Vol. 1 (Ed. J.E. DESMEDT), pp. 519–32. Karger, Basel.

KADEFORS, R., KAISER, E. & PETERSEN, I. (1968) Dynamic spectrum analysis of myo-potentials with special reference to muscle fatigue. *Electromyography*, **8**, 39–74.

KOMI, P. & VIITASALO, J.H.T. (1976) Signal characteristics of EMG at different levels of muscle tension. *Acta Physiol. Scand.*, **96**, 267–76.

LINDSTRÖM, L., MAGNUSSON, R. & PETERSEN, I. (1970) Muscular fatigue and action potential conduction velocity changes studied with frequency analysis of EMG signals. *Electromyography*, **10**, 341–56.

LINDSTRÖM, L., KADEFORS, R. & PETERSÉN, I. (1977) An electromyographic index for localized muscle fatigue. *J. Appl. Physiol.*, **43**, 750–54.

LIPPOLD, O.C.J. (1952) The relationship between integrated action potentials in a human muscle and its isometric tension. *J. Physiol.*, **177**, 492–99.

LOUHEVAARA, V., LONG, A., OWEN, P., AICKIN, C. & MCPHEE, B. (1990) Local muscle and circulatory strain in load lifting, carrying and holding tasks. *Int. J. Indust. Ergonom*, **6**, 151–62.

LYNN, P.A. (1979) Direct on-line estimation of muscle fiber conduction velocity by surface electromyography. *IEEE Trans. Biomed. Engng*, **26**, 564–71.

MCGILL, K.C. & HUYNH, A. (1988) A model of surface-recorded motor-unit action potential. *Proc. Ann. Internat. Conf. of the IEEE Engng. in Med. and Biol. Soc.*, pp. 1697–99, New Orleans.

MCGILL, S.M. & NORMAN, R.W. (1986) Partitioning of the L4–L5 dynamic moment into disc, ligamentous and muscular components during lifting. *Spine*, **11**, 666–77.

MCGILL, K.C. & DORFMAN, L.J. (1985) Automatic decomposition electromyograph (ADEMG): validation and normative data in brachial biceps. *EEG Clin. Neurophysiol.*, **61**, 453–61.

MCGILL, K.C., CUMMINS, K.L. & DORFMAN, L.J. (1985) Automatic decomposition of the clinical electromyogram. *IEEE Trans. BME*, **20**, 470–77.

MERLETTI, R. FAN, Y. & LO CONTE, R.L. (1992) Estimation of scaling factors in electrically evoked EMG signals. *Proc. 14 Ann. Conf. IEEE Engng. Med. Biol. Soc.*, pp. 1362–3, Paris.

MILNER-BROWN, H.S., STEIN, R.B. & YEMM, R. (1973) The contractile properties of human motor units during voluntary isometric contractions. *J. Physiol.*, **228**, 285–306.

MORTIMER, J.T., MAGNUSSON, R. & PETERSEN, I. (1970) Conduction velocity in ischemic muscle: effect on EMG frequency spectrum. *Am. J. Physiol.*, **219**, 1324–29.

OLNEY, S.J. & WINTER, D.A. (1985) Prediction of knee and ankle moments in walking from EMG and kinematic data. *J. Biomech.*, **18**, 9–20.

OUNPUU, S. & WINTER, D.A. (1989) Bilateral EMG analysis of lower limbs during walking in normal adults. *EEG Clin. Neurophysiol.*, **72**, 429–38.

PEAT, M., DUBO, H.I.C., WINTER, D.A. QUANBURY, A.O., STEINKE, T. & GRAHAME, R. (1976) Electromyographic temporal analysis of gait: Hemiplegic locomotion. *Arch. Physiol. Med. Rehabil.*, **57**, 421–25.

PERRY, J., EASTERDAY, C.S. & ANTONELLI, D. (1981) Surface vs. intramuscular electrodes for EMG of superficial and deep muscles. *Physical Therapy*, **61**, 7–15.

PETROFSKY, J.S. (1979) Frequency and amplitude analysis of the EMG during exercise on the bicycle ergonometer. *Eur. J. Appl. Physiol.*, **41**, 1–15.

PETROFSKY, J.S. & LIND, A.R. (1980) Frequency analysis of the surface electromyogram during sustained isometric condtions. *Eur. J. Appl. Physiol.*, **43**, 173–82.

SCOTT, S.H. & WINTER, D.A. (1990) Internal forces at chronic running injury sites. *Med. Sci. Sports Exerc.*, **22**, 357–69.

SHIAVI, R. (1985) Electromyographic patterns in adult locomotion : a comprehensive review. *J. Rehabil. Res. Devel.*, **22**, 85–98.

SOLLIE, G., HERMENS, H.J., BOON, K.L., WALLINGA-DE JONGE, W. & ZILVOLD, G. (1985) The measurement of conduction velocity of muscle fibres with surface EMG according to the cross-correlation method. *EMG Clin. Neurophysiol.*, **25**, 193–204.

VIITASALO, J.H.T. & KOMI, P.V. (1977) Signal characteristics of EMG during fatigue. *Eur. J. Physiol.*, **37**, 111–21.

VREDENBREGT, J. & RAU, G. (1973) Surface electromyography in relation to force, muscle length and endurance. In *New Developments in Electromyography and Clinical Neurophysiology*, Vol. 1 (Ed. J.E. DESMEDT), pp. 607–22. Karger, Basel.

WINTER, D.A. (1984) Pathological gait diagnosis with computer averaged electromyographic profiles. *Arch. Phys. Med. Rehabil.*, **65**, 393–98.

WINTER, D.A. (1990) *Biomechanics and Motor Control of Human Movement.* Wiley, New York.

WINTER, D.A. & YACK, H.J. (1987) EMG profiles during normal human walking: stride-to-stride and intersubject variability. *EEG. Clin. Neurophysiol.*, **67**, 402–11.

WINTER, D.A., FUGLEVAND, A.J. & ARCHER, S.E. (1994) Crosstalk in surface electromyography: theoretical and practical estimates. *J. Electromyogr. Kinesiol.*, **4**, 15–26.

YANG, J.F. & WINTER, D.A. (1984) Electromyographic amplitude normalization methods: improving their sensitivity as diagnostic tools in gait analysis. *Arch. Phys. Med. Rehabil.*, **65**, 517–21.

CHAPTER FIVE

Muscle energetics and electromyography

TOSHIO MORITANI

Kyoto University, Kyoto

5.1 INTRODUCTION

This chapter deals with myoelectric signal changes associated with muscular work with special reference to energy metabolism, excitation–contraction (EC) coupling, cardiorespiratory response, motor learning, and muscle soreness. These topics will be organized under three main topics, i.e. energy metabolism and neuromuscular regulation, electromyographic estimation of physical work capacity, and electromyographic manifestations of muscle soreness. In Section 5.2, motor unit (MU) and myoelectric signal characteristics, oxygen availability and MU activity, energy metabolism and excitation–contraction processes, and motor unit activity and cardiorespiratory response are discussed. Readers will find the use of electromyography techniques for estimating physical working capacity and motor control in Section 5.3: determination of the maximal work capacity at the neuromuscular fatigue threshold and electromyographic manifestations of neuromuscular adaptations during acquisition of motor tasks. Movement-related cortical potentials are also discussed in Section 5.3 (movement-related cortical potentials and EMG signal amplitude during motor tasks). Finally, a current view of the delayed-onset of muscle soreness (DOMS) is organized in Section 5.4 which contains a brief review of the etiology of DOMS followed by review of acute effects of static stretching on muscle soreness and effects of aerobic exercise on neuromuscular relaxation.

5.2 ENERGY METABOLISM AND NEUROMUSCULAR REGULATION

5.2.1 Motor Unit Activity and Myoelectric Signal

A motor unit (MU) consists of a motor neuron in the spinal cord and the muscle fibers it innervates. The number of MUs per muscle in humans may range from approximately 100 for a small hand muscle to 1000 or more for large limb muscles (Henneman and Mendell, 1981). It has also been shown that different MUs vary greatly in force-generating capacity, i.e. a 100-fold or more difference in twitch force (Stephens and Usherwood, 1977; Garnett *et al.*, 1979). In voluntary contractions, force is modulated by a combination of MU recruitment and changes in the MU activation frequency (rate coding) (Milner-Brown *et al.*, 1973; Kukulka and Clamann, 1981; Moritani and Muro, 1987). The greater the number of MUs recruited and their discharge frequency, the greater the force will be. During MU recruitment the muscle force, when activated at any constant discharge frequency, is approximately 2–5 $kg\,cm^{-2}$ and in general is relatively independent of species, gender, age, and training status (Ikai and Fukunaga, 1970; Alway *et al.*, 1990).

The electrical activity in a muscle is determined by the number of MUs recruited and their mean discharge frequency of excitation, i.e., the same factors that determine muscle force (Woods *et al.*, 1978; Bigland-Ritchie, 1981; Moritani and Muro, 1987). Thus, a direct proportionality between an electromyogram (EMG) and force might be expected. Under certain experimental conditions, these proportionalities can be well demonstrated by recording the smoothed rectified or integrated EMG (IEMG) (deVries, 1968a; Milner-Brown and Stein, 1975; Moritani and deVries, 1978, 1979) and reproducibility of EMG recordings are remarkably high, e.g. the test–retest correlation is in the range 0.97–0.99 (Komi and Buskirk, 1970; Moritani and deVries, 1978, 1979). However, the change in the surface EMG should not automatically be attributed to changes in either MU recruitment or excitation frequencies as the EMG signal amplitude is further influenced by the individual muscle fiber potential, the degree of MU discharge synchronization, and fatigue (Milner-Brown *et al.*, 1973; Jessop and Lippold, 1977; Bigland-Ritchie *et al.*, 1979, 1986; Bigland-Ritchie, 1981; Moritani *et al.*, 1985b, 1986b).

The surface electromyogram (EMG) frequency power spectral analysis in studies of muscle function has attained increasing attention during recent years and has been applied in assessing muscle endurance capacity (Viitasalo and Komi, 1977; Hagberg, 1981), anaerobic and lactate thresholds (Nagata *et al.*, 1981; Moritani *et al.*, 1984), neuromuscular diseases (Muro *et al.*, 1982; Mills and Edwards, 1984), motor unit activities (Komi and Viitasalo, 1977; Komi *et al.*, 1978; Moritani *et al.*, 1986b, 1988; Moritani and Muro, 1987), and skeletal muscle fatigue (Kadefors *et al.*, 1968; Lindström *et al.*, 1977; Viitasalo and Komi, 1977; Hagberg, 1981; Moritani *et al.*, 1982, 1985b, 1986b; Bigland-Ritchie *et al.*, 1983). It has been well documented that the

EMG frequency power spectrum shifts to lower frequency bands together with increased signal amplitude during development of muscle fatigue. Because of its tight link to the circulatory and metabolic processes in the muscle, the power spectrum shift can define localized muscle fatigue before the sensation of fatigue becomes apparent (Kadefors *et al.*, 1968; Lindström *et al.*, 1977).

During development of muscle fatigue, earlier electromyographic studies (deVries, 1968b; Bigland-Ritchie, 1981; deVries *et al.*, 1982; Moritani *et al.*, 1982; Bigland-Ritchie *et al.*, 1986) indicated that amplitude of myoelectric signals from the surface electrodes increased progressively as a function of time during sustained submaximal contractions. It was suggested that additional MUs were progressively recruited to compensate for the loss of contractility due to some degree of impairment of fatigued MUs. However, this increased amplitude of the surface EMG could not be demonstrated during sustained maximal voluntary contractions (MVCs) (Bigland-Ritchie *et al.*, 1979, 1983b; Moritani *et al.*, 1985b) (see Figure 5.1). Recent evidence has indicated that there is a progressive reduction in MU firing rates during sustained MVCs in

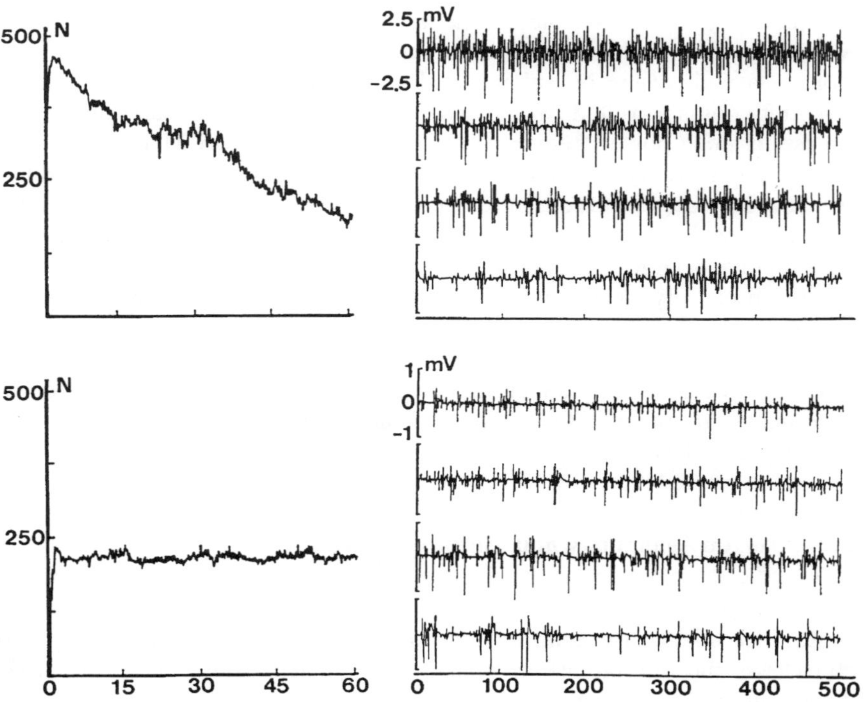

Figure 5.1 Left: averaged force data ($n = 12$) during sustained maximal voluntary contraction (MVC) (top) and 50% MVC (bottom). Right: a typical set of data showing the time course of changes in intramuscular spike activities during sustained MVC and 50% MVC. Based on Moritani et *al.* (1986b).

the absence of any measurable neuromuscular transmission failure (Bigland-Ritchie *et al.*, 1983a; Moritani *et al.*, 1986b). The apparent parallel decline in both force and MU activity could not be attributed to 'central fatigue', i.e. loss in central motor drive. Studies in humans clearly indicated that this can be overcome in sustained MVCs when performed by adequately motivated and practiced subjects, for the lost force could not be restored by supramaximal tetanic nerve stimulation (Bigland-Ritchie *et al.*, 1983b; Moritani *et al.*, 1985b). Jones *et al.* (1979) and Moritani *et al.* (1985b) also showed that prolonged stimulation at 80 Hz results in a faster rate of force loss than in a sustained MVC. This notion of 'high-frequency fatigue' (Edwards, 1981) suggests that maintenance of a high MU firing frequency may lead to impaired EC coupling, possibly associated with extracellular Na^+ depletion (Jones *et al.*, 1979; Moritani *et al.*, 1985a). Thus, it appears that a reduced MU firing rate during fatiguing MVCs may ensure the optimal force output by avoiding peripheral neuromuscular transmission failure (Bigland-Ritchie *et al.*, 1979, 1983a). This fact may also appear to be a 'fail-safe' mechanism against the muscle going into rigor through depletion of ATP (Nassar-Gentina *et al.*, 1978; Kugelberg and Lindegren, 1979). These findings clearly indicate the existence of different MU recruitment and rate-coding mechanisms during sustained maximal and submaximal voluntary contractions.

During a wide range of submaximal contractions not all of the available motor unit pool is recruited. We have demonstrated that there was a progressive decrease in mean power frequency (MPF) of the surface EMG signal during sustained contractions at 50% of MVC, but this decline was accompanied by a significant increase in the root mean square value of the EMG amplitude and a progressive MU recruitment as evidenced by an increased number of MUs with a relatively large intramuscular spike amplitude (Moritani *et al.*, 1986b). Recent muscle glycogen study by Vollestad *et al.* (1984) indicated that glycogen content of types IIa and IIb was unchanged during the first part of exercise. Later a decrease was observed, first in type IIa and finally in IIb, suggesting a decrease in the recruitment threshold force of these fibers during development of fatigue. Our intramuscular EMG studies (Moritani *et al.*, 1986a, b, 1993) have provided strong support for this notion.

5.2.2 Oxygen Availability and Motor Unit Activity

During cycling exercise an increase in the plasma lactate concentration occurs at ≈50–70% of the maximal oxygen uptake rate over the lung (VO_{2max}) and well before the aerobic capacity is fully utilized (Jorfeldt *et al.*, 1978; Wasserman *et al.*, 1985; Sahlin *et al.*, 1987). During an incremental exercise test on a bicycle ergometer, however, Sjøgaard (1978) showed that the peak force for a pedal thrust during bicycle exercise was less than 20% of the maximal voluntary contraction at work-loads eliciting VO_{2max} at 60 r.p.m. Despite these relatively low force outputs and moderate speed of contractions during cycling, the glycogen content of fast-twitch oxidative glycolytic (FOG or type IIa) and fast-twitch glycolytic (FG or type IIb) fibers was progressively decreased (first in type IIa and then in type IIb) during a prolonged submaximal exercise bout (Vollestad *et al.*, 1984). Furthermore, Gollnick *et al.* (1974) have also shown that in isometric contractions, slow-twitch units are the only ones to be depleted of glycogen at force developments of up to 15–20% MVC. Above this level, fast-twitch units are also depleted of glycogen. This suggests that with this type of muscular contraction the availability of oxygen and the developed force influences recruitment of fast-twitch units since blood flow is usually restricted during a sustained contraction of approximately 20% MVC (Humphreys and Lind, 1963). These available data thus suggest that not only the force and speed of contraction, but also the availability of oxygen may affect the recruitment of high-threshold motor units.

We have recently investigated the interrelationships between blood supply, motor unit (MU) activity, and lactate during intermittent isometric contractions of the hand grip muscles. Subjects performed repeated 2 s, 20% of maximal voluntary contractions followed by 2 s rest for 4 min under free circulation and/or arterial occlusion given between the first and second min. The constancy of both the intramuscular motor unit spikes and surface EMG activity during isometric contraction indicated no electrophysiological signs of muscular fatigue with occlusion-free conditions (see Figure 5.2). However, significant changes in the above-mentioned parameters were evident during contractions with arterial occlusion. Since the availability of oxygen and blood-borne substrates, e.g. glucose and free fatty acids, are severely reduced during occlusion, a progressive recruitment of additional motor units might have taken place so as to compensate for the deficit in force development (Bigland-Ritchie *et al.*, 1986; Moritani *et al.*, 1986b). This may occur if motor units become depleted of glycogen (Gollnick *et al.*, 1974; Vollestad *et al.*, 1984) or affected by some degree of intramuscular acidification (Wilkie, 1986; Metzger and Moss, 1987). This in turn interferes with the excitation–contraction (EC) coupling with a subsequent decrease in the developed force. Under these physiological conditions, if the force output were to be maintained, a progressive recruitment of additional MUs with possibly more glycolytic (type IIa and IIb) fibers would take place.

These results and our recent findings of significant increases in the MU firing rate and MU spike amplitude associated with arterial occlusion (Moritani *et al.*, 1992) thus suggest that not only the force and speed of contraction, but also the availability of oxygen may affect the recruitment of high-threshold motor units. In good agreement with this hypothesis, ^{31}P NMR experiments (Bylund-Fellenius *et al.*, 1981) demonstrated that the rate of recovery in phosphocreatine (PCr)/inorganic phosphate (P_i) during contraction was dependent on oxygen delivery. These results and recent evidence (Idström *et al.*, 1985; Katz and Sahlin, 1987) suggest a causal relationship between oxygen supply and energy state in the contracting as well as recovering skeletal muscles. Considering this evidence, it might be suggested that oxygen

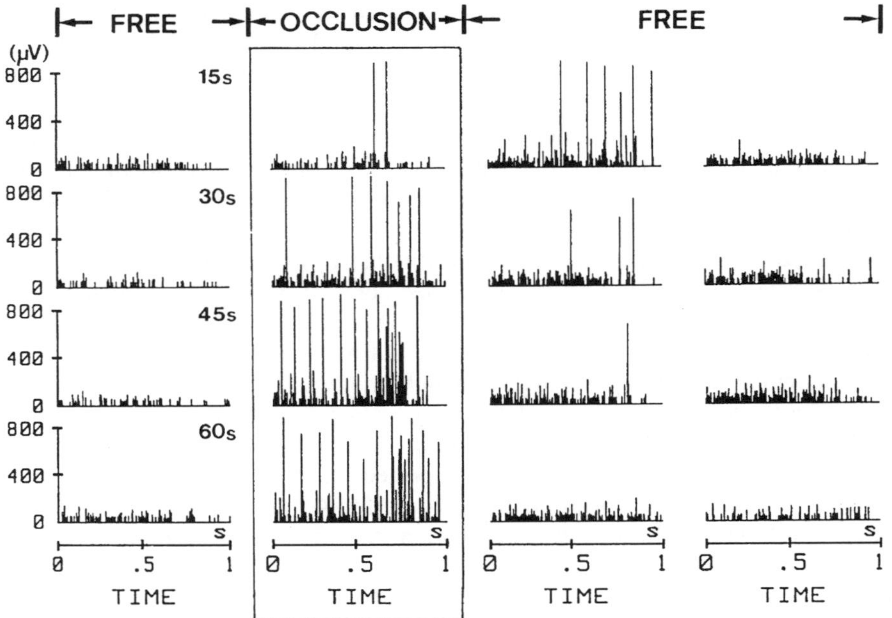

Figure 5.2 A typical set of changes in the intramuscular spike observed during experimental condition (free circulation plus occlusion between first and second minutes). In this figure only 1 s samples of MU spikes obtained at every 15 s interval (raws) are presented for clarity. Thus, in this figure each column represents 1 min of repeated intermittent isometric contraction. Based on Moritani et *al.* (1992).

availability may play an important role in regulating MU recruitment and firing frequency as there exists a close link between the state of energy supply and types of muscle fibers being recruited (Idström *et al.*, 1985; Bigland-Ritchie *et al.*, 1986; Moritani *et al.*, 1985b, 1986b, 1992).

At present, the specific messenger to which the motor control system responds with varying patterns of motor unit recruitment and firing frequency has not yet been clearly established. There is some evidence that the MU recruitment order might be modified by changes in proprioceptive afferent activity (Grimby and Hannerz, 1976; Garnett and Stephens, 1981; Kirsch and Rymer, 1987). The observed large and progressive increases in both the MU spike amplitude and firing frequency during the period of arterial occlusion suggested new MU recruitment and an increased MU discharge rate of relatively high-threshold units (Moritani *et al.*, 1992). This might thus be the result of such changes in the proprioceptive afferent inputs of the affected muscle fibers to the motor neurons as a consequence of severely reduced oxygen availability and/or a strong pressure applied during the arterial occlusion. However, the large MU spike amplitudes seen during the occlusion were still seen even after the complete release of the occlusion for more than

30 s in one study (Moritani *et al.*, 1992). Rotto and Kaufman (1988) have also shown that very large concentrations of sodium lactate (much larger than the lactate changes we observed) are required to stimulate group III and IV muscle afferents. Therefore, any effects of increased pressure or metabolic by-products on the group III or IV afferent fibers that might modify MU recruitment patterns could not be a major mechanism for the increased MU recruitment and firing frequency of relatively high-threshold units. Other and possibly more likely explanations might be that the stretch receptors in muscle spindles and Golgi tendon organs could signal the need for adding more motor units (Enoka and Stuart, 1985; Nelson and Hutton, 1985; Hutton and Nelson, 1986; Kirsch and Rymer, 1987) as a result of a fall in the contractility of some motor units affected by the reduced oxygen supply and/or the depletion of the intramuscular glycogen store (Gollnick *et al.*, 1974; Bylund-Fellenius *et al.*, 1981; Vollestad *et al.*, 1984; Idström *et al.*, 1985). Alternatively, there may be some influences on the motor neuron pool activity from the sensory afferent nerve fibers originating in the 'metaboreceptors' (Mahler, 1979; Sjøgaard, 1990).

5.2.3 Energy Metabolism and Excitation–Contraction Processes

Undoubtedly there is a close link between energy metabolism and excitation processes. Figure 5.3 summarizes possible metabolic and electrophysiologic consequences of muscular activity which might lead to fatigue based upon current studies. Prolongation and reduction in the evoked action potential have been reported during high-frequency nerve stimulation (Jones *et al.*, 1979, Moritani *et al.*, 1985a) or during ischemic contractions (Duchateau and Hainault, 1985), indicating a possible dependency on energy supply for membrane function or removal of metabolites and ions. Edwards and Wiles (1981) have shown that patients who are unable to utilize glycogen because of phosphorylase deficiency manifest a rapid decline in the surface-recorded evoked action potential amplitude and the failure of recovery during local ischemia following stimulated contractions at 20 Hz, which in normal subjects recovers rapidly. Furthermore, the depletion of extracellular Na^+ has been shown to accelerate the rate of force fatigue in an isolated curarized preparation (Jones *et al.*, 1979). This reduction of extracellular $[Na^+]$ or accumulation of K^+ may reduce the muscle membrane excitability sufficiently during high-frequency tetani to account for the excessive loss of force (Jones *et al.*,1979; Moritani *et al.*, 1985a). In addition, Moritani *et al.* (1985a) have also demonstrated that the recording of intramuscular-evoked potentials showed the predominantly 'fast-twitch' gastrocnemius muscle to have greater reductions in the potential amplitude and conduction time as compared to those of the 'slow-twitch' soleus muscle. Thus, energy metabolism clearly plays an important role influencing neural excitation and electrolyte balance within the cell.

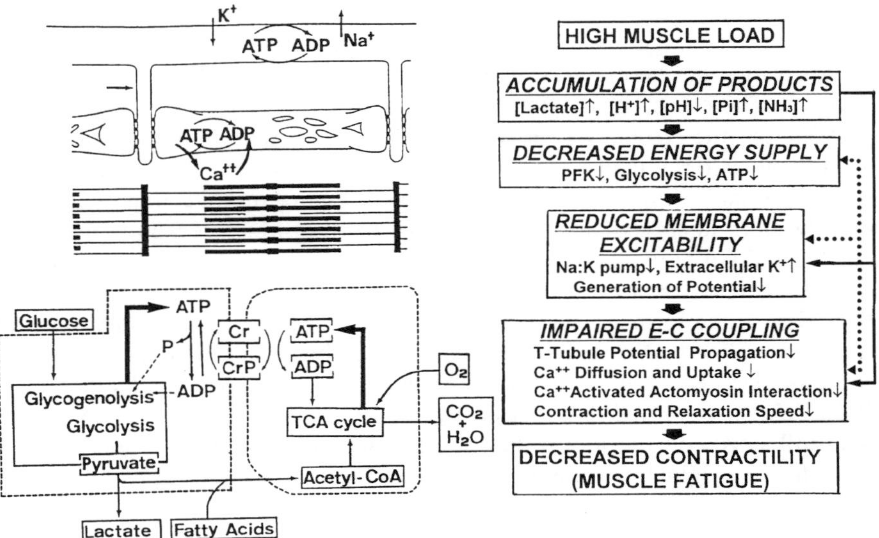

Figure 5.3 Possible metabolic and electrophysiologic consequences of muscular activity which might lead to fatigue.

There has been some evidence that a decrease in intracellular pH could interfere with muscular contractile function. For example, the increase in $[H^+]$ has been shown to interfere with Ca^{2+} binding to troponin by lowering the apparent binding constant (Fuchs *et al.*, 1970). Nakamaru and Schwartz (1972) found the affinity of the sarcoplasmic reticulum for Ca^{2+} being specifically dependent on pH, thus suggesting the possible participation of $[H^+]$ in excitation–contraction coupling with a subsequent deficit in the developed tension. The findings of Karlsson *et al.* (1975) seem to suggest that at tensions of 30–50% MVC the increase in lactate could be responsible for fatigue by direct or indirect changes in pH. However, at higher and lower tensions the possibility that lactate is directly implicated in the development of fatigue seems remote, as electrical and metabolic factors may further complicate this phenomenon (Nassar-Gentina *et al.*, 1978; Bigland-Ritchie *et al.*, 1983a, b; Moritani *et al.*, 1985b, 1986b). For example, when the cat gastrocnemius and soleus muscles were experimentally made to contract, the lactate release was the same from both muscles and yet the gastrocnemius muscle fatigued to a much greater extent than the soleus (Hudlicka, 1971). The results of NMR study (Edwards *et al.*, 1982) also demonstrated that PFK-deficient patients showed virtually no change in pH during muscle fatigue. Hence, other possible mechanisms must be considered. Accumulation of inorganic phosphate (P_i) and ammonia (NH_4^+), for example, have also been shown to occur during muscular activity as possible inhibitory metabolites contributing to fatigue (Mutch and Banister, 1983; Hibberd *et al.*, 1985). It has been suggested that P_i

may bind to myosin in such a way so as to increase the forward rate of cross-bridge cycling and thereby to reduce force output (Cooke and Pate, 1985). Other evidence of P_i-induced force reduction is that patients with McArdle's disease demonstrated greater fatiguability than normal individuals and a concomitantly larger increase in P_i accumulation (Lewis *et al.*, 1985).

5.2.4 Motor Unit Activity and Cardiorespiratory Response

The transition from aerobic to anaerobic metabolism has been the subject of special focus in human experiments during the last few years. The level of work just below that at which metabolic acidosis occurs has been called the anaerobic threshold (AT) (Wasserman *et al.*, 1973). The physiological requirements for performing work above the AT are considerably more demanding than for lower intensities. Lactic acidosis (anaerobic) threshold occurs at a metabolic rate that is specific to the individual and is usually caused by an inadequate oxygen supply (Beaver *et al.*, 1986; Katz and Sahlin, 1988; Wasserman *et al.*, 1990). Thus, the AT can be considered to be an important assessment of the ability of the cardiovascular system to supply oxygen at a rate adequate to prevent muscle anaerobiosis during work (Wasserman *et al.*, 1990).

In a series of EMG studies, we have demonstrated that an abrupt increase in integrated EMG (IEMG) representing changes in MU recruitment and/or the MU firing frequency during incremental exercise significantly correlated ($r = 0.973$, $n = 36$) with the onset of gas exchange AT VO_2 (Moritani, 1980). Two independent studies (Nagata *et al.*, 1981; Viitasalo *et al.*, 1985) have recently confirmed our findings. Figure 5.4 shows the changes in the IEMG equivalent for VO_2 (IEMG/VO_2, i.e. the slope coefficients) similar to ventilatory equivalents before and after the onset of the gas exchange AT. This indicates that the IEMG/VO_2 for this subject increased from 111 to 364 μV to achieve an O_2 uptake of 1 l after the onset of the AT. Gladden *et al.*,(1978) demonstrated that in all cases studied (before and after fatigue, partial neuromuscular block with curare or ischemia), there was no change in the O_2 uptake per unit of tension developed, indicating a constant coupling between O_2 consumption and developed tension.

Therefore, during exercise at sufficiently high intensities, e.g. above the AT, a constant coupling between O_2 uptake and developed tension can still be maintained, but may require a considerably greater IEMG (increase in MU recruitment and firing frequency) to achieve this constancy so as to compensate for a deficit in the developed tension due to the effects of lowered intracellular pH on the excitation–contraction coupling (Gevers and Dowdle, 1963; Fuchs *et al.*, 1970).

Our subsequent study (Moritani *et al.*, 1981) also demonstrated a sharp and well defined rise in the previously stable IEMG and ventilation (V_E) upon applying an arterial occlusion cuff to the leg while working at constant levels

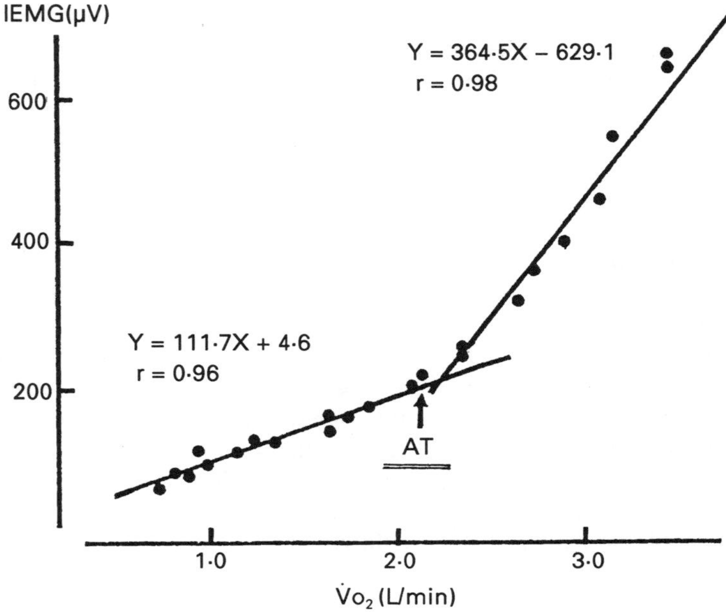

Figure 5.4 A typical set of data obtained during the analysis of EMG equivalent for O_2 (IEMG/VO_2), i.e. slope of regression line) before and after anaerobic threshold. Based on Moritani (1980).

of power output on the bicycle ergometer. In line with these results, Tibes (1977) and, more recently, Busse *et al.*, (1989) have shown some evidence that skeletal muscle group II and IV afferent nerves connected to the respiratory center could respond to local chemical stimuli including $[K^+]$, PO_2, osmolarity, pH, and PCO_2. Furthermore, Hagberg *et al.* (1982) showed that patients with McArdle's disease responded with normal hyperventilation during intense exercise despite having no H^+ or lactate release, suggesting non-humoral stimuli originating in the active muscles or in the brain during heavy exercise. Since the IEMG levels were very constant and showed no sign of fatigue prior to the occlusion, it may be concluded that some shift in the MU recruitment and/or firing frequency takes place due to local muscle hypoxia caused by the occlusion (Moritani *et al.*, 1981, 1992). On the other hand, the observed sharp increase in V_E from its steady state level after the occlusion may be mediated through some neural pathways to the respiratory center as suggested by Kao (1963), Dempsey *et al.*, (1975), and Mahler (1979), since the abrupt occlusion of the circulation to and from the exercising limb would isolate the respiratory center and chemoreceptors from the effects of chemical products of muscle metabolism. If one could assume that the sharp increases in IEMG during the occlusion represent the summation of a progressively increasing MU recruitment and firing frequency due to the compensation of reduced contractility of some fatigued MUs, a progressive increase in the extracellular K^+, for example, could be expected as a result of increased MU activities. And by this means, the ventilatory response may be stimulated via a neural pathway in the absence of circulation. In agreement with these results, recent studies (Moritani *et al.*, 1987a; Busse *et al.*, 1989) have shown some evidence that ventilatory and thereby $[H^+]$ regulation and a large proportion of variance in the gas exchange parameters can be accounted for by the plasma K^+ concentration and mechanophysiological properties of the peripheral muscles, respectively.

5.3 ELECTROMYOGRAPHIC ESTIMATION OF PHYSICAL WORK CAPACITY

5.3.1 Determination of Maximal Work Capacity at Neuromuscular Fatigue Threshold

Human muscle fatigue studies have been performed under a variety of experimental conditions and the electromyographic (EMG) signal has quite often been used as a means of assessment of muscle fatigue. In many cases, the concept of fatigue is applied to assessing deterioration of muscle performance, i.e. 'failure point' at which the muscle is no longer able to sustain the required force output level. The physiological data within the muscle shows time-dependent changes during the time course of muscle fatigue development. Such changes have been shown to be related to hydrogen ion and metabolite accumulation (Nakamaru and Schwartz, 1972; Wilkie, 1986) and to sodium and potassium ion concentration shifts (Jones *et al.*, 1979). These changes would in turn affect the muscle excitation–contraction coupling including the muscle membrane properties and muscle action potential propagation, leading to EMG manifestations of muscle fatigue distinct from mechanical manifestations (Bigland-Ritchie, 1981; Moritani *et al.*, 1985b, 1986b, 1990b; Matsumoto *et al.*, 1991).

We define the maximal power output at the neuromuscular fatigue threshold as EMG_{FT}. Our cycle ergometer is electrically braked and provides a work intensity which is independent of pedalling frequency. Surface electrodes (10 mm silver–silver chloride) in a bipolar lead system are applied, with the active electrodes (10 cm interelectrode distance) on the lateral portion of the dominant quadriceps femoris muscle and the reference electrode over the iliac crest. The large interelectrode distance is specifically chosen so as to increase sampling over a large muscle volume which would in turn allow more representative and meaningful interpretation of the mechanical involvement of large muscles (deVries *et al.*, 1990).

Figure 5.5 illustrates the IEMG method applied to determine the EMG_{FT}. This shows IEMG data for 6 s integration periods plotted as a function of time

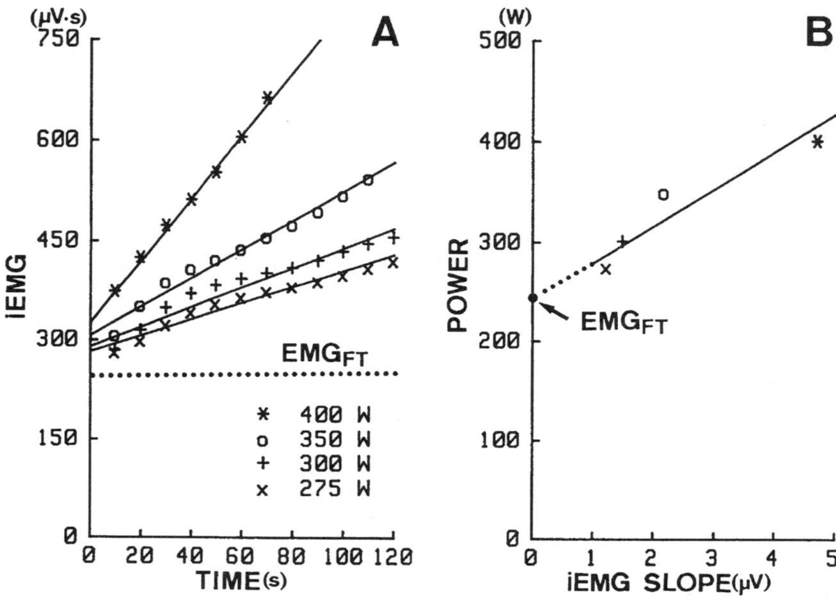

Figure 5.5 Method for estimating fatigue threshold (EMG_{FT}) from IEMG data. (a) IEMG data for each subject integrated over 6 s periods are plotted as a function of time for different exercise intensities on an ergometer. (b) IEMG slopes from (a) are plotted for each of the four exercise intensities and the zero slope (EMG_{FT}) is estimated by extrapolation to the *y* intercept. Based on Moritani *et al.* (1993).

at four different powers (400, 350, 300, and 275 W) on an ergometer to determine the rate of increase in the IEMG (IEMG slope). Note that the greater the power loading imposed, the greater the corresponding increase in the IEMG slope, indicating greater fatiguability. The IEMG slopes so obtained (Figure 5.5, left) were then plotted against the power loadings imposed, resulting in linear plots which were extrapolated to zero slope to give an intercept on the power axis (Figure 5.5, right) which was interpreted as the EMG_{FT}, i.e., the highest work intensity sustainable without evidence of neuromuscular fatigue, in this case 243 W.

We tested 20 individual EMG_{FT}s which was then expressed in terms of the VO_2 equivalent based on the individual delta mechanical efficiency calculated during a linearly increasing maximal work test on the same bicycle ergometer. Results indicated that there was a highly significant correlation ($r = 0.92$, $p < 0.01$) between EMG_{FT} VO_2 and AT VO_2 as determined by the gas exchange method. It has been suggested that the AT is the assessment of the ability of the cardiovascular system to supply oxygen at a rate which is adequate to prevent muscle anaerobiosis during exercise (Wassermann *et al.*, 1990). The magnitude of correlation ($r = 0.92$) seems to suggest the tight link between the maximal power output at the neuromuscular fatigue threshold and the AT.

The mean EMG_{FT} VO_2 (1.84 ± 0.55 l min^{-1}) was, however, significantly greater ($p < 0.01$) than that for the AT VO_2 (1.72 ± 0.54 l min^{-1}). This difference may indicate different physiological mechanisms governing these two parameters. Jenkins and Quigley (1990) showed that highly trained endurance cyclists can tolerate high blood lactate levels (8.9 mmol l^{-1}, SD 1.6) during 30 min exercise at their sustainable power asymptote value, i.e. the critical power (CP). This may be due to the fact that muscle respiratory capacity is of primary importance in determining the exercise intensity at which blood lactate begins to increase rapidly (Ivy *et al.*, 1980). Our data seem to support this notion and provided some evidence that the EMG_{FT} could represent the highest power output that can be sustained without evidence of neuromuscular fatigue. It could be speculated on the basis of the significant difference in VO_2 values at the AT and at the EMG_{FT} that the EMG_{FT} might be more closely associated with the steady state of lactate metabolism in the active muscles than with the AT. We suggest that the EMG_{FT} may provide an attractive alternative to the measurement of the highest work rate that can be sustained without evidence of neuromuscular fatigue.

5.3.2 Electromyographic Manifestations of Neuromuscular Adaptations during Acquisition of Motor Tasks

Since the 'size principle' of Henneman *et al.* (1965) was first proposed based upon results from cat motor neurones, strong evidence has been presented that in muscle contraction there is a specific sequence of recruitment in order of increasing motor neuron and motor unit (MU) size (Milner-Brown *et al.*, 1973; Freund *et al.*, 1975; Kukulka and Clamann 1981; De Luca *et al.*, 1982). Goldberg and Derfler (1977) also showed positive correlations between recruitment order, spike amplitude, and twitch tension of single MUs in human masseter muscle. It is well documented that motor unit recruitment and the firing frequency (rate coding) depend primarily upon the level of force and the speed of contraction. When low-threshold MUs are recruited, this results in a muscular contraction characterized by low force-generating capabilities and high fatigue resistance. With requirements for greater force and/or faster contraction, high-threshold-fatiguable MUs are recruited (Fruend *et al.*, 1975; Henneman and Mendell, 1981). However, Kukulka and Clamann (1981) and Moritani *et al.* (1986a) demonstrated in human adductor pollicis that for a muscle group with mainly type I fibers, rate coding plays a more prominent role in force modulation. For a muscle group composed of both type I and II fibers, MU recruitment seems to be the major mechanism for generating extra force above 40–50% of the maximal voluntary contraction(MVC) (Kukulka and Clamann, 1981; De Luca *et al.*, 1982; Moritani *et al.*, 1986a; Moritani and Muro 1987).

5.3.2.1 Neural Adaptations and Strength Gain

Earlier studies indicated that repeated testing of the strength of skeletal muscles resulted in increasing test scores in the absence of measurable muscle hypertrophy (deVries, 1968a; Coleman, 1969). In some cases, several weeks of intensive weight training resulted in a significant improvement in strength without a measurable change in girth (deVries, 1968a). It has also been shown that when only one limb is trained, the paired untrained limb improves significantly in subsequent retests of strength but without evidence of hypertrophy (Coleman, 1969; Ikai and Fukunaga, 1970; Moritani and deVries, 1979, 1980). Rasch and Morehouse (1957) demonstrated strength gains from 6 weeks training in tests when muscles were employed in a familiar way, but little or no gain in strength was observed when unfamiliar test procedures were employed. These data suggest that the higher scores in strength tests resulting from the training programs reflected largely the acquisition of skill and training-induced alterations in antagonist muscle activity, i.e. an enhanced reciprocal inhibition that contributes to greater net force production, reduced energy expenditure, and more efficient coordination (Davies *et al.*, 1988; Kamen and Gormley, 1968; Rutherford and Jones, 1986).

All of the above findings support the importance of 'neural factors', which although not yet well defined, certainly contribute to the display of maximal muscle force which we call strength. On the other hand, a strong relationship has been demonstrated between both absolute strength and the cross-sectional area of the muscle (Rodahl and Horvath, 1962) and between strength gain and increase in muscle girth or cross-sectional area (Ikai and Fukunaga, 1970). It is quite clear, therefore, that human voluntary strength is determined not only by the quantity (muscle cross-sectional area) and quality (muscle fiber types) of the involved muscle mass, but also by the extent to which the muscle mass has been activated (neural factors) (see Moritani (1992, 1993) for a review).

Earlier studies (Kawakami, 1955; Cracraft and Petajan, 1977) regarding the neural factors involved in muscle training demonstrated that specific exercise programs (high-intensity, short-duration static exercise versus low-intensity, long-duration dynamic exercise) can effectively produce changes in the firing patterns of single motor units and the expected direction of that change can be predicted based on the type of exercise (static or dynamic). We have demonstrated that the neural factors play a major role in strength development at early stages of strength gain for both young and old men and then hypertrophic factors gradually dominate over the neural factors for the young subjects in the contribution to further strength gain (see Moritani and deVries (1979, 1980) for more details). The strength gain seen for the untrained contralateral arm flexors provides further support for the concept of cross-education. It is reasonable to assume that the nature of this cross-education effect may rest entirely on the neural factors presumably acting at various levels of the nervous system which could result in increasing the maximal level of muscle activation. Subsequent studies (Häkkinen *et al.*, 1981; Houston *et al.*, 1983; Davies *et al.*, 1985; Jones and Rutherford, 1987; Narici *et al.*, 1989; Ishida *et al.*, 1990); have confirmed these observations and provided evidence for the concept that in strength training the increase in voluntary neural drive accounts for the larger proportion of the initial strength increment and thereafter both neural adaptation and hypertrophy take place for a further increase in strength, with hypertrophy becoming the dominant factor after the first 3–5 weeks (Moritani and deVries, 1979; Häkkinen *et al.*, 1981).

5.3.2.2 *Electromyographic Manifestations of Neuromuscular Adaptations During Acquisition of Motor Tasks*

The study of movement has been partitioned into several specialized fields, including neurophysiology and motor behavior. Recently, investigators in these two fields have been interested in a common issue: the regulation of force over time. Unfortunately, research in the two fields has been directed toward the issue as it relates specifically to one field or the other, rather than incorporating the knowledge of both fields. The physiologists, on the other hand, have emphasized the production of muscular control through motor unit (MU) recruitment and firing frequency patterns, while the researchers in motor behavior have investigated motor output, force, and force variability, with respect to theoretical control mechanisms such as Fitts' speed–accuracy trade-off (Fitts, 1954) and Schmidt's impulse-variability theory (Schmidt *et al.* 1979).

Researchers in motor behavior have attempted to establish mathematical relationships between movement parameters, but have had little success. One of the few to be established and to withstand the scrutiny of further delineating study, is Fitts' law. In 1954, Fitts proposed a mathematical model representing the relationship between speed and accuracy in aimed tasks. He found a relationship between the amplitude of a movement (A), the target width (W), and the resultant average movement time (MT), such that $MT = a + b(\log_2(2A/W))$. Thus, he found a linear relationship between the average movement time and $log_2(2A/W)$ which corresponds to the index of difficulty for the movement. Fitts' law implies an inverse relationship between the difficulty of a movement and the speed with which it can be performed. Thus, it has been called the speed–accuracy trade-off (Pachella and Pew, 1968).

Since the advent of Fitts' law, may researchers have investigated its applicability to different movement tasks and have found it to be robust (e.g. Fitts and Peterson, 1964; Schmidt *et al.*, 1978). Subsequently, attention turned toward an explanation of the mechanisms behind Fitts' law. Keele and Posner (1968) proposed that movement toward a target involves a series of corrections, each requiring a constant amount of time (approximately 190 ms), and having an error proportional to the remaining distance to the target. The movement time would then be a function of the number of corrections necessary to reach the target. Thus, movements with a large amplitude or small target width, would require longer movement times.

More recently, Schmidt *et al.* (1979) have proposed an alternative explanation, as a result of a large amount of evidence indicating limitations in the feedback control theory. Examples of these limitations were the existence of rapid movements such as the rapid target shooting, which appeared to be the result of motor program utilization, and the evidence that in some tasks, error corrections were almost impossible due to time constraints. Schmidt *et al.*, (1979) therefore modified the classical definition of a motor program to include the possibility of peripheral feedback in the correction of minor errors in execution, such as reflex control loops, or alpha–gamma co-activation (Schmidt, 1976; Smith, 1976) in addition to the idea of a general motor program, with variable parameters to specify the exact execution of the program (Schmidt, 1975, 1976; Schmidt *et al.*, 1978).

Thus, given the above framework, Schmidt *et al.* (1979) proposed the impulse-variability theory as a possible explanation for the speed–accuracy trade-off, applicable for rapid movements. According to the proposed theory, as the amount of force necessary to produce a movement is increased, there is a parallel increase in neurological noise, resulting in variability in the movement. A large movement amplitude would require a large amount of force production, causing a greater variability in the movement, thus, a longer movement time.

Schmidt *et al.*, (1979) and Sherwood and Schmidt (1980) found a linear relationship between force and force variability in the low and mid-range force levels, providing support for the impulse-variability explanation of the speed–accuracy trade-off. In 1980, however, Sherwood and Schmidt found that at a force requirement greater than 60% of MVC, the force variability reached a plateau and then decreased with increasing force levels. Their results seem to contradict the speed–accuracy trade-off, i.e. at near-maximal levels, force output appears to be more consistent, rather than more variable.

We have recently conducted a series of studies in an attempt to shed some light on this unsolved issue and to investigate the possible neurophysiological adaptations during a variety of different motor tasks (Yamashita and Moritani, 1989; Moritani and Mimasa, 1990; Moritani *et al.*, 1990a, 1991a, b; Yamashita *et al.*, 1990; Moritani and Shibata, 1994). In our earlier attempt, we studied the effects of extended practice on the parameters of motor output variability such as force variability, maximal rate of force development, contraction time interval, and accuracy during force-varying isometric muscular contractions with respect to the variability in neural outputs as determined by surface EMG power spectral characteristics. Subjects were instructed to produce 'shots' of force-varying isometric contractions corresponding to 20 and 60% of the maximal voluntary contraction of the biceps brachii muscle. They attempted ten shots for each trial as rhythmically as possible as the dot crossed the screen of the oscilloscope. All the subjects returned to the laboratory for 1500 extended practice trials (a total of 15 000 shots) over a 1 week period of time. The force data were processed by computer so as to determine motor output variability such as force variability, maximal rate of force development (dF/dt), contraction time interval, and accuracy (constant error – CE (average algebraic error), absolute error – AE (average absolute error), and variable error – VE (standard deviation of error)) (for more details, see Poulton, 1981).

Results indicated that all of the motor output parameters showed significant improvements after the extended practice in terms of accuracy (AE, CE, and VE) and variability in dF/dt and contraction time interval for both 20 and 60% MVC trials (e.g. see Figures 5.6 and 5.7). These changes were accompanied by significant reductions in the neural output variability as evidenced by significantly smaller coefficients of variation in the mean power frequency (MPF) and root mean square amplitude (RMS). Interestingly, when 20 and 60% MVC trials were compared after the extended practice, significantly greater improvements in accuracy and less variability in the neural output parameters were found for the 60% MVC trials. These data strongly support the recent findings of Sherwood and Schmidt (1980) who have demonstrated the limitation of Fitts' law for rapid movements. Our data and those reported by Sherwood and Schmidt (1980) seem to be consistent with well-established neurophysiological evidence that motor unit (MU) recruitment is the primary factor in increasing muscular force at low force levels, while rate coding (MU firing frequency modulation) becomes significant and predominant at

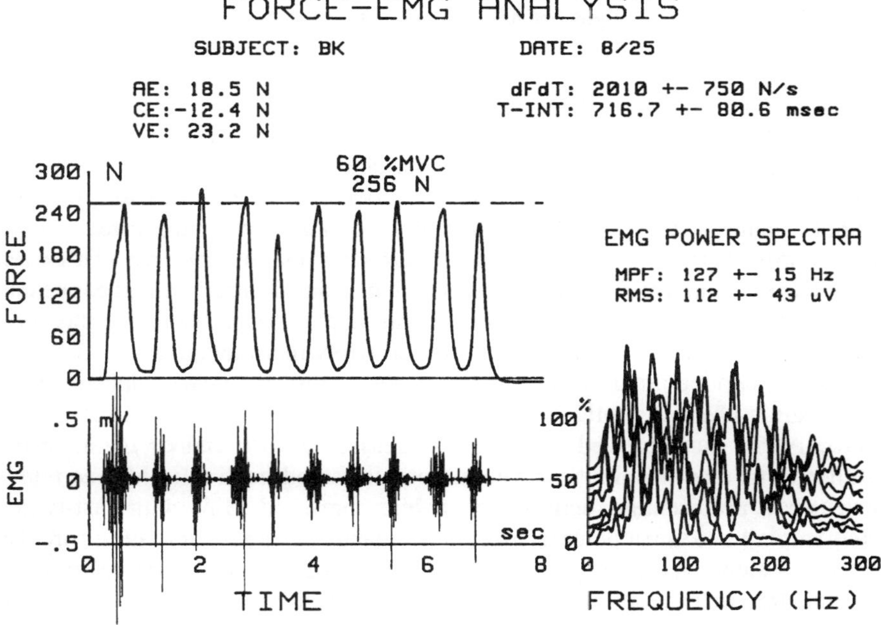

Figure 5.6 Computer analysis results showing force curve and corresponding EMG signals together with mechanical (error, dF/dt, and contraction time interval) as well as neural (MPF and RMS) parameters obtained at the beginning of practice session.

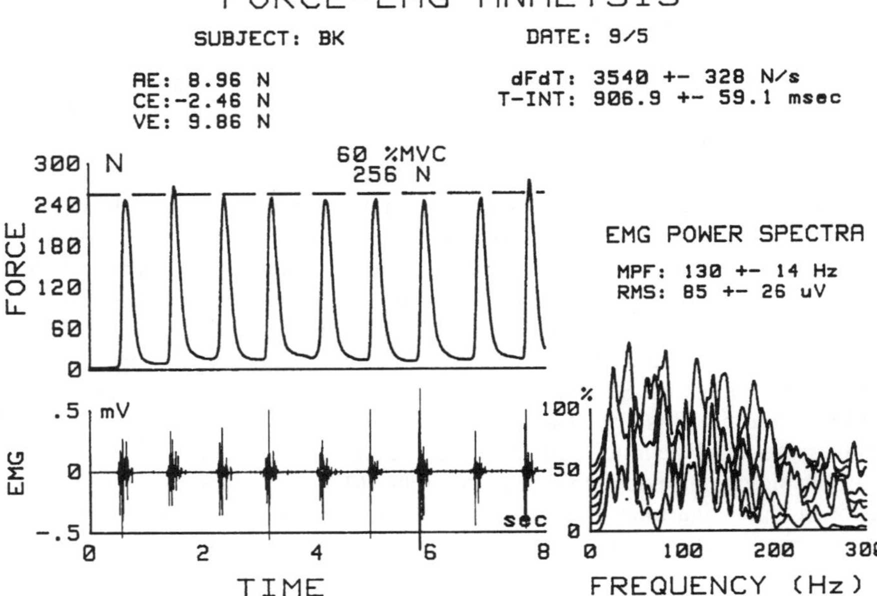

Figure 5.7 Computer analysis results obtained after the end of 1500 practice session (total of 15 000 shots).

intermediate and high force levels (Milner-Brown *et al.*, 1973; Kukulka and Clamann, 1981; Moritani *et al.*, 1986a; Moritani and Muro, 1987). Because the rate coding would bring about much smoother force regulation through temporal summation than MU recruitment, in which a small 'error' would cause recruitment of 'high-threshold motor neurons' innervating fast-twitch fibers (types IIa and IIb) capable of producing strong contractile force, one can thus expect much less mechanical and neural output variability during the 60% MVC trials as most of the motor units are probably recruited.

Considering the relationship between surface EMG power spectra (e.g. MPF) and underlying MU activities (Moritani *et al.*, 1986a; Moritani and Muro, 1987), the observed significant increases in MPF and dF/dt after the extended practice thus may indicate the possible modification in MU activities such that a preferential recruitment of high-threshold MUs with fast-twitch fibers might have taken place to meet the demands of rapid alternating forceful motor activities. Available experimental results suggest that MU recruitment patterns are not stereotyped motor patterns, but can be specifically modulated for different functional requirements in animals (Smith *et al.*, 1980; Hodgson, 1983) and in humans (Nardone and Schieppati, 1988; Nardone *et al.*, 1988; Moritani *et al.*, 1990a). This modulation may occur in relation to different phases of the motor learning process as well as to a varying degree in fast- and slow-twitch fibers, depending on the demands of the force and speed of the motor activity (Moritani and Mimasa, 1990; Moritani *et al.*, 1990a, 1991a, b). Recent evidence also suggests that even short-latency, largely monosynaptic reflexes show a high degree of modulation during simple human motor activities and that the pattern of modulation can be specifically altered for the different functional requirements of each activity (Capaday and Stein, 1987; Crenna and Frigo, 1987; Yamashita and Moritani, 1989; Moritani *et al.*, 1990a).

5.3.2.3 *Movement-related Cortical Potentials and EMG Signal Amplitude During Motor Tasks*

Potentials occurring immediately preceding and following a voluntary movement have been defined as movement-related cortical potentials (MRPs) (Neshige *et al.*, 1988b). Human MRPs have been studied in normal people (Deecke *et al.*, 1969; Gerbrandt *et al.*, 1973; Taylor, 1978; Shibasaki *et al.*, 1980; Neshige *et al.*, 1988a, b) and in patients (Neshige *et al.*, 1988b, Singh and Knight, 1990). The fact that MRPs onset up to 1 s prior to movements suggest that these potentials are generated by neural circuits involved in motor preparation and initiation.

MRPs recorded from chronically implanted subdural electrodes in patients have indicated discrete MRP sources in the pre- and postcentral gyrus with additional contributions from supplementary and premotor cortices (Neshige *et al.*, 1988a, b). Most investigators agree that the motor potential (MP) has its major neural source in the primary motor area (Deecke *et al.*, 1969; Gerbrabdt *et al.*, 1973; Shibasaki *et al.*, 1980; Singh and Knight, 1990). Recent intracranial data in humans (Neshige *et al.*, 1988a, b) also indicate that the sensorimotor cortex is a major contributor to the earlier MRP components (readiness potential, RP, and late portion of the RP, termed as negative shift, NS').

Figure 5.8 (overleaf) represents examples of MRPs recorded from scalp electrodes using the international 10–20 system. In this figure, only Fz, C3, Cz, and C4 together with force and EMG recordings were shown for clarity. In order to observe the EEG activity before and after isometric left arm flexion, an average EEG, time locked to the onset of force production, was prepared using special computer programs developed in our laboratory. The RP, which corresponds to the previously described Bereitschafts potential (Deecke *et al.*, 1969), started at least 1000 ms prior to the force output and slowly increased in amplitude. Approximately 500 ms before the force onset, the slope of this negative potential became steeper (negative slope NS' – according to the terminology used by Shibasaki *et al.*, 1980). Isopotential maps for each MRP component were also constructed and shown in this figure (see Figure 5.8). These data clearly indicate that NS' and MP are maximal over scalp sites contralateral to movements which in turn suggest that sensorimotor areas and the supplementary motor area participate in the preparation of

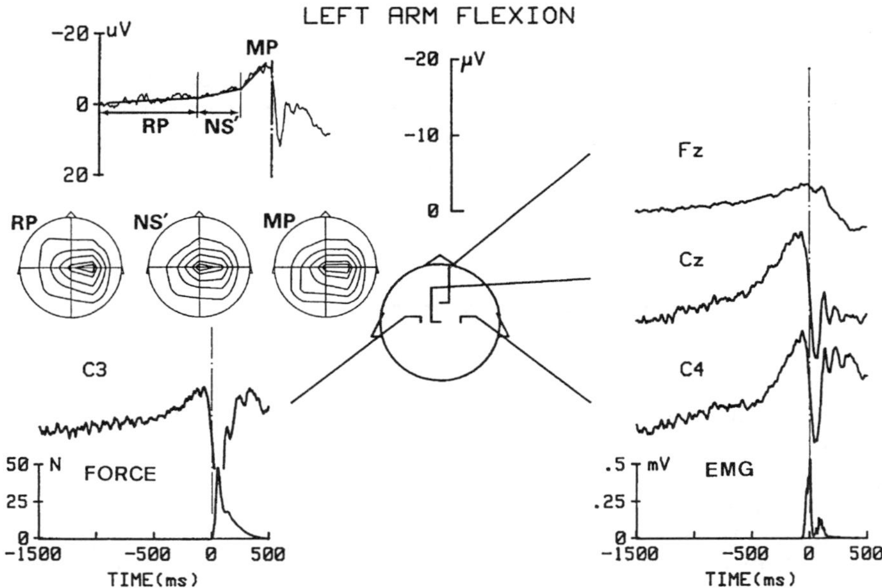

Figure 5.8 Movement-related cortical potentials (MRPs) during left arm flexion. Isopotential maps for readiness potential (RP), negative slope (NS') and motor potential (MP) are also constructed in this figure.

movements, but that mainly the contralateral cortex generates the discharges necessary to produce the actual movement (Neshige *et al.*, 1988b; Singh and Knight, 1990).

Figure 5.9 shows the relationship between MRPs and the force output amplitude during execution of target shooting, i.e. force-varying isometric contractions corresponding to 10 and 50% MVC. The significant increase of MP corresponding to the exerted force level (10 versus 50% MVC) might indicate the relative increase of the pyramidal tract cell discharge. This notion has been tested in the subsequent studies in which the exerted isometric target force (20% MVC) was held for 2 s after its initiation under free circulation (CON) or intermittent arterial occlusion (EXP). Figure 5.10 summarizes the results (averaged data, $n = 10$) indicating that MRPs remain elevated for nearly 2 s after movement initiation under both experimental conditions. This may suggest that MRPs reflect underlying pyramidal tract cell activity responsible for the execution of a motor task. The marked increase in MRPs during force maintenance period under ischemic conditions seems to support this notion as greater MU activities are expected during movement with ischemia (see Figure 5.10).

Further studies are obviously necessary to draw any conclusions from these attempts, but MRP analyses may provide an attractive tool in movement-related field of studies.

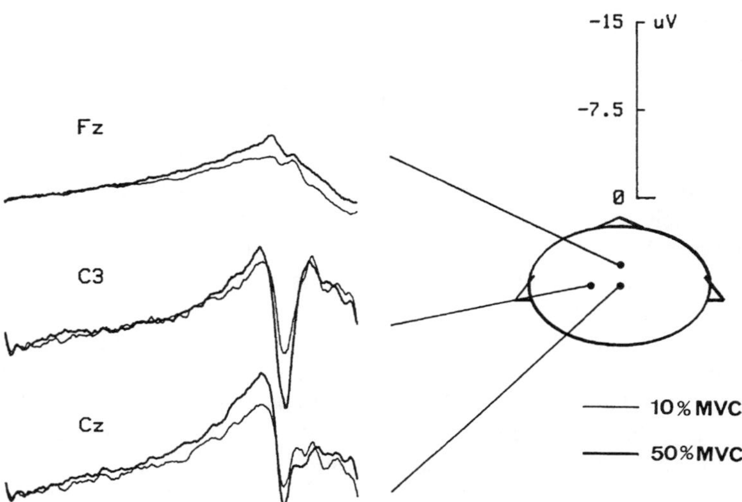

Figure 5.9 Changes in movement-related cortical potentials (MRPs) during right arm flexion (force-varying isometric contractions) corresponding to 10 and 50% MVC.

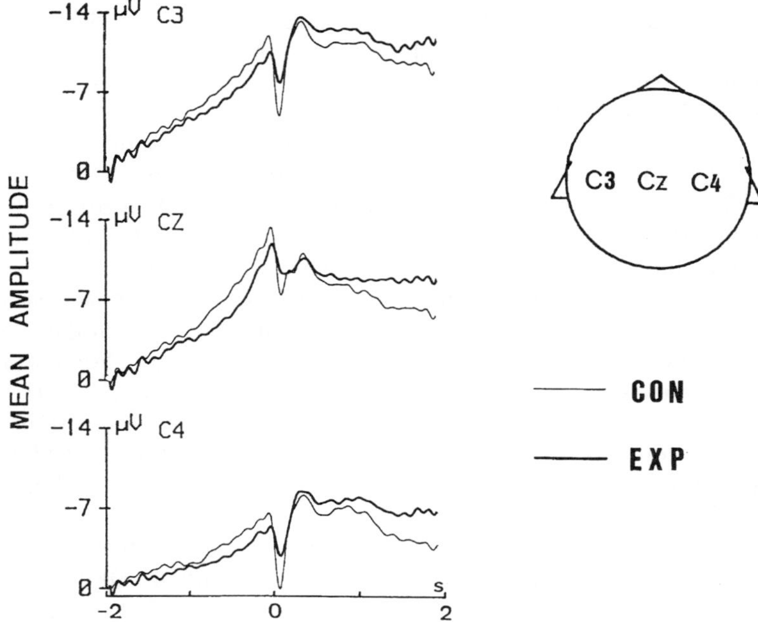

Figure 5.10 Changes in movement-related cortical potentials (MRPs) during right arm flexion (force-varying isometric contractions held for 2 s at 20% MVC) under free circulation and muscle ischemia conditions.

5.4 ELECTROMYOGRAPHIC MANIFESTATIONS OF MUSCLE SORENESS

Nearly everyone has experienced delayed-onset muscle soreness (DOMS) at some time; many have suffered from this common ailment on numerous occasions. DOMS is characterized by stiffness, tenderness, and pain during active movements and weakness of the affected musculature. A number of investigators have demonstrated that the eccentric component of dynamic work plays a critical role in determining the occurrence and severity of exercise-induced muscle soreness (Friden *et al.*, 1983; Newham *et al.*, 1983; McCully and Faulkner, 1986; Berry *et al.*, 1990). It has been also demonstrated that type II fibers are predominantly affected by this type of muscular contraction (Friden *et al.*, 1983; Jones *et al.*, 1986).

It is well established that eccentric (lengthening) muscle action requires less oxygen and a lower amount of ATP than concentric muscle action (Infante *et al.*, 1964; Davies and Barnes, 1972). Both surface (Komi and Viitasalo, 1977) and intramuscular (Moritani *et al.*, 1988) EMG studies have demonstrated that MU recruitment patterns are qualitatively similar in both types of contractions, but for a given MU, the force at which MU recruitment occurs is greater in eccentric muscle action than in either isometric or concentric (shortening) muscle actions.

Based on these findings and the results of EMG studies cited earlier, it is most likely that DOMS associated with the eccentric component of dynamic exercise might be in part due to high mechanical forces produced by relatively few numbers of active MUs which may in turn result in some degree of disturbance in structural proteins in muscle fibers, particularly those of high recruitment threshold MUs.

Despite the fact that DOMS is a well-known phenomenon in the sphere of sports as well as working life, the exact pathophysiological mechanisms underlying this are still not well understood. According to Armstrong (1984), a number of hypotheses may exist to explain the etiology and cellular mechanisms of DOMS. The following model may be proposed: (1) high tension, particularly associated with eccentric muscle action in the contractile and elastic system of the muscle result in structural damage, (2) muscle cell membrane damage leads to disruption of Ca^{2+} homeostasis in the injured muscle fibers resulting in necrosis that peaks approximately 2 days post-exercise, and (3) products of macrophage activity and intracellular contents accumulate in the interstitium, which stimulate free nerve endings of group IV sensory neurons in the muscles leading to the sensation of DOMS.

Earlier EMG work by deVries (1961a, b) demonstrated that symptomatic soreness and tenderness seemed to parallel EMG amplitude. DeVries (1961a, b) thus proposed the spasm theory: DOMS is caused by tonic, localized spasm of motor units as a result of a vicious cycle in which the activity induced ischemia which in turn led to further pain and reflex activity. Later workers have been unable to demonstrate any EMG activity in resting painful muscles

(Abraham, 1977). Berry *et al.*, (1990) showed that muscle soreness only occurred in the muscles that had contracted eccentrically and did not occur at the time of greatest myoelectric signal changes. Although elevated EMG activity (Newham *et al.*, 1983; Moritani *et al.*, 1987b; Berry *et al.*, 1990) was accompanied by eccentric muscle action, a dissociation in the time course between these two parameters seem to exist. The data of Newham *et al.* (1983) have also demonstrated that eccentric muscle action has a long lasting effect on the muscle's ability to generate force after exercise. When muscle was stimulated at low frequencies, it was not able to develop the same force as it had under similar conditions before eccentric exercise. On the other hand, high-frequency stimulations elicited similar force before and after such exercise. The underlying mechanism of this so-called 'low-frequency fatigue' has been postulated to be impaired excitation–contraction coupling (Edwards, 1981) due to a reduced release of calcium or possibly because of impaired transmission in the transverse tubular system, as a result of muscle damage in the period of ischemic activity.

5.4.1 Acute Effects of Static Stretching on Muscle Soreness

We conducted a series of studies to determine the physiological effects of static muscle stretching upon DOMS which was induced experimentally by heel rase (10 repetitions, 10 sets) with a 70% MVC equivalent weight attached on a universal shoulder press equipment or step test (Thigpen *et al.*, 1985; Moritani *et al.*, 1987b; Berry *et al.*, 1990). Electrophysiological parameters, e.g. the maximal mass action potential (M-wave), H-wave, and H/M ratio for determination of α motor neuron excitability were measured during standing rest (control), 24 h post-experimental fatigue, and immediate post-static stretching. Changes in the resting EMG signal up to 48 h post-experimental fatigue were subjected for frequency power spectral analysis in order to determine the degree of muscle fatigue and resting action potential amplitude.

Figure 5.11 (overleaf) represents a typical set of evoked potentials and muscle twitch data obtained before and after 24 h post-experimental fatigue. Group data revealed that there were significant decreases in peak force, maximal dF/dt, and relaxation dF/dt. Note also the existence of tonic EMG activity 24 h post-experimentally induced muscle soreness immediately prior to and after the electrical stimulation. The EMG frequency power spectral analyses revealed that (1) the experimentally induced DOMS was associated with a significantly higher resting action potential amplitude and lower mean power frequency, suggesting the existence of some degree of muscle spasm and a possible synchronization of tonic motor neurons (Moritani *et al.*, 1987; Berry *et al.*, 1990) and (2) static muscle stretching (three sets of 20 s duration) showed immediate and quite noticeable effects in restoring these electrophysiological parameters back towards the control level (see Figure 5.12).

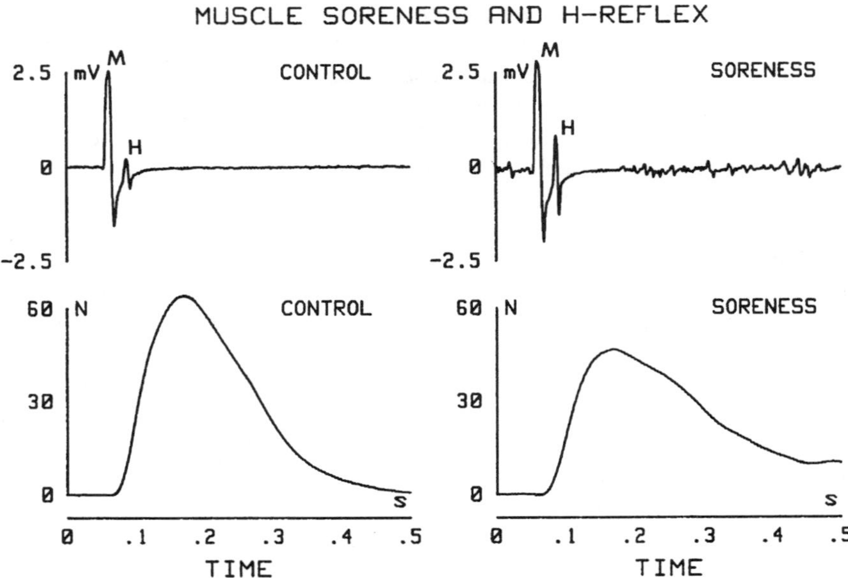

Figure 5.11 A typical set of evoked potentials and muscle twitch data obtained before and after 24 h post-experimental fatigue.

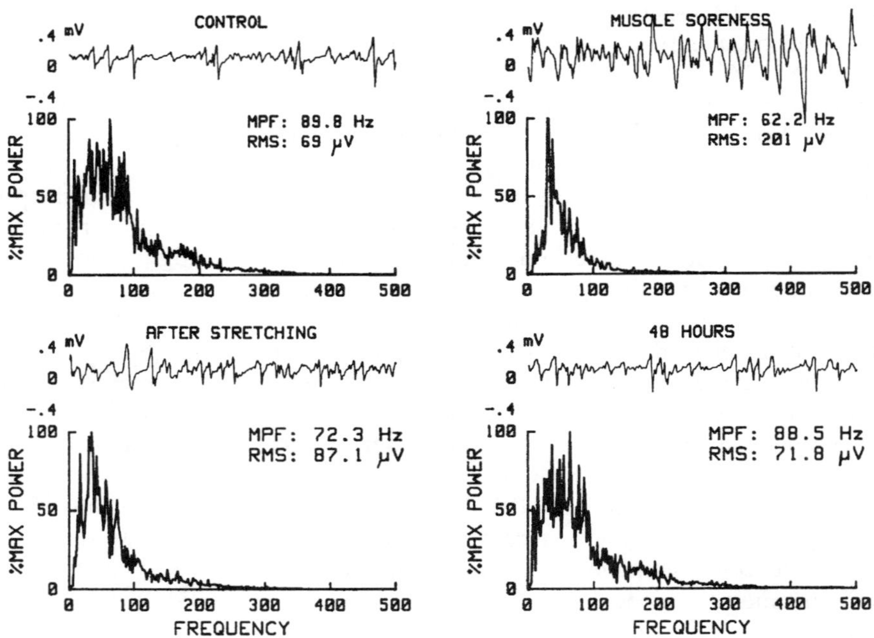

Figure 5.12 Changes in the resting EMG frequency power spectra during the development of DOMS and after static muscle stretching.

Results on α motor neuron excitability indicated that there was very little change, if any, in the maximum amplitude of the H-waves for the control leg while the experimental leg post-H-wave was markedly reduced by static stretching. The mean relative reduction in the H/M ratio from pre- to post-test for the control and experimental legs were 0.63 ($p > 0.05$) and 21.5% ($p < 0.01$), respectively (see Figure 5.13). These results are entirely consistent with earlier studies (deVries 1961a, b) and further suggest that the inverse myotatic reflex which originates in the Golgi tendon organs (GTO) may be the basis for the relief of DOMS by static stretching. Since the H-reflex involves tonic MUs (McIlwain and Hayes, 1977), it is likely that the Ib afferent inhibitory effects from the GTO could be mediated through the tonic MUs, thus reducing the evoked H-wave amplitude.

5.4.2 Effects of Aerobic Exercise on Neuromuscular Relaxation

A tranquilizer effect of aerobic exercise has been demonstrated by several investigators (deVries and Adams, 1972; deVries *et al.*, 1981; Bulbulian and Darabos, 1986; Mimasa *et al.*, 1990). To further shed some light on this tranquilizer effect of aerobic exercise, changes in the spinal motor neuron

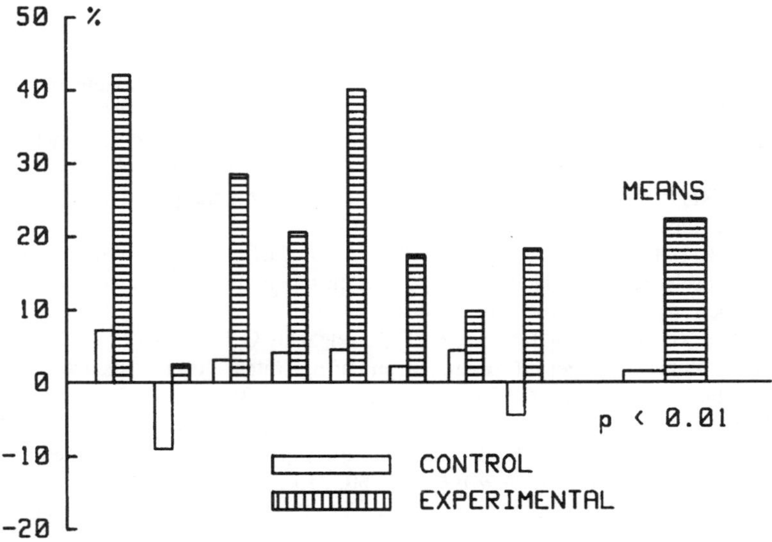

Figure 5.13 Percent reduction in the H/M ratio following treatment interventions, i.e., resting control and static muscle stretching. Based on Moritani *et al.* (1987b).

excitability and electroencephalogram (EEG) were studied in our laboratory (Mimasa *et al.*, 1990). The eight subjects acted as their own controls in an experimental design where each subject was tested twice before and after exercise and twice before and after control (quiet sitting). Exercise consisted of 20 min cycling on an ergometer with the load adjusted to elicit a heart rate rise of 60% between resting and maximal values. Results indicated that there was no significant difference in the H/M ratio on the control day (−0.27%, $p>0.05$), but showed a highly significant decrease after exercise (−15.9%, $p<0.01$).

EEG spectral analysis demonstrated significant increases ($p<0.05$) in relatively low frequency components, e.g. delta (+23.8%), theta (+6.8%), and alpha wave (+18.1%), after exercise treatment while no such significant changes were observed during the control rest condition. These results confirm earlier resting EMG results and further suggest that tranquilizer effect of exercise may reside not only in the peripheral muscle motor neuron excitability, but also in the higher central nervous system.

References

ABRAHAM, W.M. (1977) Factors in delayed muscle soreness. *Med. Sci. Sports*, **9**, 11–26.

ALWAY, S.E., STRAY-GUNDERSEN, J, GRUMBT, W.H. & GONYEAR, W.J. (1990) Muscle cross sectional area and torque in resistance-trained subjects. *Eur. J. Appl. Physiol.*, **60**, 86–90.

ARMSTRONG, R.B. (1984) Mechanisms of exercise-induced delayed onset muscular soreness: a brief review. *Med. Sci. Sports Exercise*, **16**, 529–38.

BEAVER, W.L., WASSERMAN, K. & WHIPP, B.J. (1986) Bicarbonate buffering of lactic acid generated during exercise. *J. Appl. Physiol.*, **60**, 472–8.

BERRY, C.B., MORITANI, T.M. & TOLSON, H. (1990) Electrical activity and soreness in muscles after exercise. *Am. J. Phys. Med. Rehabil.*, **69**, 60–6.

BIGLAND-RITCHIE, B., (1981) EMG/force relations and fatigue of human voluntary contractions. *Exercise Sports Sci.*, **9**, 75–117.

BIGLAND-RITCHIE, B., JONES, D.A. & WOODS, J.J. (1979) Excitation frequency and muscle fatigue: electrical responses during human voluntary and stimulated contractions. *Exp. Neurol.*, **64**, 414–27.

BIGLAND-RITCHIE, B., JOHANSSON, R., LIPPOLD, O.J.C., SMITH, S. & WOODS, J.J. (1983a) Changes in motoneuron firing rates during sustained maximal voluntary contractions. *J. Physiol. (London)*, **340**, 335–46.

BIGLAND-RITCHIE, B., JOHANSSON, R., LIPPOLD, O.J.C. & WOODS, J.J. (1983b) Contractile speed and EMG changes during fatigue of sustained maximal voluntary contraction. *J. Neurophysiol.*, **50**, 313–24.

BIGLAND-RITCHIE, B., CAFARELLI, E. & VOLLESTAD, N.K. (1986) Fatigue of submaximal static contractions. *Acta Physiol. Scand.*, **128** (Suppl. 556), 137–48.

BULBULIAN, R. & DARABOS, B.L. (1986) Motor neuron excitability: the Hoffmann reflex following exercise of low and high intensity. *Med. Sci. Sports Exercise*, **18**, 697–702.

BUSSE, M.W., MAASSEN, N., KONRA, D.H. & BONING, D. (1989) Interrelationship between pH, plasma potassium concentration and ventilation during intense continuous exercise in man. *Eur. J. Appl. Physiol.*, **59**, 256–61.

BYLUND-FELLENIUS, A.C., WALKER, P.M., ELANDER, A., HOLM, J. & SCHERSTEN, T. (1981) Energy metabolism in relation to oxygen partial pressure in human skeletal muscle during exercise. *Biochem. J.*, **200**, 247–55.

CAPADAY, C. & STEIN, R.B. (1987) Difference in the amplitude of the human soleus H reflex during walking and running. *J. Physiol. (London)*, **392**, 513–22.

COLEMAN, E.A. (1969) Effect of unilateral isometric and isotonic contractions on the strength of the contralateral limb. *Res. Quart.*, **40**, 490–95.

COOKE, R. & PATE, E. (1985) Inhibition of muscle contraction by the products of ATP hydrolysis: ADP and phosphate. *Biophys. J.*, **47**, 25a.

CRACRAFT, J.D. & PETAJAN, J.H. (1977) Effect of muscle training on the pattern of firing of single motor units. *Am. J. Phys. Med.*, **56**, 183–94.

CRENNA, P. & FRIGO, C. (1987) Excitability of the soleus H-reflex arc during walking and stepping in man. *Exp. Brain Res.*, **66**, 49–60.

DAVIES, C.T.M. & BARNES, C. (1972) Negative (eccentric) work. II. Physiological responses to walking uphill and downhill on a motor-driven treadmill. *Ergonomics*, **15**, 121–31.

DAVIES, C.T., DOOLEY, P., MCDONAGH, M.J.N. & WHITE, M. (1985) Adaptation of mechanical properties of muscle to high force training in man. *J. Physiol. (London)*, **365**, 277–84.

DAVIES, J., PARKER, D.F., RUTHERFORD, O.M. & JONES, D.A. (1988) Changes in strength and cross sectional area of the elbow flexors as a result of isometric strength training. *Eur. J. Appl. Physiol.*, **57**, 667–70.

DEECKE, L., SCHEID, P. & KORNHUBER, H.H. (1969) Distribution of readiness potential, pre-motion positivity, and motor potential of the human cerebral cortex preceding voluntary finger movements. *Exp. Brain Res.*, **7**, 158–68.

DE LUCA, C.J., LEFEVER, R.S., MCCUE, M.P. & XENAKIS, A.P. (1982) Behavior of human motor units in different muscles during linearly varying contractions. *J. Physiol. (London)*, **329**, 113–28.

DEMPSEY, J.A., FORSTER, H.V., DLEDHILL, N. & DEPICO, G.A. (1975) Effects of moderate hypoxemia and hypocapnia on CSF [H^+] and ventilation in man. *J. Appl. Physiol.*, **38**, 665–74.

DEVRIES, H.A. (1961a) Electromyographic observations of the effects of static stretching upon muscular distress. *Res. Quart.*, **32**, 468–79.

DEVRIES, H.A. (1961b) Prevention of muscular distress after exercise. *Res. Quart.*, **32**, 177–85.

DEVRIES, H.A. (1968a) Efficiency of electrical activity as a measure of the functional state of muscle tissue. *Am. J. Phys. Med.*, **47**, 10–22.

DEVRIES, H.A. (1968b) Method for evaluation of muscle fatigue and endurance from electromyographic fatigue curves. *Am. J. Phys. Med.*, **47**, 125–35.

DEVRIES, H.A. & ADAMS, G.M. (1972) Electromyographic comparison of single doses of exercise and meprobamate as to effects on muscular relaxation. *Am. J. Phys. Med.*, **51**, 130–41.

DEVRIES, H.A., WISWELL, R.A., BULBULIAN, R. & MORITANI, T. (1981) Tranquilizer effect of exercise: acute effects of moderate aerobic exercise on spinal reflex activation level. *Am. J. Phys. Med.*, **60**, 57–66.

DEVRIES, H.A., MORITANI, T., NAGATA, A. & MAGNUSSEN, K. (1982) The relation between critical power and neuromuscular fatigue as estimated from electromyographic data. *Ergonomics*, **25**, 783–91.

DEVRIES, H.A., HOUSH, T.J., JOHNSON, G.O., EVANS, S.A., THARP, G.D., HOUSH, D.J. & HUGHES, R.A. (1990) Factors affecting the estimation of physical working capacity at the fatigue threshold. *Ergonomics*, **33**, 25–33.

DUCHATEAU, J. & HAINAULT, K. (1985) Electrical and mechanical failures during sustained and intermittent contractions. *J. Appl. Physiol.*, **58**, 942–47.

EDWARDS, R.H.T. (1981) Human muscle function and fatigue. In: *Human Muscle Fatigue: Physiological Mechanisms* (Ed. R. PORTER & J. WHELAN), pp. 1–18. Pitman, London.

EDWARDS, R.H.T. & WILES, C.M. (1981) Energy exchange in human skeletal muscle during isometric contraction. *Circ. Res.*, **48** (Suppl.), I11–I17.

EDWARDS, R.H.T., DAWSON, M.J., WILKIE, D.R., GORDON, R.E. & SHAW, D. (1982) Clinical use of nuclear magnetic resonance in the investigation of myopathy. *Lancet*, **1**, 725–31.

ENOKA, R.G. & STUART, D.G. (1985) The contribution of neuroscience to exercise studies. *Fed. Proc.*, **44**, 2279–85.

FITTS, P.M. (1954) The information capacity of the human motor system in controlling the amplitude of movement. *J. Exp. Psychol.*, **47**, 381–91.

FITTS, P.M. & PETERSON, J.R. (1964) Information capacity of discrete motor responses. *J. Exp. Psychol.*, **67**, 103–12.

FREUND, H.J., BUDINGEN, H.J. & DIETZ, V. (1975) Activity of single motor units from human forearm muscles during voluntary isometric contractions. *J. Neurophysiol.*, **38**, 993–46.

FRIDEN, J., SJOSTROM, M. & EKBLOM, B. (1983) Myofibrillar damage following intense eccentric exercise in man. *Int. J. Sports Med.*, **4**, 170–6.

FUCHS, F., REDDY, Y. & BRIGGS, F.M. (1970) The interaction of cations with the calcium-binding site of troponin. *Biochim. Biophys. Acta*, **221**, 407–9.

GARNETT, R.A.F., O'DONOVAN, M.J., STEPHENS, J.A. & TAYLOR, A. (1979) Motor unit organization of human medial gastrocnemius. *J. Physiol.* (London), **263**, 33–43.

GARNETT, R. & STEPHENS, J.A. (1981) Changes in the recruitment threshold of motor units produced by cutaneous stimulation in man. *J. Physiol. (London)*, **311**, 463–73.

GERBRANDT, L.K., GOFF, W.R. & SMITH, D.B. (1973) Distribution of the human average movement potential. *ECG Clin. Neurophysiol.*, **34**, 461–74.

GEVERS, W. & DOWDLE, E. (1963) The effect of pH on glycolysis *in vivo*. *Clin. Sci.*, **25**, 343–9.

GLADDEN, L.B., MACINTOSH, B.R. & STAINSBY, W.W. (1978) O_2 uptake and developed tension during and after fatigue, curare, and ischemia. *J. Appl. Physiol.*, **45**, 751–5.

GOLDBERG, L.J. & DERFLER, B. (1977) Relationship among recruitment order, spike amplitude, and twitch tension of single motor units in human masseter muscle. *J. Neurophysiol.*, **40**, 879–90.

GOLLNICK, P.D., KARLSSON, J., PIEHL, K. & SALTIN, B. (1974) Selective glycogen depletion in skeletal muscle fibers of man following sustained contractions. *J. Physiol. (London)*, **241**, 59–66.

GRIMBY, L. & HANNERZ, J. (1976) Disturbances in voluntary recruitment order of low and high frequency motor units on blockades of proprioceptive afferent activity. *Acta Physiol. Scand.*, **96**, 207–16.

HAGBERG, J.A., COYLE, E.F., CARROLL, J.E., MILLER, J.M., MARTIN, W.H. & BROOKE, M.H. (1982) Exercise hyperventilation in patients with McArdle's disease. *J. Appl. Physiol.*, **52**, 991–4.

HAGBERG, M. (1981) Muscular endurance and surface electromyogram in isometric and dynamic exercise. *J. Appl. Physiol.*, **51**, 1–7.

HÄKKINEN, K., KOMI, P.V. & TESCH, P. (1981) Effect of combined concentric and eccentric strength training and detraining on force-time, muscle fiber and metabolic characteristics of leg extensor muscles. *Scand. J. Sports Sci.*, **3**, 50–8.

HENNEMAN, E. & MENDELL, L.M. (1981) Functional organization of the motoneuron pool and its inputs. In *Handbook of Physiology. The Nervous System* (Ed. V.B. BROOKS), pp. 423–507. American Physiological Society, Bethesda.

HENNEMAN, E., SOMJEM, G. & CARPENTER, D.O. (1965) Functional significance of cell size in spinal motoneurons. *J. Neurophysiol.*, **28**, 560–80.

HIBBERD, M.G., DANTZING, J.A., TRENTHAM, D.R. & GOLDMAN, Y.E. (1985) Phosphate release and force generation in skeletal muscle fibers. *Science*, **228**, 1317–19.

HODGSON, J.A. (1983) The relationship between soleus and gastrocnemius muscle activity in conscious cats – a model for motor unit recruitment *J. Physiol. (London)*, **337**, 553–62.

HOUSTON, M.E., FROESE, E.A., VALERIOTE, St P. & GREEN, H.J. (1983) Muscle performance, morphology and metabolic capacity during strength training and detraining: a one leg model. *Eur. J. Appl. Physiol.*, **51**, 25–35.

HUDLICKA, O. (1971) Differences in development of fatigue in slow and fast muscles. In *Limiting Factors of Physical Performance* (Ed. J. KEUL), pp. 36–41. Thieme, Stuttgart.

HUMPHREYS, P.W. & LIND, R.A. (1963) The blood flow through active and inactive muscles of the forearm during sustained hand-grip contractions. *J. Physiol. (London)*, **166**, 120–35.

HUTTON, R.S. & NELSON, D.L. (1986) Stretch sensitivity of Golgi tendon organs in fatigued gastrocnemius muscle. *Med. Sci. Sports Exercise*, **18**, 69–74.

IDSTRÖM, J.P., SUBRAMANIAN, V.H., CHANCE, B., SCHERSTEN, T. & BYLUND-FELLENIUS, A.C. (1985) Energy metabolism in relation to oxygen supply in contracting rate skeletal muscle. *Fed. Proc.*, **45**, 2937–41.

IKAI, M. & FUKUNAGA, T. (1970) A study on training effect on strength per unit cross-sectional area of muscle by means of ultrasonic measurements. *Internationale Zeitschrift Fur Angewandte Physiologie Einschliesslich Arbeitsphysiologie*, **28**, 173–80.

INFANTE, A.A., KLAUPIKS, D. & DAVIES, R.E. (1964) Adenosine triphosphate changes in muscle doing negative work. *Science*, **62**, 595–604.

ISHIDA, K., MORITANI, T. & ITOH, K. (1990) Changes in voluntary and electrically induced contractions duringstrength training and detraining. *Eur. J. Appl. Physiol.*, **60**, 244–8.

IVY, J.L., WITHERS, R.T., VAN HANDEL, P.J., ELGER, D.H. & COSTILL, D.L. (1980) Muscle respiratory capacity and fiber type as determinants of the lactate threshold. *J. Appl. Physiol.*, **48**, 523–7.

JENKINS, D.G. & QUIGLEY, B.M. (1990) Blood lactate in trained cyclists during cycle ergometry at critical power. *Eur. J. Appl. Physiol.*, **61**, 278–83.

JESSOP, J. & LIPPOLD, O.C.J. (1977) Altered synchronization of motor unit firing as a mechanism for long-lasting increases in the tremor of human hand muscles following brief, strong effort. *J. Physiol. (London)*, **269**, 29P–30P.

JONES, D.A. & RUTHERFORD, O.M. (1987) Human muscle strength training: the effects of three different regimes and the nature of the resultant changes. *J. Physiol. (London)*, **391**, 1–11.

JONES, D.A., BIGLAND-RITCHIE, B. & EDWARDS, R.H.T. (1979) Excitation frequency and muscle fatigue: mechanical responses during voluntary and stimulated contractions. *Exp. Neurol.*, **64**, 401–13.

JONES, D.A., NEWHAM, D.J., ROUND, J.M. & TOLFREE, E.J. (1986) Experimental human muscle damage: morphological changes in relation to other indices of damage. *J. Physiol. (London)*, **375**, 435–48.

JORFELDT, L., JUHLIN-DANNFELT, A. & KARLSSON, J. (1978) Lactate release in relation to tissue lactate in human skeletal muscle during exercise. *J. Appl. Physiol.*, **44**, 350–2.

KADEFORS, R., KAISER, E. & PETERSEN, I. (1968) Dynamic spectrum analysis of myopotentials with special reference to muscle fatigue. *Electromyography*, **8**, 39–74.

KAMEN, G. & GORMLEY, J. (1968) Muscular activity pattern for skilled performance and during learning of a horizontal bar exercise. *Ergonomics*, **22**, 345–57.

KAO, F.F. (1963) Experimental study of the pathways involved in exercise hyperpnea employing cross-circulation techniques. In *Regulation of Human Respiration* (Ed. D.J.C. CONNINGHAM & B.B. LLOYD) pp. 461–501. Blackwell Scientific Publication, Oxford.

KARLSSON, J., FUNDERBURK, C.F., ESSEN, B. & LIND, A.R. (1975) Constituents of human muscle in isometric fatigue. *J. Appl. Physiol.*, **38**, 208–11.

KATZ, A. & SAHLIN, K. (1987) Effect of decreased oxygen availability on NADH and lactate contents in human skeletal muscle during exercise. *Acta Physiol. Scand.*, **131**, 119–27.

KATZ, A. & SAHLIN, K. (1988) Regulation of lactic acid production during exercise. *J. Appl. Physiol.*, **65**, 509–18.

KAWAKAMI, M. (1955) Training effect and electromyogram. I. Spatial distribution of spike potentials. *Jap. J. Physiol.*, **5**, 1–8.

KEELE, S.W. & POSNER, M.I. (1986) Processing of visual feedback in rapid movement. *J. Exp. Psychol.*, **77**, 155–8.

KIRSCH, R.F. & RYMER, W.Z. (1987) Neural compensation for muscular fatigue: evidence for significant force regulation in man. *J. Neurophysiol.*, **57**, 1893–910.

KOMI, P.V. & BUSKIRK, E.R. (1970) Reproducibility of electromyographic measurements with inserted wire electrodes and surface electrodes. *Electromyography*, **4**, 357–67.

KOMI, P.V. & VIITASALO, J.T. (1977) Changes in motor unit activity and metabolism in human skeletalmuscle during and after repeated eccentric and concentric contractions. *Acta Physiol. Scand.*, **100**, 246–56.

KOMI, P.V., VIITASALO, J.T., RAURAMAA, R. & VIHKO, V. (1978) Effect of isometric strength training on mechanical, electrical and metabolic aspects of muscle function. *Eur. J. Appl. Physiol.*, **40**, 45–55.

KUGELBERG, E. & LINDEGREN, B. (1979) Transmission and contraction fatigue of rat motor units in relation to succinate dehydrogenase activity of motor unit fibers. *J. Physiol. (London)*, **288**, 285–300.

KUKULKA, C.G. & CLAMANN, H.P. (1981) Comparison of the recruitment and discharge properties of motor units in human brachial biceps and adductor pollicis during isometric contractions. *Brain Res.*, **219**, 45–55.

LEWIS, S.F., HALLER, R.G., COOK, J.D. & NUNNALLY, R.L. (1985) Muscle fatigue in McArdle's disease studied by ^{31}P-NMR: effect of glucose infusion. *J. Appl. Physiol.*, **59** 1991–4.

LINDSTRÖM, L., KADEFORS, R. & PETERSEN, I. (1977) An electromyographic index for localized muscle fatigue. *J. Appl. Physiol.*, **43**, 750–4.

MCCULLY, K.K. & FAULKNER, A. (1986) Characteristics of lengthening contractions associated with injury to skeletal muscle fibers. *J. Appl. Physiol.*, **61**, 293–9.

MCILWAIN, J.S. & HAYES, K.C. (1977) Dynamic properties of human motor units in the Hoffmann-reflex and M response. *Am. J. Phys. Med.*, **56**, 704–10.

MAHLER, M. (1979), Neural and humoral signals for pulmonary ventilation arising in exercising muscle. *Med.Sci. Sports Exercise*, **11**, 191–7.

MATSUMOTO, T., ITO, K. & MORITANI, T. (1991) The relationship between anaerobic threshold and electromyographic fatigue threshold in college women. *Eur. J. Appl. Physiol.*, **63**, 1–5.

METZGER, J.M. & MOSS, R.L. (1987) Greater hydrogen ion-induced depression of tension and velocity in skinned single fibers of rat fast than slow muscles. *J. Physiol. (London)*, **393**, 724–42.

MILLS, K. & EDWARDS, R.T.H. (1984) Muscle fatigue in myophosphorylase deficiency: power spectral analysis of electromyogram. *ECG Clin. Neurophysiol.*, **57**, 330–5.

MILNER-BROWN, H.S. & STEIN, R.B. (1975) The relation between the surface electromyogram and muscular force. *J. Physiol. (London)*, **246**, 549–69.

MILNER-BROWN, H.S., STEIN, R.B. & YEMM, R. (1973) Changes in firing rate of human motor units during linearly changing voluntary contractions. *J. Physiol. (London)*, **230**, 371–90.

MIMASA, F., MATSUMOTO, T. & MORITANI, T. (1990) Acute effects of aerobic exercise on spinal reflex excitability and electroencephalogram. *Jap. J. Phys. Educat.*, **35**, 261–9.

MORITANI, T. (1980) Anaerobic threshold determination by surface electromyography. PhD dissertation, University of Southern California, California.

MORITANI, T. (1992) Time course of adaptations during strength and power training. In *Strength and Power in Sport* (Ed. P.V. KOMI), pp. 266–78. Blackwell Scientific Publications, Oxford.

MORITANI, T. (1993) Neuromuscular adaptations during the acquisition of muscle strength, power and motor tasks. *J. Biomechan.*, **26**, 95–107.

MORITANI, T. & DEVRIES, H.A. (1978) Reexamination of the relationship between the surface integrated electromyogram (IEMG) and force of isometric contraction. *Am. J. Phys. Med.*, **57**, 263–77.

MORITANI, T. & DEVRIES, H.A. (1979) Neural factors versus hypertrophy in the time course of muscle strength gain. *Am. J. Phys. Med.*, **58**, 115–30.

MORITANI, T. & DEVRIES, H.A. (1980) Potential for gross muscle hypertrophy in older men. *J. Gerontol.*, **35**, 672–82.

MORITANI, T. & MIMASA, F, (1990) An electromyographic analysis of neuromuscular control during the acquisition of a motor task. *J. Sports Med. Sci. (Japan)*, **4**, 35–43.

MORITANI, T. & MURO, M. (1987) Motor unit activity and surface electromyogram power spectrum during increasing force of contraction. *Eur. J. Appl. Physiol.*, **56**, 260–5.

MORITANI, T., NAGATA, A. & MURO, M. (1981) Electromyographic manifestations of neuromuscular fatigue of different muscle groups during exercise and arterial occlusion. *J. Phys. Fitness Japan*, **30**, 183–92.

MORITANI, T. & SHIBATA, M. (1994) Premovement electromyographic silent period and alpha motoneuron excitability. *J. Electromyogr. Kinesiol.*, **4**, 27–36.

MORITANI, T., NAGATA, A. & MURO, M. (1982) Electromyographic manifestations of muscular fatigue. *Med. Sci. Sports Exercise*, **14**, 198–202.

MORITANI, T., TANAKA, H., YOSHIDA, T., ISHII, C., YOSHIDA, T. & SHINDO, M. (1984) Relationship between myoelectric signals and blood lactate during incremental forearm exercise. *Am. J. Phys. Med.*, **63**, 122–32.

MORITANI, T., MURO, M. & KIJIMA, A. (1985a) Electromechanical changes during electrically induced and maximal voluntary contractions: electrophysiologic responses of different muscle fiber types during stimulated contractions. *Exp. Neurol.*, **88**, 471–83.

MORITANI, T., MURO, M., KIJIMA, A., GAFFNEY, F.A. & PERSONS, A. (1985b) Electromechanical changes during electrically induced and maximal voluntary contractions: surface andintramuscular EMG responses during sustained maximal voluntary contraction. *Exp. Neurol.*, **88**, 484–99.

MORITANI, T., MURO, M., KIJIMA, A. & BERRY, M.J. (1986a) Intramuscular spike analysis during ramp force and muscle fatigue. *EMG Clin. Neurophysiol.*, **26**, 147–60.

MORITANI, T., MURO, M. & NAGATA, A. (1986b) Intramuscular and surface electromyogram changes during muscle fatigue. *J. Appl. Physiol.*, **60**, 1179–85.

MORITANI, T., BERRY, M.J., BACHARACH, D.W. & NAKAMURA, E. (1987a) Gas exchange parameters, muscle blood flow and electromechanical properties of the plantar flexors. *Eur. J. Appl. Physiol.*, **56**, 30–7.

MORITANI, T., ISHIDA, K. & TAGUCHI, S. (1987b) Physiological effects of stretching upon DOMS: electrophysiological analyses. *Descente Sports Sci. (Japan)*, **8**, 212–20.

MORITANI, T., MURAMATSU, S. & MURO, M. (1988) Activity of motor units during concentric and eccentric contractions. *Am. J. Phys. Med.*, **66**, 338–50.

MORITANI, T., ODDSSON, L. & THORSTENSSON, A. (1990a) Differences in modulation of the gastrocnemius and soleus H-reflexes during hopping in man. *Acta Physiol. Scand.*, **138**, 575–6.

MORITANI, T., ODDSSON, L. & THORSTENSSON, A. (1990b) Electromyographic evidence of selective fatigue during the eccentric phase of stretch/shortening cycles in man. *Eur. J. Appl. Physiol.*, **60**, 425–9.

MORITANI, T., ODDSSON, L. & THORSTENSSON, A. (1991a) Phase dependent preferential activation of the soleus and gastrocnemius muscles during hopping in humans. *J. Electromyogr. Kinesiol.*, **1**, 34–40.

MORITANI, T., ODDSSON, L. & THORSTENSSON, A. (1991b) Activation patterns of the soleus and gastrocnemius muscles during different motor tasks. *J. Electromyogr. Kinesiol.*, **1**, 81–8.

MORITANI, T., SHERMAN, W.M., SHIBATA, M., MATSUMOTO, T. & SHINOHARA, M. (1992) Oxygen availability and motor unit activity in humans. *Eur. J. Appl. Physiol.*, **64**, 552–6.

MORITANI, T., TAKAISHI, T. & MATSUMOTO, T. (1993) Determination of maximal power output at neuromuscular fatigue threshold. *J. Appl. Physiol.*, **74**, 1729–34.

MURO, M., NAGATA, A., MURAKAMI, Y. & MORITANI, T. (1982) Surface EMG power spectral analysis of neuromuscular disorder patients during isometric and isotonic contractions. *Am. J. Phys. Med.*, **61**, 244–54.

MUTCH, B.J.C. & BANISTER, E.W. (1983) Ammonia metabolism in exercise and fatigue: a review. *Med. Sci. Sports Exercise*, **15**, 41–50.

NAGATA, A., MURO, M., MORITANI, T. & YOSHIDA, T. (1981) Anaerobic threshold determination by blood lactate and myoelectric signals. *Jap. J. Physiol.*, **31**, 585–97.

NAKAMARU, Y. & SCHWARTZ, A. (1972) The influence of hydrogen ion concentration on calcium binding and release by skeletal muscle sarcoplasmic reticulum. *J. Gen. Physiol.*, **59**, 22–32.

NARDONE, A. & SCHIEPPATI, M. (1988) Shift of activity from slow to fast muscle during voluntary lengthening contractions of the triceps surae muscles in humans. *J. Physiol. (London)*, **395**, 363–81.

NARDONE, A., ROMANO, C. & SCHIEPPATI, M. (1989) Selective recruitment of high-threshold human motor units during voluntary isotonic lengthening of active muscles. *J. Physiol. (London)*, **409**, 451–71.

NARICI, M.V., ROI, G.S., LANDONI, L., MINETTI, A.E. & CERRETELLI, P. (1989) Changes in force, cross-sectional area and neural activation during strength training and detraining of the human quadriceps. *Eur. J. Appl. Physiol.*, **59**, 310–19.

NASSAR-GENTINA, V., PASSONNEAU, J.V., VERGARA, J.L. & RAPAPORT, S.I. (1978) Metabolic correlates of fatigue and recovery from fatigue in single frog muscle fibers. *J. Gen. Physiol.*, **72**, 593–606.

NELSON, D.L. & HUTTON, R.S. (1985) Dynamic and static stretch responses in muscle spindle receptors in fatigued muscle. *Med. Sci. Sports Exercise*, **17**, 445–50.

NESHIGE, R., LUDERS, H., FRIEDMAN, L. & SHIBASAKI, H. (1988a) Recording of movement-related potentials from the human cortex. *Ann. Neurol.*, **24**, 439–45.

NESHIGE, R., LUDERS, H. & SHIBASAKI, H. (1988b) Recording of movement-related potentials from scalp and cortex in man. *Brain*, **111**, 719–36.

NEWHAM, D.J., MILLS, K,R,, QUIGLEY, B.M. & EDWARDS, R.H.T. (1983) Pain and fatigue after concentric and eccentric muscle contractions. *Clin. Sci.*, **64**, 55–62.

PACHELLA, R.C. & PEW, R.W. (1968) Speed–accuracy trade-off in reaction time: effect of discrete criterion times. *J. Exp. Psychol.*, **76**, 19–24.

POULTON, E.C. (1981) Human manual control. In *Handbook of Physiology, Sect. 1. The Nervous System. Volume II. Motor Control.* (Ed. V.B. BROOKS), pp. 1337–89. American Physiological Society, Maryland.

RASCH, P.J. & MOREHOUSE, L.E. (1957) Effect of static and dynamic exercise on muscular strength and hypertrophy. *J. Appl. Physiol.*, **11**, 29–34.

RODAHL, K. & HORVATH, S.M. (1962) *Muscle as a Tissue*. McGraw-Hill Book Company Inc., New York.

ROTTO, D.M. & KAUFMAN, M.P. (1988) Effect of metabolic products of muscular contraction on discharge of group III and IV afferents. *J. Appl. Physiol.*, **64**, 2306–13.

RUTHERFORD, O.M. & JONES, D.A. (1986) The role of learning and coordination in strength training. *Eur. J. Appl. Physiol.*, **55**, 100–5.

SAHLIN, K., KATZ, A. & HENRIKSSON, J. (1987) Redox state and lactate accumulation in human skeletal muscle during dynamic exercise. *Biochem. J.*, **245**, 551–6.

SCHMIDT, R.A. (1975) A schema theory of discrete motor skill learning. *Psychol. Rev.*, **82**, 225–6.

SCHMIDT, R.A. (1976) Control processes in motor skills. *Exercise Sport Sci. Rev.*, **4**, 229–61.

SCHMIDT, R.A., ZELAZNIK, H.N. & FRANK, J.S. (1978) Sources of inaccuracy in rapid movement. In *Information Processing in Motor Control and Learning*, (Ed. G.E. STELMACH), Academic Press, New York.

SCHMIDT, R.A., ZELAZNIK, H.N., HAWKINS, B., FRANK, J.S. & QUINN, J.T. (1979) Motor output variability: a theory for the accuracy of rapid motor acts. *Psychol.Rev.*, **86**, 416–49.

SHERWOOD, D.E. & SCHMIDT, R.A. (1980) The relationship between force and force variability in minimal and near-maximal static and dynamic contractions. *J. Motor Behav.*, **12**, 75–89.

SHIBASAKI, H., BARRETT, G., HALLIDAY, E. & HALLIDAY, A.M. (1980) Components of the movement-related cortical potential and the scalp topography. *ECG Clin. Neurophysiol.*, **49**, 213–26.

SINGH, J. & KNIGHT, R.T. (1990) Frontal lobe contribution to voluntary movements in humans. *Brain Res.*, **531**, 45–54.

SJØGAARD, G. (1978) Force–velocity curve for bicycle work. In *Biomechanics VI-A*. (Ed. E. ASMUSSEN), pp. 93–9. University Press, Baltimore.

SJØGAARD, G. (1990) Exercise-induced muscle fatigue: the significance of potassium. *Acta Physiol. Scand.*, **140** (Suppl 593), 1–63.

SMITH, J.L. (1976) Fusimotor loop properties and involvement during voluntary movement. *Exercise Sport Sci. Rev.*, **4**, 297–333.

SMITH, J.L., BETTS, B., EDGERTON, V.R. & ZERNICKE, R.F. (1980) Rapid ankle extension during paw shakes: selective recruitment of fast ankle extensors. *J. Neurophysiol.*, **43**, 612–20.

STEPHENS, J.A. & USHERWOOD, T.P. (1977) The mechanical properties of human motor units with special reference to their fatigability and recruitment threshold. *Brain Res.*, **125**, 91–7.

TAYLOR, M.G. (1978) Bereitschaftspotential during the acquisition of a skilled motor task. *ECG Clin. Neurophysiol.*, **45**, 568–76.

THIGPEN, L.K., MORITANI, T., THIEBAUD, R. & HARGIS, J.L. (1985) The acute effects of static stretching on alpha motoneuron excitability. In *Biomechanics IX-A* (Ed. D.A. WINTER, R.W. NORMAN, R.P. WELLS, K.C. HAYES & A.E. PATLA), pp. 352–7. Human Kinetics, Champaign, Illinois.

TIBES, U. (1977) Reflex inputs to the cardiovascular and respiratory centers from dynamically working canine muscles: some evidence for involvement of Group II and IV nerve fibers. *Circ. Res.*, **41**, 332–41.

VIITASALO, J.T. & KOMI, P.V. (1977) Signal characteristics of EMG during fatigue. *Eur. J. Appl. Physiol.*, **37**, 111–21.

VIITASALO, J.T., LUHTANEN, P., RAHKILA, P. & RUSKO, H. (1985) Electromyographic activity related to aerobic and anaerobic threshold in ergometer bicycling. *Acta Physiol. Scand.*, **124**, 287–93.

VOLLESTAD, N.K., VAAGE, O. & HERMANSEN, L. (1984) Muscle glycogen depletion patterns in type I and subgroups of type II fibers during prolonged severe exercise in man. *Acta Physiol. Scand.*, **122**, 433–41.

WASSERMAN, K., WHIPP, B.J. KOYAL, S.N. & BEAVER, W.L. (1973) Anaerobic threshold and respiratory gas exchange during exercise. *J. Appl. Physiol.*, **35**, 236–43.

WASSERMAN, K., BEAVER, W.L., DAVIS, J.A., PU, J.Z., HERBER, D. & WHIPP,

B., (1985) Lactate, pyruvate, and lactate-to-pyruvate ratio during exercise and recovery. *J. Appl. Physiol.*, **59**, 935–40.

WASSERMAN, K., BEAVER, W.L. & WHIPP, B.J. (1990) Gas exchange theory and the lactic acidosis (anaerobic threshold). *Circulation*, **81** (Suppl. II), 1114–30.

WILKIE, D.R. (1986) Muscular fatigue: effects of hydrogen ions and inorganic phosphate. *Fed. Proc.*, **45**, 2921–3.

WOODS, J.J., JONES, D.A. & BIGLAND-RITCHIE, B. (1978) Components of the surface EMG during stimulated and voluntary contractions. *Med. Sci. Sports Exercise*, **10**, 67.

YAMASHITA, N. & MORITANI, T. (1989) Anticipatory changes of soleus H-reflex amplitude during execution process for heel raise from standing position. *Brain Res.*, **490**, 148–51.

YAMASHITA, N., NAKABAYASHI, T. & MORITANI, T. (1990) Interrelationships among anticipatory EMG activity, Hoffmann reflex amplitude and EMG reaction time during voluntary standing movement. *Eur. J. Appl. Physiol.*, **60**, 98–103.

CHAPTER SIX

EMG alterations at sustained contractions with special emphasis on applications in ergonomics

GÖRAN M. HÄGG and ROLAND KADEFORS

National Institute for Working Life, Solna, Lindholmen Development and Chalmers University of Technology, Gothenburg

6.1 INTRODUCTION

Evaluation of the effect of sustained muscular contractions is of interest in a multitude of situations. Examples in clinical medicine include pain syndromes linked to the involuntary building up of muscular tension, for instance localized to the trapezius or the masseter muscles. In ergonomics the problems associated with sustained muscular load play a profound role. This is due to the fact that prolonged muscular effort, particularly at high levels of voluntary activation, evokes pain and discomfort, that the biomechanical loading on body structures may cause tissue damage in the long run, and that muscular fatigue may impair the possibility to carry out the working task as required. Muscle fatigue also enhances tremor and reduces fine motor control.

6.1.1 Occupational Health Aspects

The link between physical loading inflicted during work and emerging pain syndromes on the musculoskeletal system is of different strength, depending upon the precise character of the exposure and the clinical symptoms observed. The terminology developed, 'repetitive strain injuries' and 'cumulative trauma disorders', indicates the current view on causality and the nature of progress. Recent reviews dealing with this family of syndromes and possible causative factors include the ones published by Kilbom (1988), Putz-Anderson (1988),

Keyserling *et al.* (1993), and Armstrong *et al.* (1993). Strain caused by prolonged repetitive or sustained static work plays a significant role in all models developed to link exposure to disease.

For instance with respect to hand and wrist problems, risk factors behind carpal tunnel syndrome include combined wrist flexion and high muscular force, which increases the pressure in the median nerve channel (Rempel, 1994). Such loading may occur in industrial work while using improperly designed hand tools (Tichauer, 1968; Sperling *et al.*, 1993). Tanaka and McGlothlin (1993) proposed a risk evaluation model for hand and wrist pain considering force, wrist position, and time.

Static loading on the shoulder, particularly in elevated arm positions, necessitates recruitment of all muscles of the rotator cuff. Epidemiological and clinical studies have linked shoulder tendinitis to exposure to work with elevated arms (Hagberg, 1982; Järvholm 1990).

In evaluation of material handling tasks, loads on the low back have been in focus. In the 'Revised NIOSH equation for the design and evaluation of lifting tasks' (Waters *et al.*, 1993), evaluation is in part based on force prediction and time.

In the illustrative examples given, it should be noted that the exposure to biomechanical loading while transferred by muscles, does not explicitly give rise to pathological processes in the muscles themselves: the principal problem sites are in nerves, tendons, ligaments, and cartilage. This is an important observation in the context of this paper, since electromyographic studies in the ergonomics field do not in general focus on revealing risk for muscular disorders; they rather use information made available and using electromyogram (EMG) as an indicator of muscular load and of physiological effects of sustained load, to assess the risk of pathological processes in other types of tissues.

There is one (or perhaps two) exception to the observation made in the above. There are observations that, in cases of clinical myalgia syndrome localized to the descending part of the trapezius muscle, so-called 'ragged red fibers' are found (Larsson *et al.*, 1988). These findings indicate that only a limited number of muscle fibres in the muscle are affected. The 'raggedness' is likely not caused by a mechanical trauma but rather by a metabolic disturbance and it is only found in red (type I) muscle fibers mainly active during long-time static work. It is hypothesized that it is related to a too high activation/rest ratio in these fibers ('cinderella fibers') (Hügg, 1991b). These fibres are likely not relieved by load reduction but rather by introducing a more dynamic load pattern. Prevention of these kinds of disorders should be accomplished by studies of EMG amplitude as a direct expression of load pattern (Veiersted *et al.*, 1993).

6.1.2 Ergonomic Aspects

As touched upon, muscular force exertion and fatigue is a valid subject for ergonomic study, even despite aspects related to the risk of development of cumulative trauma disorders. The sensation of fatigue contributes to the stress induced during work.

In the process of acquisition of skill, muscular tension gradually decreases, entailing increased physiological economy and improving fine motor control. For instance in welding work, which combines high demands on motor skills and force exertion, it was found that experienced welders were less fatigue prone in the shoulder muscles than those who were relatively new to the job (Kadefors *et al.*, 1976).

It should be noted that there is a complex relationship between work stress and the development of muscular tension. Waersted and Bjørklund (1991) have shown that under certain circumstances, a stressful task in data entry work may induce muscular strain in the shoulder muscles. Lundberg *et al.* (1994) demonstrated that women exposed to laboratory stress responded by increased muscle tension in the descending part of the trapezius muscle. These observations show that sustained muscle contraction may be at hand even if the work situation under study may not indicate such at first glance. Occurrence of occasional muscular rest plays a significant role for endurance and for the development of fatigue.

6.1.3 Purpose of the Present Paper

The background given here indicates the need for monitoring sustained contractions and the physiological effects thereof in ergonomic studies. It has been found that electromyography provides extremely valuable tools for this purpose. However, it should be noted that there is a considerable diversity in the precise hypotheses to be tested in ergonomic applications, making necessary the adoption of methods which are best suited for the purpose. Such methodological streamlining means that different dimensions of the EMG signal have to be considered.

The goal of the present paper is to provide a comprehensive review of the role of electromyography in the study of sustained contractions. Since various types of statistical changes in the properties of the EMG signal are of significance in this context, we consider changes in the amplitude domain, in the frequency domain, and in the time domain as well. Again, it should be noted that the chief aim is primarily applications in the field of ergonomics, rater than in clinical electromyography or other medical fields outside the general area of occupational health.

6.2 DIMENSIONS OF THE EMG SIGNAL

6.2.1 Action Potential Conduction Velocity Changes

The depolarization wave propagates from the motor end-plate along the muscle fiber with a velocity in the range of 4–6 $m\,s^{-1}$ (Arendt-Nielsen and Mills,

1988). This velocity is a function of fiber diameter (Håkansson, 1956). The level of velocity is a result of complex electrochemical phenomena localized to the muscle cell membrane. At sustained contractions, the internal physiological environment of the muscle changes gradually, and this affects virtually all aspects associated with the contractile process, including the propagation velocity. In particular, these effects are pronounced at high muscle activation levels, as the metabolism goes into the anaerobic state, and the composition of the intracellular and intercellular fluids is changed rapidly. Propagation velocity change is one of the most important mechanisms behind the EMG spectral compression observed in fatiguing contractions (see below).

Action potential conduction velocity (APCV) changes may be studied readily by means of electromyography. Such evaluations may be carried out in the time or in the frequency domain. For instance, using a microelectrode, Stålberg (1966) was able to evaluate the delay time needed for the motor unit action potentials (MUAPs) to travel the minute distance between the electrode surfaces. In sustained(several hours) contraction at very low activation levels, the propagation velocity was reduced significantly.

Intramuscular evaluations of sustained contractions are valuable in providing reference data. However, in ergonomic applications it is also important to be able to carry out field studies without penetrating the skin. Such studies of propagation velocity have been carried out using various types of differential surface electrode techniques and cross-correlation methods. For a review the reader is referred to Arendt-Nielsen and Zwarts (1989). These methods have mainly been used for basic muscle research but attempts have been made to apply these methods in an ergonomic context (Caffier *et al.*, 1993; Hägg, 1993).

It is interesting to note that provided a pair of electrodes are aligned in the fiber direction, and they are located on the same side of the point of innervation, the differential lead constitutes a filter which produces predicted 'dips' in the spectrum for frequencies such that

$$\sin(d/v) = 0 \tag{1}$$

where $2d$ is the interelectrode distance and v is the propagation velocity along the muscle fibers of the motor unit considered (Lindström, 1973). Using these spectral 'dip' locations, it is possible to calculate the conduction velocity. Using this method, Broman (1977) was able to monitor the decrease in the conduction velocity during a fatiguing contraction in the biceps muscle and the subsequent recovery.

According to the model presented by Lindström (e.g. Lindström *et al.*, 1970), the average propagation velocity has an overall influence on the shape of the spectrum. The spectrum of the SEMG may be expressed as

$$\psi(j\omega) = G(j\omega/v)\, A_1(j\omega/v) \tag{2}$$

where $A_1(j\omega/v)$ marks the Fourier amplitude of the action potential and $G(j\omega/v)$ describes the spectral changes due to, for example geometrical parameters. It is clear here that changes in the propagation velocity v are always accompanied by a compression along a linear frequency axis or a translation with logarithmic axis as demonstrated in Figure 6.1.

Generally, detection of motor unit action potentials with surface electrodes is influenced by electrode geometry (Fuglevand *et al.*, 1992). Hence, when comparing results from different laboratories, such aspects should be kept in mind.

6.2.2 Synchronization and Firing Rate

The timing of firings of motor units (MUs) contributing to a surface EMG (SEMG) is of great importance for the spectral properties. Data describing intervals between firings are generally denoted as firing statistics. The firing rates of MUs show large intraindividual as well as interindividual differences. The intraindividual variation is one of two possibilities to increase the muscular force output. The firing rate of a MU is determined by the type of MU and type of muscle and actual force output. Typical values are 5–50 firings per second.

The firing of a MU is seldom strictly periodic. Small temporal displacements makes it relevant to study the variation of the duration between two firings

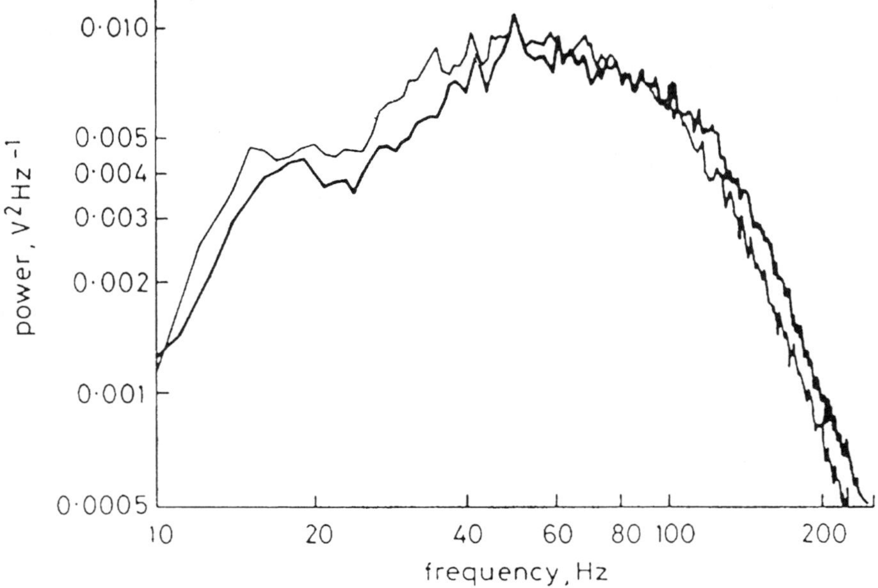

Figure 6.1 Illustration of average power spectrum parallel shift due to action potential velocity decrease (logarithmic frequency axis). Thick line – unfatigued. Thin line – fatigued. From Hägg (1991a).

denoted the interpulse interval (IPI). If the IPI variance is small, the power spectrum of a MUAP train is well defined, with peaks at the firing rate and its harmonics. A larger IPI variance yields more diffuse power spectra contributions.The theoretical basis for these matters is complex and for a thorough review, the reader is referred to some of the major contributions (Lago and Jones, 1977; Weytjens and van Steenberghe, 1984a, b; Pan *et al.*, 1989; Hermens *et al.*, 1992; Fuglevand *et al.*, 1993).

The concept of synchronization of MU firings as a phenomenon accompanying fatigue is old (Buchthal and Madsen, 1950; Lippold *et al.*, 1960; Bigland-Ritchie *et al.*, 1981). According to the theory, synchronization implies a specific increase of the low frequency components of the SEMG spectrum. Even if attempts with intramuscular electrodes to demonstrate the existence of progressive synchronization have failed (De Luca *et al.*, 1993), there are several results from fatigue experiments indicating such a low frequency increase in addition to the compression caused by an APCV decrease (Blinowska *et al.*, 1980; Krantz *et al.*, 1983; Hägg, 1991a; Krogh-Lund and Jørgensen, 1991, 1992). Hence, when discussing spectral EMG alterations at sustained contraction, these types of effects should be considered as a significant factor, in particular close to exhaustion.

6.2.3 Recruitment

Additional recruitment of MUs is the major way to increase muscular force output. The strategies of the central nervous system for work allocation among available MUs are of great importance for our subject but are only partly known. Two basic principles are in focus here. The first one concerns the organization of motor units into task groups. These functional groups sometimes involve MUs from several adjacent anatomically defined muscles (Loeb, 1985; Enoka *et al.*, 1989; Wolf *et al.*, 1993). Hence, from a CNS point of view, MUs are not grouped following gross anatomical structures but according to functional principles.

Another basic principle is ordered recruitment (and reversed order derecruitment) according to motor neuron size as discovered by Henneman *et al.* (1965). This principle has been verified in short laboratory experiments during strict static isometric conditions (Milner-Brown *et al.*, 1973; Stashuk and De Luca, 1989; Masuda and De Luca, 1991). Type I MUs which are slow but persistent and fatigue resistant are recruited first while fast type II MU:s with short endurance have higher recruitment thresholds. However, little is known about the validity of this principle in more dynamic tasks. It has been hypothesized that other groups of MUs, adapted to dynamic activity (mainly type II), are recruited when a faster response is needed. In a recent thesis by Søgaard (1994) it was shown that a considerable number of MUs in m. biceps brachii are active during static as well as slow dynamic (10°/s) tasks.

A key issue for our topic is what happens during sustained contraction. The problem can be separated into two parts. Does additional recruitment occur when already active MUs due to fatigue do not give the required force contribution? Also do all MUs recruited at the start of a contraction remain active to the end or is there any recruitment rotation among MUs?

The first question is probably the easiest to answer. In spite of a lack of definite scientific proof, there is today concensus that additional recruitment does exist. Part of the EMG amplitude increase seen during sustained contraction may be contributed to the low-pass filtering properties of the tissue in conjunction with decreased frequency content. However, this effect alone cannot explain the large amplitude increase seen during sustained contraction. Additional recruitment is the most feasible explanation (Moritani *et al.*, 1986; Krogh-Lund and Jørgensen, 1991, 1992). This view is supported by findings of amplitude increase even when APCV is uneffected (Krogh-Lund and Jørgensen, 1992). This amplitude increase has been applied as a simple fatigue index (Laurig *et al.*, 1987). The newly recruited MUs have a higher conduction velocity (CV) due to larger fibres. An important consequence for the SEMG power spectrum is that a moderate decrease in CV in earlier recruited MUs due to fatigue may be cancelled by a higher CV in MUs recruited later (Arendt-Nielsen *et al.*, 1989; Hägg, 1991a).

The second question is harder to answer. Some indications of recruitment rotation have been reported (Sjøgaard *et al.*, 1986; Fallentin *et al.*, 1993). However, no definite conclusions can be drawn.

6.2.4 Muscle Temperature

Muscle temperature is another important source for EMG signal alteration which usually is a confounding factor. Generally, muscular CV decreases when the temperature decreases (Buchthal and Engbæk, 1963; Morimoto *et al.*, 1980; Schneider *et al.*, 1988). Spectrum indexes are also influenced (Petrofsky and Lind, 1980) according to relations between such indexes and the CV (see below). This effect is estimated to be 3–3.5% $°C^{-1}$ (Merletti *et al.*, 1984). The relation between CV and EMG signal characteristics is discussed above. The EMG amplitude at constant mechanical output is increased at decreased temperature (Winkel and Jørgensen, 1991). This is likely due to impaired mechanical efficiency at decreased muscle temperature. More MUs have to be recruited to accomplish the same mechanical output which implies a higher amplitude. The CV decrease also implies an increased EMG amplitude (Lindström *et al.*, 1970). It should be noted that the increased metabolism during muscular activity tends to raise the muscle temperature (Saltin *et al.*, 1968).

A schematic summary of the impact on SEMGPS of different physiological phenomena is shown in Figure 6.2.

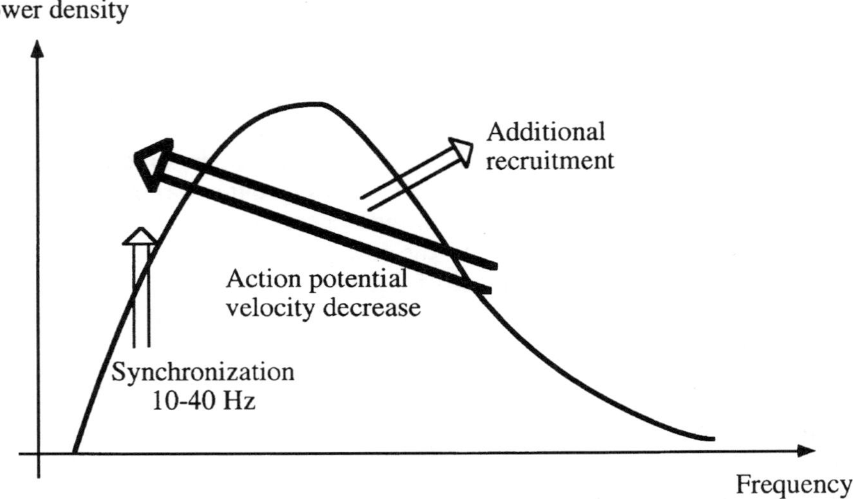

Figure 6.2 Schematic illustration of power spectrum impact from different physiological phenomena (figure outline from de Luca 1985)).

6.2.5 Fatigue Indexes

Usually when discussing fatigue effects on the EMG signal, power spectrum properties are addressed. The SEMG power spectrum contains a large amount of data. Hence, studies of spectral alterations needs reduction of spectral data to a single index like the median frequency (MF) (Stulen and De Luca, 1981) or the mean power frequency (MPF) (Kwatny *et al.*, 1970). The median frequency is defined as the frequency which divides the power spectrum in two parts with equal areas. The mean power frequency is defined according to the standard definition of a mean in statistics, based on a continuous distribution, in this case the power spectrum. Conventionally, the power spectrum is calculated and thereafter the MF or the MPF. However, hardware implementations have also been suggested (Broman and Kadefors, 1979; Stulen and De Luca, 1979). An alternative approach is the so far fairly untested autoregressive modeling technique suggested by Paiss and Inbar (1987).

Another much simpler method for monitoring surface EMG power spectrum (SEMGPS) alterations is to count the number of zero crossings (ZC) of the SEMG signal per time unit (Hägg, 1981; Inbar *et al.*, 1986). Even if the ZC method is not based on the SEMGPS, the number of expected ZCs per time unit can be expressed in terms of the power spectrum (Rice, 1945).

Some other methods such as the mode frequency (frequency at the highest spectrum peak) (Schweitzer *et al.*, 1979) or the power ratio between a high- and a low-frequency band (Kogi and Hakamada, 1962; Bigland-Ritchie *et al.*, 1981; Moxham *et al.*, 1982) have been suggested. The mode estimate is unstable (Schweitzer *et al.*, 1979) and shows a poor relationship with APCV decrease (Sadoyama *et al.*, 1983). The high-to-low ratio technique is not related to any specific fatigue phenomenon by any known model. Furthermore, the band limits for the high and low bands are arbitrarily set without standardization.

It is well documented that any spectrum index (MF, MPF, or ZC) decreases at a sustained strong contraction (Lindström *et al.*, 1977; Hägg, 1981; Sadoyama *et al.*, 1983; De Luca *et al.*, 1986). Chaffin (1973) was the first to suggest SEMGPS alterations as indicators of local muscular fatigue. The physiological background for these alterations has been a matter of controversy over decades and a complete explanation of this phenomenon is still lacking. However, three major events may be pointed out here. APCV decrease (see above) is one of them (Lindström *et al.*, 1977). The average APCV decrease is equivalent to a scaling of the frequency axis or a spectral compression. The second is synchronization of motor unit firings (Buchthal and Madsen, 1950; Lippold *et al.*, 1960; Bigland-Ritchie *et al.*, 1981) which imply increased peaks in the low frequency band of the SEMGPS. Finally, additional recruitment of new MUs most likely also influence the alteration during a sustained contraction. MUs with higher initial APCV are recruited, increasing index readings. However, when more fatiguable type II MUs are recruited the APCV decrease is accelerated.

At sustained contraction, these different factors are likely to coincide. Hence, an index alteration represents the compound action of phenomena with different physiological origin. The APCV decrease is a genuine local muscular event while the others are related to CNS factors. At medium to high contractions levels, which have been investigated extensively, the APCV decreases. Towards exhaustion synchronization dominates, and a typical exponential index response is seen with decreasing time constants at increasing load levels (Lindström *et al.*, 1977; Hagberg, 1981a; De Luca *et al.*, 1986; Kilbom *et al.*, 1992).

At low contraction levels which are frequent in working life, the APCV decrease is low (mainly type I MUs) (Krogh-Lund and Jørgensen, 1992; Caffier *et al.*, 1993). Synchronization or kindered effects related to firing statistics presumably have little influence even if some signs of increased low frequency band peaks have been reported during occupational work (Hägg, 1991a). Under these circumstances, additional recruitment of MUs (mainly type I) with a higher APCV can counteract the decrease or even produce an increase in the fatigue index (Arendt-Nielsen and Mills, 1988; Hägg, 1991a).

When choosing between MF, MPF, or ZC it has been argued that MF should be preferred due to superior high-frequency noise immunity (Stulen and De Luca, 1981). This is true but there are other systematic differences between the responses of these indexes. All three exhibit the same relative response to APCV decrease (Hägg, 1991a). However, synchronization and kindered effects related to firing statistics are mainly seen in the low-frequency band of the SEMGPS. Due to the different spectral moments involved, MF is most sensitive to these phenomena, and ZC least sensitive with MPF in between (Hägg, 1991c). A consequence of this is that MF most often exhibits the largest compound fatigue response, while ZC is more true as an index of specific APCV decrease. These relationships are summarized in Table 6.1.

Finally, the amplitude increase index originally suggested by deVries (1968) should be mentioned. This has mainly been applied by Laurig and his group (Laurig, 1976; Laurig *et al.*, 1987; Luttman *et al.*, 1990, 1994). If a work task is performed repeatedly under constant biomechanical conditions, an increase of the mean EMG amplitude is an indicator of fatigue. Theoretically, this can be caused either by APCV decrease (Lindström *et al.*, 1977) and/or by additional recruitment. However, when low-level, long-lasting contractions are studied, APCV decrease is small or absent (Krogh-Lund and Jørgensen, 1992; Caffier *et al.*, 1993). In these cases, the amplitude increase should be interpreted as a sign of additional recruitment to maintain the same force output. Even at higher load levels, the major contribution to the amplitude increase is likely to come from additional recruitment.

Another important aspect is the relation between EMG fatigue indexes and fatigue defined in other ways. One basic definition is loss of force generation capacity. Comparison between these two aspects of fatigue made during recovery show that MPF after a fatiguing experiment recovered within a couple of minutes while reduced force-generating capacity remains for a much longer time (Jonsson and Nilsson, 1979; Kroon and Naeije, 1988). Another interesting aspect is the relations between subjectively experienced fatigue and electromyographic signs of fatigue. At moderate to high load levels, correlation is quite high while at low contraction levels correlation is poor (Hasson *et al.*, 1989; Öberg *et al.*, 1994).

For an extensive general review of this field, the reader is referred to De Luca (1985) and Hägg (1992).

Table 6.1 A summary of different physiological phenomena and their effects on different spectrum alteration estimates (from Hägg 1991c)

	Effect on spectrum shift estimate		
Physiological effect	Median frequency	Mean power frequency	Zero crossing rate
Average APV	Proportional		
Firing statistics alteration (10–40 Hz increase)	– – –[a]	– –[a]	–[a]
Additional recruitment	+	+	+
Muscle temperature	Approximately proportional, 3–4% $°C^{-1}$		

[a] The number of negative signs indicates mutual magnitude of the effect.

6.3 APPLICATIONS

6.3.1 Laboratory

A basic principle in ergonomics is that fatigue should be avoided. A natural application of EMG in ergonomics is therefore to identify load situations which cause electromyographic signs of fatigue (ESF) and then try to modify them to minimize or eliminate ESF. Almost all such studies have focused on static work-loads and a large majority of these studies have been performed on medium to high load levels in load situations which to a varying degree simulate real occupational work-loads. Hence, studies have addressed repetitive arm elevations (Hagberg, 1981b; Sundelin and Hagberg, 1992; Sundelin, 1993), arm positioning related to welding (Herberts *et al.*, 1980, 1981), forearm exertion during carrying of heavy loads (Kilbom *et al.*, 1992), forearm load at gripping work (Byström and Kilbom, 1990; Roy *et al.*, 1991), and seated assembly work in the laboratory (Mathiassen, 1993). Generally, at higher load levels, the index response is related to load level while at low load levels results are ambiguous.

The validity of extrapolating laboratory data on static loads to occupational work has recently been questioned by Mathiassen (1994). When comparing the ESF response in a laboratory experiment with similar measurements from occupational work on the same static level with superimposed dynamic activity he found a higher ESF in the laboratory setting in spite of the lower muscular load.

6.3.2 Field

In some early studies, the fatigue index technique was applied on an EMG detected during ordinary work (Örtengren *et al.*, 1975; Kadefors *et al.*, 1976; Malmqvist *et al.*, 1981; Winkel *et al.*, 1983). A basic problem in these studies was the direct influence of activity level on any ESF index which has been documented later to be most pronounced up to 30% MVC (Hagberg and Ericson, 1982; Moritani and Muro, 1987; Gerdle *et al.*, 1990). A possible solution is to introduce a test contraction (TC) at a well-known load level as originally suggested by Kadefors (1978). This idea was further developed by Hägg *et al.* (1987) as short TCs (10 s) interrupting ordinary work every tenth minute, every TC yielding a single ESF estimate. This approach was applied by Suurküla and Hägg (1987) on seated assembly work and coil-winding. An alternative was suggested and applied by Christensen (1986a, b) using a sustained TC before and after work. The results from this approach indicated that the initial reading of the analysis gave significant results while the further index development was of minor significance. A drawback with sustained TCs is that fatigue, not associated with the ordinary job, is added. This makes multiple TCs impossible. The concept with short test contractions interrupting ordinary work was further applied on cable bundle assembly work (Hägg and Suurküla, 1991). In this study which is the only longitudinal study of relations between the ESF during work and later contracted muscular shoulder disorders it was not possible to demonstrate any significant relations between the ESF and later disorders. However, a relation was found between the ESF during work and disorders when they were established which confirms earlier cross-sectional findings (Suurküla and Hägg, 1987). These findings in interplay with the Henneman size principle and the recognition of 'ragged red fibers' (Larsson *et al.*, 1988) triggered the formulation of 'the cinderella hypothesis' mentioned in the introduction (Hägg, 1991b).

The approach using amplitude increase as a fatigue indicator was applied by Laurig and co-workers in heavy industrial work (Laurig *et al.*, 1987), check-out work in a shop (Luttman *et al.*, 1990) and in a study of urology surgeons (Luttman *et al.*, 1994). A prerequisite is that the biomechanical conditions are not changed. When this condition is fulfilled the interpretation of the results is straightforward.

6.4 DISCUSSION

It is evident that the set of electromyographic methods presented in this paper provide valuable tools in ergonomic studies of sustained muscle contractions. It is possible to judge aspects such as load level and accumulation of fatigue regarding muscular involvement during work, in various muscles. This means that it is possible to trace effects which are highly relevant in ergonomic studies and where no other methods of high specificity exists. Psychophysical rating techniques (Borg, 1982) may provide valuable information, but do not make available muscle-specific data. Biomechanical modeling on the other hand, tends to present conservative estimates of muscular tension due to aspects associated with joint stabilization which are difficult to take into account. Physiological effects of sustained contractions may be monitored using statistical signal processing methods in the time or spectral domains, providing information on development of 'localized muscular fatigue' (Chaffin, 1973).

On the other hand, EMG methods have many shortcomings. To a considerable extent, this is due to our imperfect knowledge of basic mechanisms controlling motor behavior, such as recruitment, temporal organization, and reactions to changes in the internal physiological state of muscle. It is admittedly a problem that virtually all ergonomic studies employing EMG methods are in one way or another, open to discussion on the validity of the results and conclusions drawn. In particular, it should be noted that at loads exceeding approximately 20% MVC, EMG fatigue indices provide fairly reliable measures of localized muscle fatigue in isometric contractions at constant effort, whereas at dynamic loads or loads below this approximate threshold indices become much more unreliable in spite of obvious fatigue reactions. In the choice between methods, the user must be guided by the sensitivity and specificity as well as the feasibility aspects. In many situations, the test contraction method may serve as a solution. This experimental approach may be applied using frequency as well as amplitude domain indexes.

It is believed that carefully designed laboratory studies using EMG methodology should be encouraged, not only because of their ergonomic significance, but also since the results may help to shed light on mechanisms of motor control. Advances in this area are needed in order to improve methods for analysis and interpretation of data collected in the study of sustained contractions.

References

ARENDT-NIELSEN, L. & MILLS, R.K. (1988) Muscle fibre conduction velocity, mean power frequency, mean EMG voltage and force during submaximal fatiguing contractions of human quadriceps, *Eur. J. Appl. Physiol.*, **58**, 20–5.

ARENDT-NIELSEN, L. & ZWARTS, M.J. (1989) The measurement of muscle fibre conduction velocity in human: techniques and applications. *J. Clin. Neurophysiol.*, **6**, 173–90.

ARENDT-NIELSEN, L., MILLS, R.K. & FORSTER, A. (1989) Changes in muscle fiber conduction velocity, mean power frequency, and mean EMG voltage during prolonged submaximal contractions. *Muscle and Nerve*, **12**, 493–7.

ARMSTRONG, T.J., BUCKLE, P., FINE, L.J., HAGBERG, M., JONSSON, B., KILBOM, Å. *et al.* (1993) A conceptual model for work-related neck and upper-limb muskuloskeletal disorders. *Scand. J. Work, Environ. Health*, **19**, 73–84.

BIGLAND-RITCHIE, B., DONOVAN, E.F. & ROUSSOS, C.S. (1981) Conduction velocity and EMG power spectrum changes in fatigue of sustained maximal efforts. *J. Appl. Physiol.*, **51**, 1300–05.

BLINOWSKA, A.J., VERROUST, J. & CANNET, G. (1980) An analysis of synchronization and double discharge effects on low frequency electromyographic power spectra. *EMG Clin. Neurophysiol.*, **20**, 465–80.

BORG, G. (1982) Psychophysical bases of perceived exertion. *Med. Sci. Sports Exec.*, **14**, 377–81.

BROMAN, H. (1977) An investigation on the influence of a sustained contraction on the succession of action potentials from a single motor unit. *EMG Clin. Neurophysiol.*, **17**, 341–58.

BROMAN, H. & KADEFORS, R. (1979) A spectral moment analyzer for quantification of electromyograms. In *Fourth Congress of the International Society of Electrophysiological Kinesiology* (Ed. C.J. DE LUCA), pp. 90 – 1. ISEK, Boston.

BUCHTHAL, F. & ENGBÆK, L. (1963) Refractory period and conduction velocity of the striated muscle fibre. *Acta Physiol. Scand.*, **59**, 199–220.

BUCHTHAL, F. & MADSEN, A. (1950) Synchronous activity in normal and atrophic muscle. *ECG Clin. Neurophysiol.*, **2**, 425–44.

BYSTRÖM, S. & KILBOM, Å. (1990) Physiological response in the forearm during and after isometric intermittent handgrip. *Eur. J. Appl. Physiol.*, **60**, 457–66.

CAFFIER, G., HEINECKE, D. & HINTERTHAN, R. (1993) Surface EMG and load level during long-lasting static contractions of low intensity. *Int. J. Indust. Ergonom.*, **12**, 77–83.

CHAFFIN, D. (1973) Localized muscle fatigue-definition and measurement *J. Occupat. Med.*, **15**, 346–54.

CHRISTENSEN, H. (1986a) Muscle activity and fatigue in the shoulder muscles during repetitive work. *Eur. J. Appl. Physiol.*, **54**, 596–601.

CHRISTENSEN, H. (1986b) Muscle activity and fatigue in the shoulder muscles of assembly plant employees. *Scand. J. Work, Environ. Health*, **12**, 582–7.

DE LUCA, C. (1985) Myoelectric manifestations of localized muscle fatigue in humans. *CRC Crit. Rev. Biomed. Engng.*, **11**, 251–79.

DE LUCA, C.J., SABBAHI, M.A. & ROY, S.H. (1986) Median frequency of the myoelectric signal. *Eur. J. Appl. Physiol.*, **55**, 457–64.

DE LUCA, C.J., ROY, A.M. & ERIM, Z. (1993) Synchronization of motor-unit firings in several muscles. *J. Neurophysiol.*, **70**, 2010–23.

DE VRIES, H.A. (1968) Method for evaluation of muscle fatigue and endurance from electromyographic fatigue curves. *Am. J. Phys. Med.*, **47**, 125–35.

ENOKA, R.M., ROBINSON, G.A. & KOSSEV, A.R. (1989) Task and fatigue effects on low-threshold motor units in human hand muscle. *J. Neurophysiol.*, **62**, 1344–59.

FALLENTIN, N., JØRGENSEN, K. & SIMONSEN, E.B. (1993) Motor unit recruitment during prolonged isometric contractions. *Eur. J. Appl. Physiol.*, **67**, 335–41.

FUGLEVAND, A.J., WINTER, D.A., PATLA, A.E. & STASHUK, D. (1992) Detection of motor unit action potentials with surface electrodes: influence of electrode size and spacing. *Biol. Cyberne.*, **67**, 143–53.

FUGLEVAND, A.J., WINTER, D.A. & PATLA, A.E. (1993) Models of recruitment and rate coding organization in motor-unit pools. *J. Neurophysiol.*, **70**, 2470–88.

GERDLE, B., ERIKSSON, N.E. & BRUNDIN, L. (1990) The behaviour of the mean power frequency of the surface electromyogram in biceps brachii with increasing force during fatigue. With special regard to the electrode distance. *EMG Clin. Neurophysiol.*, **30**, 483–89.

HAGBERG, M. (1981a) Muscular endurance and surface electromyogram in isometric and dynamic exercise. *J. Appl. Physiol.: Respirat. Environ. Exercise Physiol.*, **51**, 1–7.

HAGBERG, M. (1981b) Work load and fatigue in repetitive arm elevations *Ergonomics*, **24**, 543–55.

HAGBERG, M. (1982) Local shoulder muscular strain-symptoms and disorders. *J. Human Ergol.*, **11**, 99–108.

HAGBERG, M. & ERICSON, B. (1982) Myoelectric power spectrum dependence on muscular contraction level of elbow flexors. *Eur. J. Appl. Physiol.*, **42**, 147–56.

HÄGG, G. (1981) Electromyographic fatigue analysis based on the number of zero crossings. *Pflügers Arch.*, **391**, 78–80.

HÄGG, G.M. (1991a) Comparison of different estimators of electromyographic spectral shifts during work when applied on short test contractions. *Med. Biol. Engng. Comput.*, **29**, 511–16.

HÄGG, G.M. (1991b) Static work load and occupational myalgia – a new explanation model. In *Electromyographical Kinesiology* (Ed. P. ANDERSON, D. HOBART & J. DANOFF), pp. 141–4.

HÄGG, G.M. (1991c) Zero crossing rate as an index of electromyographic spectral alterations and its applications to ergonomics. Doctoral thesis, Chalmers University of Technology.

HÄGG, G.M. (1992) Interpretation of EMG spectral alterations and alteration indexes at sustained contraction. *J. Appl. Physiol.*, **73**, 1211–17.

HÄGG, G.M. (1993) Action potential velocity measurements in the upper trapezius muscle. *J. Electromyogr. Kinesiol.*, **3**, 231–5.

HÄGG, G. M. & SUURKÜLA, J. (1991) Zero crossing rate of electromyograms during occupational work and endurance test as predictors for work related myalgia in the shoulder/neck region. *Eur. J. Appl. Physiol.*, **62**, 436–44.

HÄGG, G. M., SUURKÜLA, J. & LIEW, M. (1987) A worksite method for shoulder muscle fatigue measurements using EMG. test contractions and zero crossing technique. *Ergonomics*, **30**, 1541–51.

HÅKANSSON, C.H. (1956) Conduction velocity and amplitude of the action potential as related to circumference in the isolated fibre of frog muscle. *Acta Physiol. Scand.*, **37**, 14–34.

HASSON, S.M., SIGNORILE, J.F. & WILLIAMS, J.H. (1989) Fatigue-induced changes in myoelectric signal characteristics and perceived exertion. *Can. J. Spt. Sci.*, **14**, 99–102.

HENNEMAN, E., SOMJEN, G. & CARPENTER, D.O. (1965) Excitability and inhibitability of motoneurons of different sizes. *J. Neurophysiol.*, **28**, 599–620.

HERBERTS, P., KADEFORS, R. & BROMAN, H. (1980) Arm positioning in manual tasks. An electromyographic study of localized fatigue *Ergonomics*, **23**, 655–65.

HERBERTS, P., KADEFORS, R., ANDERSSON, G.B.J. & PETERSÉN, I. (1981) Shoulder pain in industry: an epidemiological study on welders. *Acta Orthopaed. Scand.*, **59**, 299–306.

HERMENS, H.J, BRUGGEN, T.A.M., BATEN, C.T.M., RUTTEN, W.L.C. & BOOM, H.B.K. (1992) The median frequency of the surface EMG power spectrum in relation to motor unit firing and action potential properties. *J. Electromyogr. Kinesiol.*, **2**, 15–25.

INBAR, G.F., ALLIN, J., PAISS, O. & KRANZ, H. (1986) Monitoring surface EMG spectral changes by the zero crossing rate *Med. Biol. Engng Comput.*, **24**, 10–18.

JÄRVHOLM, U. (1990) On shoulder muscle load. Doctoral thesis, University of Göteborg.

JONSSON, B. & NILSSON, T. (1979) Electromyographic fatigue effects and recovery of endurance in forearm muscles. In *Fourth Congress of the International Society of Electrophysiological Kinesiology*, (Ed. C.J. DE LUCA), pp. 98–9. ISEK, Boston.

KADEFORS, R. (1978) Application of electromyography in ergonomics: New vistas. *Scand. J. Rehab. Med.*, **10**, 127–33.

KADEFORS, R., PETERSÉN, I. & HERBERTS, P. (1976) Muscular reaction to welding work: an electromyographic investigation. *Ergonomics*, **19**, 543–58.

KEYSERLING, W.M., STETSON, D.S., SILVERSTEIN, B.A. & BROUWER, M.L. (1993) A checklist for evaluating ergonomic risk factors associated with upper extremity cumulative trauma disorders. *Ergonomics*, **36**, 807–31.

KILBOM, Å. (1988) Intervention programmes for work-related neck and upper limb disorders: strategies and evaluation. *Ergonomics*, **31**, 735–47.

KILBOM, Å., HÄGG, G.M. & KÄLL, C. (1992) One-handed load carrying – cardiovascular, muscular, and subjective indices of endurance and fatigue. *Eur. J. Appl. Physiol.*, **65**, 52–8.

KOGI, K. & HAKAMADA, T. (1962) Slowing of surface electromyogram and muscle strength in muscle fatigue. *Rep. Inst. Sci. Lab.*, **60**, 27–41.

KRANTZ, H., WILLIAMS, A.M., CASSEL, J., CADDY, D.J. & SILBERSTEN, R.B. (1983)

Factors determining the frequency content of the electromyogram. *J. Appl. Physiol.*, **55**, 392–9.

KROGH-LUND, C. & JØRGENSEN, K. (1991) Changes in conduction velocity, median frequency and root mean square-amplitude of the electromyogram during 25% maximal voluntary contraction of the triceps brachii muscle, to limit of endurance. *Eur. J. Appl. Physiol.*, **63**, 60–9.

KROGH-LUND, C. & JØRGENSEN, K. (1992) Modification of myo-electric power spectrum in fatigue from 15% maximal voluntary contraction of human elbow flexors, to limit of endurance: reflexion of conduction velocity variation and/or centrally mediated mechanisms? *Eur. J. Appl. Physiol.*, **64**, 359–70.

KROON, G.W. & NAEIJE, M. (1988) Recovery following exhaustive dynamic exercise in the human biceps muscle. *Eur. J. Appl. Physiol.*, **58**, 228–32.

KWATNY, E., THOMAS, D.H. & KWATNY, H.G. (1970) An application of signal processing techniques to the study of myoelectric signals. *IEEE Trans. Biomed. Engng*, **17**, 303–12.

LAGO, P. & JONES, N.B. (1977) Effect of motor-unit firing time statistics on e.m.g. spectra. *Med. Biol. Engng Comput.*, **15**, 648–55.

LARSSON, S.E., BENGTSSON, A., BODEGÅRD, L., HENRIKSSON, K.G. & LARSSON, J. (1988) Muscle changes in work-related chronic myalgia. *Acta Orthopaed. Scand.*, **59**, 552–6.

LAURIG, W. (1976) Methodological and physiological aspects of electromyographic investigations. In *Fifth International Congress on Biomechanics* (Ed. P. KOMI), pp. 219–30. Human Kinetics Publishers, Champaign, Il.

LAURIG, W., LUTTMANN, A. & JÄGER, M. (1987) Evaluation of strain in shop-floor situations by means of electromyographic investigations. In *Trends in Ergonomics* (Ed. S.S. ASFOUR), pp. 685–92. Elsevier Science Publishers, Amsterdam.

LINDSTRÖM, L. (1973) A model describing the power spectrum of myoelectric signals. Part 1: single fiber signals, Research Laboratory of Medical Electronics, Chalmers University of Technology, 5:73.

LINDSTRÖM, L.R., MAGNUSSON, R. & PETERSÉN, I. (1970) Muscular fatigue and action potential conduction velocity changes studied with frequency analysis of EMG signals. *Electromyography*, **4** 341–53.

LINDSTRÖM, L., KADEFORS, R. & PETERSÉN, I. (1977) An electromyographic index for localized muscle fatigue. *J. Appl. Physiol.*, **43**, 750–4.

LIPPOLD, O.C.J., REDFEARN, J.W.T. & VUCO, J. (1960) The electromyography of fatigue. *Ergonomics*, **3**, 121–31.

LOEB, G.E. (1985) Motorneurone task groups: coping with kinematic heterogeneity. *J. Exp. Biol.*, **115**, 137–46.

LUNDBERG, U., KADEFORS, R., MELIN, B., PALMERUD, G., HASSMÉN, P., ENGSTRÖM, M. & ELFSBERG-DOHNS, I. (1994) Psychophysiological stress and EMG activity of the trapezius muscle. *Int. J. Behav. Med.*, **1**, 354–70.

LUTTMAN, A., JÄGER, M. & LAURIG, W. (1990) Electromyographic studies of check-out work. In *Eighth International Congress of International Society of Electrophysiological Kinesiology*, (Eds P.A. ANDERSSON, D.J. HOBART & J.V. DANOFF), pp. 145–8. Elsevier Science Publishers, Baltimore.

LUTTMAN, A., SÖKELAND, J. & LAURIG, W. (1994) Muscular strain and fatigue among surgeons in urology. In *12th Triennial Congress of the International Ergonomics Association* (Ed. S. MCFADDEN, L. INNES & M. HILL), pp. 315–17. Human Factors Association of Canada, Toronto.

MALMQVIST, R., EKHOLM, I., LINDSTRÖM, L., PETERSEN, I., ÖRTENGREN, R., BJURÖ, T. *et al.* (1981) Measurement of localized muscle fatigue in building work. *Ergonomics*, **24**, 695–709.

MASUDA, T. & DELUCA, C.J. (1991) Recruitment threshold and muscle fibre conduction velocity of single motor units. *J Electromyogr. Kinesiol.*, **1**, 116–23.

MATHIASSEN, S.E. (1993) Variation in shoulder-neck activity. Doctoral thesis, Karolinska Institute, Stockholm.

MATHIASSEN, S.E. (1994) On the validity of isometric exercise as a model of occupational shoulder-neck activity. In Proceedings of *IEA-94* Vol. 2 (Eds S. MCFADDEN, L. INNES & M. HILL) pp. 189–91. Human Factors Association of Canada, Toronto.

MERLETTI, R., SABBAHI, M.A. & DELUCA, C.J. (1984) Median frequency of the myoelectric signal. Effects of muscle ischemia and cooling. *Eur. J. Appl. Physiol.*, **52**, 258–65.

MILNER-BROWN, H.S., STEIN, R.B. & YEMM, R. (1973) The orderly recruitment of human motor units during voluntary isometric contractions. *J. Physiol.*, **230**, 359–70.

MORIMOTO, S., UMAZUME, Y., & MASUDA, M. (1980) Properties of spike potentials detected by surface electrode in intact human muscle. *Jap. J. Physiol.*, **30**, 71–80.

MORITANI, T. & MURO, M. (1987) Motor unit activity and surface electromyogram power spectrum during increasing force of contraction. *Eur. J. Appl. Physiol.*, **56**, 260–65.

MORITANI, T., MURO, M. & NAGATA, A. (1986) Intramuscular and surface electromyogram changes during muscle fatigue. *Eur. J. Appl. Physiol.*, **60**, 1179–85.

MOXHAM, J. EDWARDS, R.H.T., AUBIER, M., DETROYER, A., FARKAS, G., MACKLEM, P.T. & ROUSSOS, C. (1982) Changes in EMG power spectrum (high-to-low ratio) with force fatigue in humans. *J. Appl. Physiol.*, **53**, 1094–99.

ÖBERG, T., SANDSJÖ, L. & KADEFORS, R. (1994) Subjective and objective evaluation of shoulder muscle fatigue. *Ergonomics*, **37**, 1323–33.

ÖRTENGREN, R., ANDERSSON, G., BROMAN, H., MAGNUSSON, R. & PETERSÉN, I. (1975) Vocational electromyography: studies of localized muscle fatigue at the assembly line. *Ergonomics*, **18**, 157–174.

PAISS, O. & INBAR, G. (1987) Autoregressive modelling of surface EMG and its spectrum with application to fatigue. *IEEE Trans. Biomed. Engng*, **34**, 760–70.

PAN, Z.S., ZHANG, Y. & PARKER, P.A. (1989) Motor unit power spectrum and firing rate. *Med. Biol. Engng Comput.*, **27**, 14–18.

PETROFSKY, J.S. & LIND, A.R. (1980) The influence of temperature on the amplitude and frequency components of the EMG during brief and sustained isometric contractions. *Eur. J. Appl. Physiol.*, **44**, 189–200.

PUTZ-ANDERSON, V. (1988) Cumulative trauma disorders. A manual for musculoskeletal diseases of the upper limb. Taylor & Francis, London.

REMPEL, D.M. (1994) Carpal tunnel pressure studies: implications for prevention and rehabilitation. Proceedings of *IEA-94* Vol. 3 (Ed. S. MCFADDEN, L. INNES & M. HILL) pp. 244–6. Human Factors Association of Canada, Toronto.

RICE, R.O. (1945) Mathematical analysis of random noise. In *Selected Papers on Noise and Stochastic Processes* (Ed. N. WAX), pp. 133–45. Dover Publications, New York.

ROY, S.H. O'HARA, J.M. & BRIGANTI, M. (1991) Use of EMG spectral parameters to evaluate fatigue associated with pressure glove work. In *Electromyographical*

Kinesiology (Ed. P. ANDERSON, D. HOBART & J. DANOFF) pp. 283–6. Elsevier Science Publishers, Amsterdam.

SADOYAMA, T., MASUDA, T. & MIYANO, H. (1983) Relationships between muscle fibre conduction velocity and frequency parameters of surface EMG during sustained contraction. *Eur. J. Appl. Physiol.*, **51**, 247–56.

SALTIN, B., GAGGE, A.P. & STOLWIJK, J.A.J. (1986) Muscle temperature during submaximal exercise in man. *J. Appl. Physiol.*, **26**, 679–88.

SCHNEIDER, J., SILNY, J. & RAU, G. (1988) Noninvasive measurement of conduction velocity in motor units influenced by temperature and excitation pattern. In *Seventh International Congress of International Society of Electrophysiological Kinesiology* (Ed. W. WALLINGA, H.B.K. BOOM & J. DE VRIES), pp. 251–4. Elsevier Science Publishers, Enschede.

SCHWEITZER, T.W., FITZGERALD, J.W., BOWDEN, J.A. & LYNNE-DAVIES, P. (1979) Spectral analysis of human inspiratory diaphragmatic electromyogram. *J. Appl. Physiol.*, **46**, 152–65.

SJØGAARD, G., KIENS, B., JØRGENSEN, K. & SALTIN, B. (1986) Intramuscular pressure, EMG and blood flow during low-level prolonged static contraction in man. *Acta Physiol. Scand.*, **128**, 475–84.

SØGAARD, K. (1994) Biomechanics and motor control during repetitive work, Doctoral thesis, University of Copenhagen.

SPERLING, L., DAHLMAN, S., WIKSTRÖM, L., KILBOM, Å & KADEFORS, R. (1993) A cube model for the classification of work with hand tools and the formulation of functional requirements. *Appl. Ergonom.*, **24**, 212–20.

STÅLBERG, E. (1966) Propagation velocity in human muscle fibres *in situ*, Doctoral thesis, University of Uppsala, Uppsala.

STASHUK, D. & DE LUCA, C. (1989) Update on the decomposition and analysis of EMG signals. In *Computer Aided Electromyography and Expert Systems* (Ed. J.E. DESMEDT), pp. 39–53. Elsevier Science Publishers, New York.

STULEN, F.B. & DELUCA, C.J. (1979) Median frequency of the myoelectric signal as a measure of localized muscular fatigue. In *Fourth Congress of the International Society of Electrophysiological Kinesiology*, (Ed. C.J. DE LUCA), pp. 92–3. ISEK, Boston.

STULEN, F.B. & DELUCA, C.J. (1981) Frequency parameters of the myoelectric signal as a measure of muscle conduction velocity. *IEEE Trans. Biomed. Engng*, **28**, 515–23.

SUNDELIN, G. (1993) Patterns of electromyographic shoulder muscle fatigue during MTM-paced repetitive arm work with and without pauses *Int. Arch. Occupat. Environ. Health*, **64**, 485–93.

SUNDELIN, G. & HAGBERG, M. (1992) Electromyographic signs of shoulder muscle fatigue in repetitive arm work paced by the methods-time-measurement system. *Scand. J. Work, Environ. Health*, **18**, 262–8.

SUURKÜLA, J. & HÄGG, G.M. (1987) Relations between shoulder/neck disorders and EMG zero crossing shifts in female assembly workers using the test contraction method. *Ergonomics*, **30**, 1553–64.

TANAKA, S. & MCGLOTHLIN, J.D. (1993) A conceptual model for prevention of work related carpal tunnel syndrome (CTS). *Int. J. Indust. Ergonomics*, **11**, 181–93.

TICHAUER, E. (1968) Electromyographic kinesiology in the analysis of work situations and hand tools. *Electromyography*, **8** Suppl 1, 197–211.

VEIERSTED, K.B., WESTGAARD, R.H. & ANDERSEN, P. (1993) Electromyographic evaluation of muscular work pattern as a predictor of trapezius myalgia. *Scand. J. Work, Environ. Health*, **19**, 284–90.

WÆRSTED, M. & BJØRKLUND, R.A. (1991) Shoulder muscle tension induced by two VDU-based tasks of different complexity. *Ergonomics*, **34**, 137–50.

WATERS, T.R., PUTZ-ANDERSON, V., GARG, A. & FINE, L.J. (1993) Revised NIOSH equation for the design and evaluation of manual lifting tasks. *Ergonomics*, **36**, 749–76.

WEYTJENS, J.L.F. & VAN STEENBERGHE, D. (1984a) The effects of motor unit synchronization on the power spectrum of the electromyogram. *Biol. Cybernet.*, **51**, 71–7.

WEYTJENS, J.L.F. & VAN STEENBERGHE, D. (1984b) Spectral analysis of the surface electromyogram as a tool for studying rate modulation: a comparison between theory, simulation, and experiment. *Biol. Cybernet.*, **50**, 95–103.

WINKEL, J. & JØRGENSEN, K. (1991) Significance of skin temperature changes in surface electromyography. *Eur. J. Appl. Physiol.*, **63**, 345–8.

WINKEL, J. EKBLOM, B., HAGBERG, M. & JONSSON, B. (1983) The working environment of cleaners. Evaluation of physical strain in mopping and swabbing as a basis for job redesign. In *Ergonomics of Workstation Design* (Ed. KVALSETH, T.O.), pp. 35–44. Butterworths, London.

WOLF, S.L., SEGAL, R.L. & ENGLISH, A.W. (1993) Task oriented electromyographic activity from human lateral gatrocnemius muscle. *J. Electromyogr. Kinesiol.*, **3**, 87–94.

CHAPTER SEVEN

Electromyography of upper extremity muscles and ergonomic applications

HELMUT STRASSER

University of Siegen, Siegen

7.1 INTRODUCTION – TASKS AND GOALS OF ERGONOMIC METHODS FOR STRAIN MEASUREMENTS

Although the advance of technology in many work areas has led to mechanization and rationalization, it still seems unfeasible to ever do completely without manpower (with 'power' meant in a literal sense). That means that humans will always be more than only an information processing element in a man–machine system. A work system, however, can only lead to optimal results if technical elements and human factors are coordinated. The realization of ergonomic knowledge in work design is of great importance, as unattractive jobs do not pay in the long run under either economic or human aspects.

However, the layout of a work-place should not only observe various measurements of the human body to ensure the necessary leg space and reaching space (to the centimeter), thus being based upon static optimizing efforts. Rather, it is important to also take biomechanical and motion technical rules of the human body into account. Forced positions, on the one hand and apparent power resulting from changes in the body posture due to work procedure, on the other hand, have to be minimized. Favorable directions of movement and beneficial angles in the kinematic chain of the hand–arm system have to be prescribed, for example during repetitive assembly work, actuating element control, and tool operating. That is the only way to avoid muscle cramping (myogelosis), damage to joints, and physiological pains which lead to uneconomical rates of absenteeism.

The EC Basic Guideline for Occupational Safety not only aims at the prevention of accidents and occupational diseases, it also aims at the avoidance of exposure to danger. Paragraphs 100a and 118a force producers of machinery, equipment, and working tools to design their products ergonomically. In order to achieve that, production engineering methods that check things such as the ergonomic quality of work-places and tools are necessary. Nowadays, it is often no longer hard physical work, but supposedly light, one-sided stress on the support and motor apparatus which still has negative consequences. Thus, methods that give insight into the local strain during work become more and more important.

In industrial work, not only heavy load carriage – for instance involved in vertical manual material handling – implies risks for occupational health; in addition, light physical work in awkward postures can be associated with at least complaints and troubles and, finally, with cumulative trauma disorders (CTD) or repetitive strain injuries (RSI). In particular, in Scandinavia, the USA, and Australia, such and other musculoskeletal disorders are assumed to be the long-term issues of light manual work which must be executed by the hand–arm–shoulder system at higher working frequencies and in unfavorable postures.

In this context, electromyographic measurement of muscular strain of the upper extremities and the body trunk during simulated work can offer the possibility of detecting bottle-necks of stress and strain. Therefore, according to Figure 7.1, experiments were carried out in the course of which – at seated work – external loads had to be handled repetitively from starting points S within the reach to a fixed point Z, a destination near the body. The following important work-related parameters were varied.

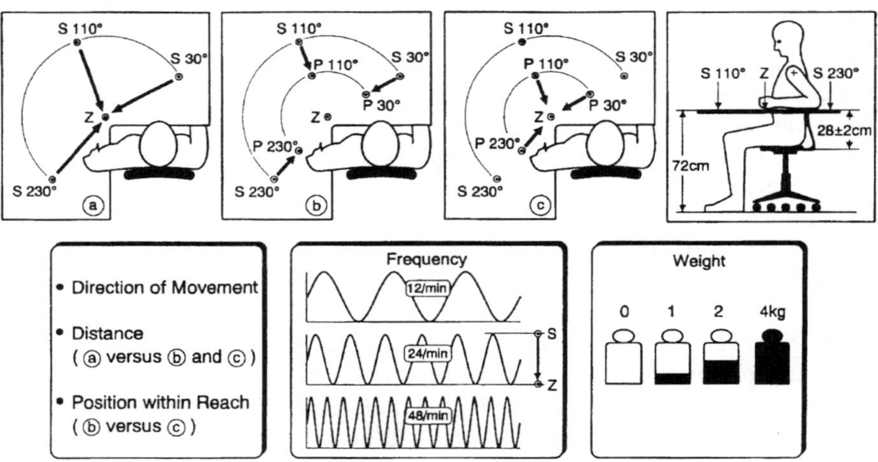

Figure 7.1 Parameters varied during a series of investigations into physiological cost of repetitive manual materials handling.

1 The direction of movements in some practically relevant working directions, measured from the frontal plane of the subject.
2 Their distance and position within the outer or inner reach.
3 The frequency of manual movements between 12 and 48 lifts per minute.
4 The external loads to be handled, varying between 0 and 4 kg.

7.2 COMPUTER-AIDED ELECTROMYOGRAPHIC EVALUATION OF DYNAMIC MUSCLE WORK OF THE UPPER EXTREMITY DURING REPETITIVE MANUAL MOVEMENTS

Since the early 1970s, electromyography has been successfully used in Germany to assess local static muscle work (shown by publications such as Laurig 1970, 1974, Rau 1977, and Einars 1979). Surface electrodes placed on the concerning muscle groups collect data on bioelectric processes and make them graphically depictable. They give insight into innervation, i.e. the muscles' state of activation. The regulation of innervation originates in the cerebrum's pyramidal cells and is controlled via the motoric system of α motor neurons in the spinal cord with branching neurites and motoric end-plates in the muscle fibers. The activation of the motoric units, varying in number and timing, leads – according to a spatial and timewise summation or superposition, respectively, of single contractions of the motoric units – to summing potentials at the skin's surface (in the range of milli- or microvolt) which can be derived via electric amplifiers.

When activating a muscle group, e.g. the biceps, in a static isometric way, myoelectric signals like those in Figure 7.2 can be picked up with the help of bipolar silver–silver chloride electrodes (note the bursts of positive and negative potential changes). According to common methods, the amplified rough electromyogram (EMG(t)) is normally rectified and smoothed (integrated over T) so that according to the formula

$$\mathrm{EA} = \frac{1}{T}\int_0^T \left|\mathrm{EMG}(t)/\mathrm{d}t\right. \tag{1}$$

the electromyographic activity EA (the envelope of the rectified rough EMG) correlates well with the exerted static strength of the muscle, measured e.g. with the help of an isometric force apparatus.

But even investigations into simple static exertions and particularly those into dynamic manual maneuvers, necessitate a multichannel registration and a more detailed work-specific evaluation of myoelectric signals in order to assess the physiological cost of the most relevant muscles and to detect possible bottle-necks with regard to strain and the interaction of the different muscles which participate as active members of a kinematic chain during the performance of a manual task.

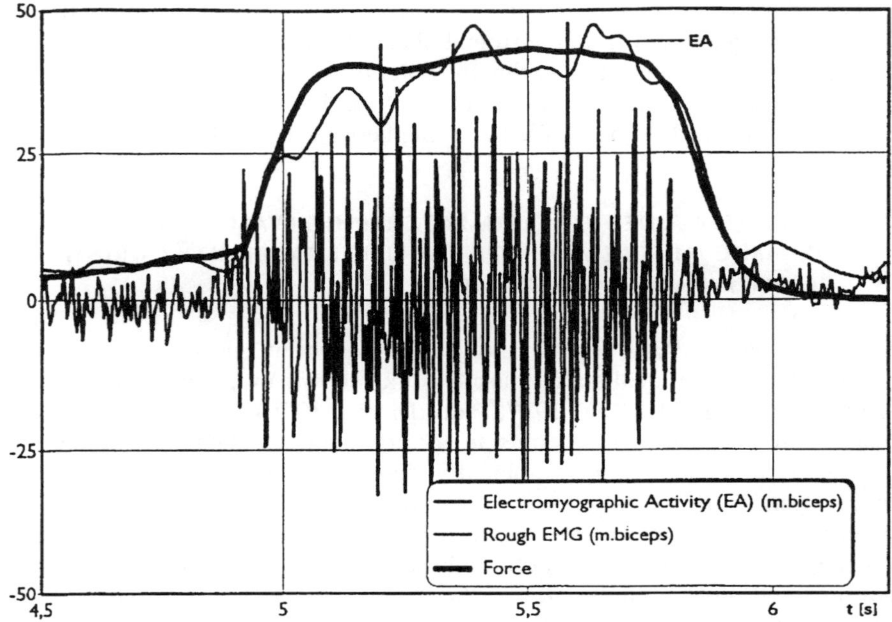

Figure 7.2 Rough myoelectric signal associated with a single exertion of the biceps within approximately 1 s. The rectified and smoothed signal (electromyographic activity) represents the envelope of the bursts and correlates well with the force, which can be measured by a dynamometer.

7.2.1 Calculation of Work-specific Characteristics of EA Data by Splitting-up Electromyographic Time Series into Static and Dynamic Components

In particular when analyzing electromyographic responses from the upper extremity, which is more involved in dynamic movements whereas the muscles of the trunk are mostly involved in static work, averaging or integration of myoelectric activity over a working period may not yield a reliable indicator for the specific underlying muscular strain.

Responses of some muscles involved in manual materials handling in the horizontal plane are shown in the left part of Figure 7.3, as they can be registered by amplifiers and a polygraph. The time series of EA over a period of approximately 50 s, of course, suggest consecutive phases of muscle contractions and relaxations. But despite identical work cycles (during which an external load weighing 1 kg had to be manipulated over a distinct distance in a defined direction), the responses were not always the same with respect to the values of EA and, above all, they were superimposed by higher frequency components. These irrelevant 'high frequency' signals can be filtered out by a

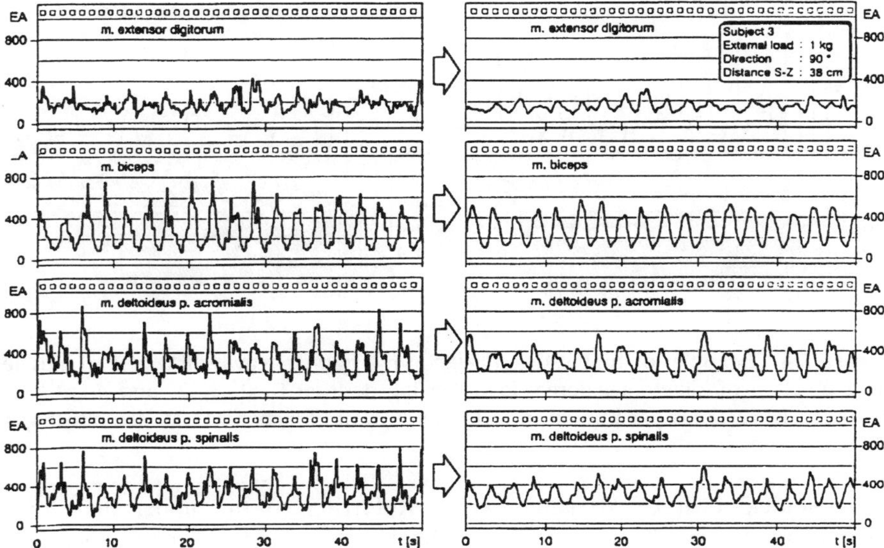

Figure 7.3 Time series of electromyographic activity from four muscles involved in repetitive arm movements (unsmoothed left and smoothed right) by computer-aided filtering processes.

computer-aided smoothening procedure. After analog-to-digital conversion of the time series, a 'sliding' mean value was calculated from each actual value and from the immediately preceding and succeeding value via applying this procedure twice; 'low-frequency' time series (like those on the right side of Figure 7.3) whose extremes can be attributed more reliably to the repetitive exertions and reliefs of the muscles result. When the signals were analog–digital converted with a frequency of 16 2/3 Hz (1000 values per minute), a cut-off frequency of approximately 3 Hz resulted from the sliding averaging processes, which was high enough to correctly reflect physiological responses to repetitive manual movements.

Figure 7.4 again shows some typical already smoothed patterns of the electromyographic activity (EA) from two muscles dependent on external load as a function of time. Above the resting level (at the bottom), the time series of myoelectric activities from the biceps and extensor digitorum in the case of unloaded movements (0 kg), as well as in the case of movements with external loads between 1 and 4 kg to be handled, are represented.

Associated with repetitive movements, curves of EA were generated, clearly reflecting contractions and relaxations within the work cycles (12 per minute) as well as a general increase of EA when heavier weights had to be handled. Traditional mean electromyographic activity, i.e. the averaging of EA over an analyzed period of time of, for example, 50 s and the leveling of the summits and dales would not be adequate processes for characterizing specific

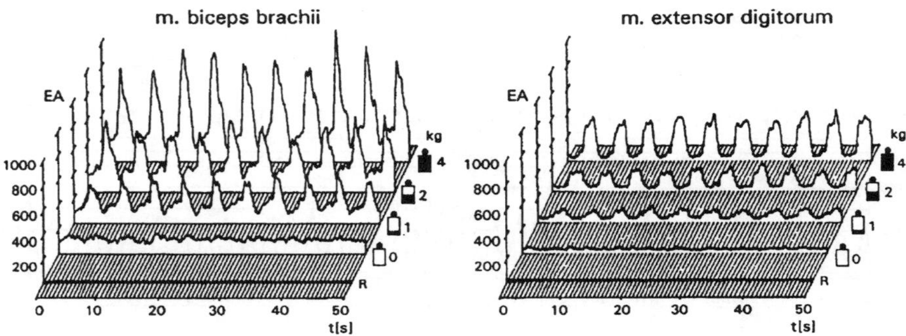

Figure 7.4 Examples of time series of electromyographic activity (EA) of two muscles as a function of different external load (resting position R, 0, 1, 2, and 4 kg) lifted horizontally 12 times per minute in a distinct moving direction (110°, see also Figure 7.1).

physiological cost of dynamic work. Referring to the upper and lower inversion points of the time series, i.e. the maximum and minimum EA values, additional information about the changing exertions and reliefs of muscle strain can be extracted, whose neglect would be a substantial loss of information. Figure 7.5 shows how work-specific characteristics from a time course of electromyographic activity were calculated (for details see Müller *et al.*, 1988a; Strasser *et al.*, 1989).

First of all, in order to avoid an overestimation of (casual) accidental short extreme values and artefacts in physiological data, a further smoothening process was applied. As can be seen in Figure 7.5, maximum and minimum ranges were defined for each 'peak' and 'valley', the limits of which, for example were set at 10% of the whole range between two successive extremes of a work cycle. The calculated mean values of the activity in the maximum

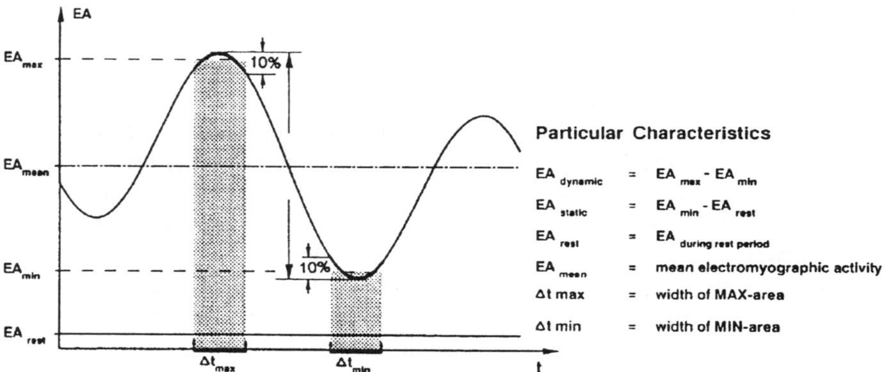

Figure 7.5 Calculation of work-specific characteristics from the time course of electromyographic activity (EA).

and minimum ranges Δt max and Δt min were named maximum activity EA_{max} and minimum activity EA_{min}. The numeric quantification, however, presupposes the localization of these characteristics in their temporal sequence. This is rather simple, if – as in this case – work is done in a precise and predetermined time course. Otherwise, plausibility testing is an inevitable prerequisite.

Because an isolated muscle response to a single manual movement can never be representative, it is advisable to average all the EA_{max} and EA_{min} values within a longer lasting working period. These (more reliable) mean responses can be calculated from a preset number of work cycles (e.g. mean extreme from working periods of 1 or 5 min).

Figure 7.6 shows the conversion of an EMG course associated with ten consecutive work cycles and the splitting-up of the EA into corresponding components. As the height of the minimum values determines the permanent contraction of a muscle, which cannot be fallen below during work, the mean of the averaged minimum values – after having subtracted the resting EA which should be measured in working posture before or after a test – can be defined as the static component of the EA. On the contrary, the range between the peak and the minimum values characterizes the part caused by limb movements, the dynamic component.

Figure 7.7 presents a survey of the experimental set-up (upper left) and the computer-aided data evaluation as shown before (lower part), and it visualizes exemplary results of static and dynamic components of the EMG from one single muscle in a circle diagram (upper right). The very small inner white arc in the graph (upper right) represents the resting activity, measured during time intervals of at least 1 min before and after a test session. The black area, associated with the static component, as well as the outer white part, the dynamic component, clearly demonstrate the increase of the EMG activity

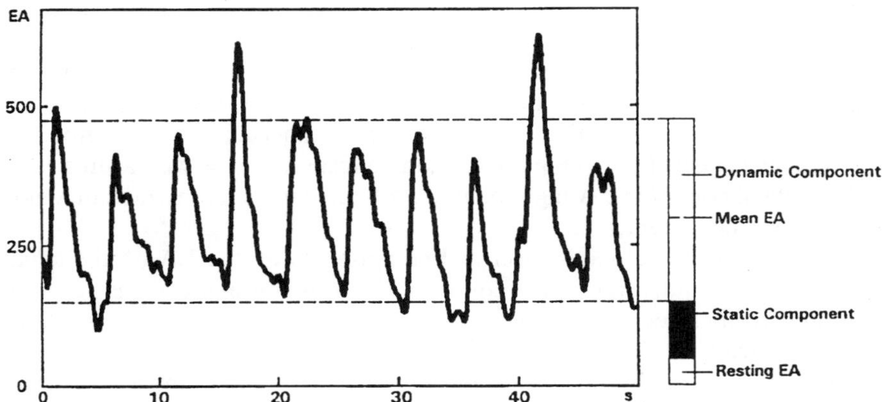

Figure 7.6 Example for splitting up electromyographic activity (EA) into static and dynamic components (from records of repetitive horizontal arm movements).

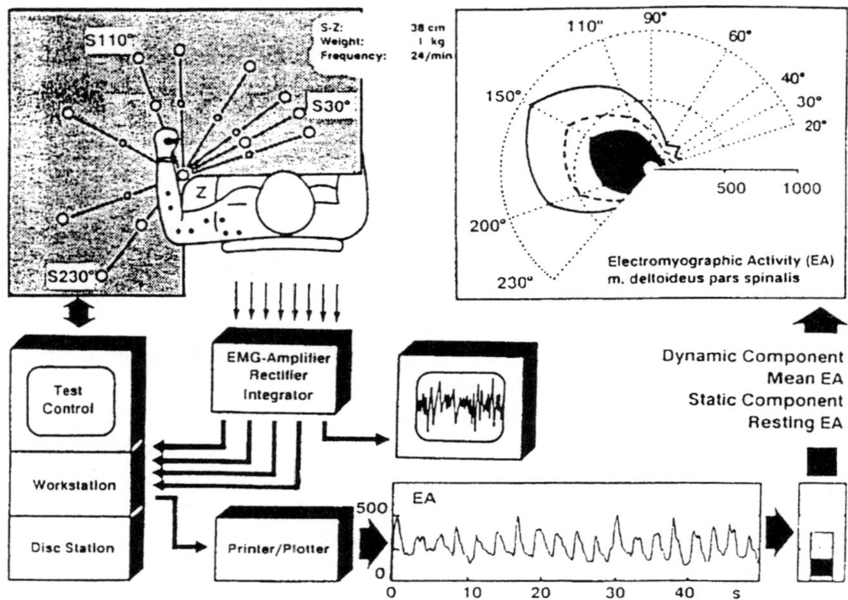

Figure 7.7 Test design and exemplary representation of electromyographic evaluation of horizontal manual materials handling (details see also Figure 7.1).

from the direction range between 20° and 60°, which is favorable (in this individual case) in comparison to the working direction of 150°, which apparently is most stressful. On the other hand, movements out of the range from behind the frontal plane – directions exceeding 180° – again turn out to have relatively low expense in terms of muscular strain.

These results for one single subject, of course, would not be especially important if they could not be replicated. Yet, the characteristics of all myograms measured were repeatable and reliable, and changes in the amplitude due to the working direction proved to be objectively assessed. For example, the diagrams in Figure 7.8 represent mean values from a group of five female subjects (Ss) who participated in a test session as well as in a retest with identical conditions. A high reliability of the data is represented in almost exactly congruent graphs.

In addition to the already described components of the EA of the spinal part of the deltoid, Figure 7.9 contains electromyographic signs in circle diagrams from three further muscles of the upper arm and the shoulder region as well as from two muscles implicated in movements of the forearm and the fingers. Before interpreting the results, it must be pointed out that there is a typical methodical lack with regard to traditional EMG measurements. That means, it is impossible to compare absolute EMG values resulting from different lead positions, due to uncontrollable

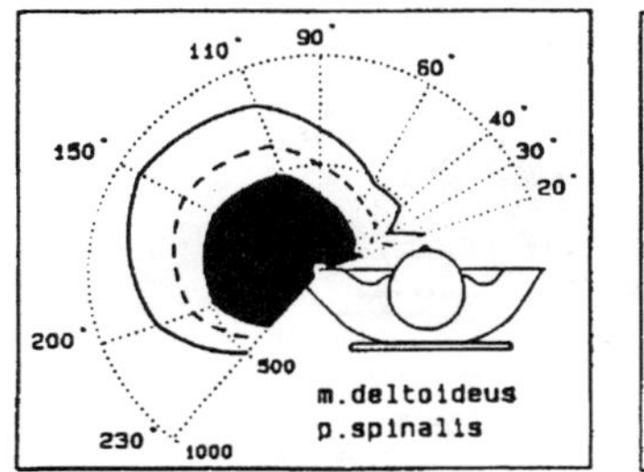

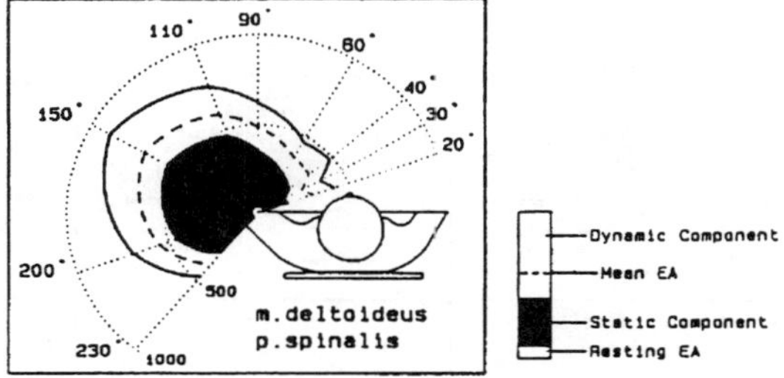

Figure 7.8 Components of electromyographic activity of one muscle part during a test and a retest with identical conditions (means from five subjects (Ss)).

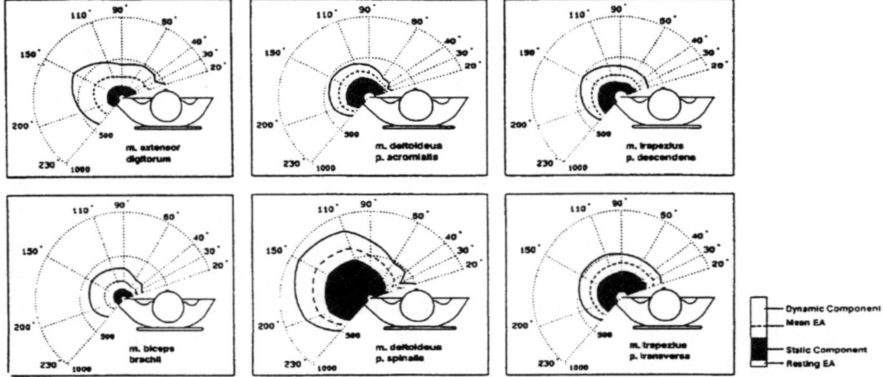

Figure 7.9 Components of EA (absolute values) of six muscles of the left upper limb (means from five Ss) dependent on the direction of horizontal arm movements (external load, 1 kg; working frequency, 24 cpm; distance S–Z, 38 cm).

recording conditions. For instance, it cannot be concluded from the diagrams in Figure 7.9 that the acromial part of the deltoid (whose results are shown in the upper middle part of the six graphs) was activated only approximately half as much as the spinal part (shown in the middle lower part of Figure 7.9).

Different amplitudes of the EMG can already result from the specific characteristics of the muscle, for example a small or a large size muscle, from lead placements, and from interelectrode distances.

7.2.2 Simplified Standardization of EA Data by Relating Actual Values to a Maximum Within a Test Series as a Reference Basis

To facilitate at least the comparison regarding the effects of the direction of arm movements on EMG, all values of a myoelectric signal within one lead

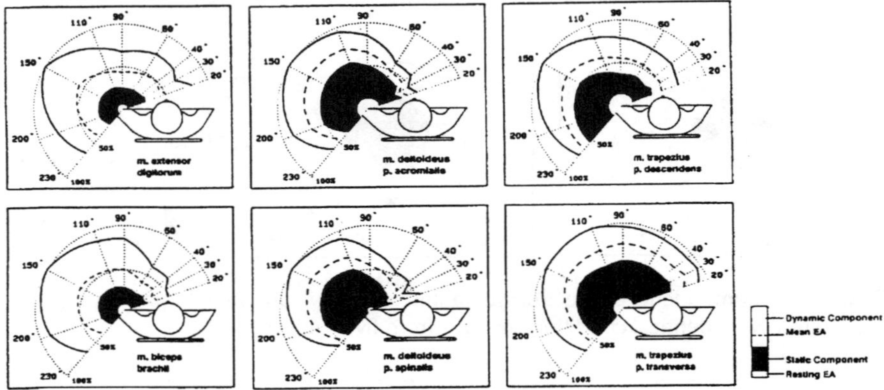

Figure 7.10 Components of EA (values related to the maximum EA at 150 ° within the test series) of six muscles of the left upper limb, based on data represented in Figure 7.9.

position had to be registered under the condition that, firstly, all the other parameters were held constant and did not disturb the electrodes once they were attached and, secondly, that the data can be related to a standardized value, for example an absolute maximum within a test series. As can be seen from the plots in Figure 7.9, that maximum was the direction of 150 ° in these investigations. When applying this procedure, at least intermuscular comparisons can be made.

A survey of such related EA values is given in Figure 7.10, which, for example generally shows relatively small static components in the muscles acting on the fingers and on the forearm. Therefore, it can be concluded that their involvement in repetitive manual tasks is much more efficient than that of the muscles of the upper arm and shoulder, which more likely could represent a bottle-neck.

Although a uniform and concordant effect of the different working directions on EA in all the monitored muscles can be shown, and though it can be concluded that there is much more static strain in the upper arm and shoulder, this effect cannot be quantified in absolute numbers and figures.

7.2.3 Dependence of EA on External Load

Before describing advanced methods of standardization of EA data, some important results of the influence of external load on myoelectric activity of the upper extremity shall be discussed.

Figure 7.11 shows the experimental set-up for the investigation of physiological cost associated with two contrary working directions in manual materials handling. Situations where the left arm is involved in

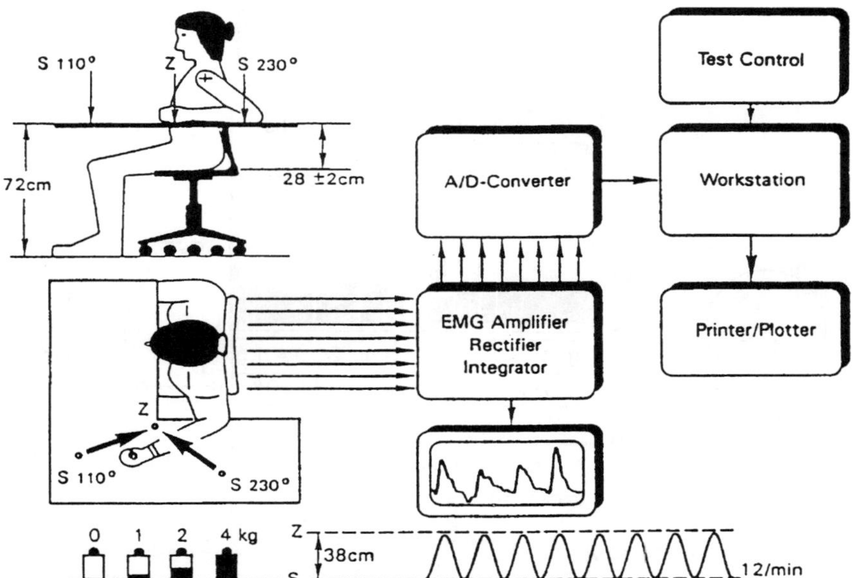

Figure 7.11 Analysis of the influence of repetitive horizontal arm movements on EMG activity in two working directions and with four different external loads.

work from the front to the rear (110 °) as well as from the rear to the front (230 °) can be found, for example in cashier check-outs in self-service shops in Germany.

The influence of repetitive horizontal arm movements in these two working directions at a working speed of 12 lifts per minute have been analyzed. Four different external loads (0, 1, 2, and 4 kg) had to be lifted over a distance of 38 cm each (S 110 °–Z; S 230 °–Z).

Figure 7.12 shows static and dynamic components as well as the mean EA from the middle and the spinal part of the deltoid. These muscle parts – which are involved in abduction and retroversion of the upper arm – reflect essential static components of the EA already without any external load. The increase of the external load is almost only decisive for the component of the EA associated with the movements.

Figure 7.13 gives a survey of the responses of altogether eight muscle groups of the upper extremity.

Muscle groups such as the extensor digitorum and the pronator teres which act on the hand as well as muscle groups which act on the forearm (biceps and triceps) show very low static components of the EA. With the exception of the triceps, a very clear dependency on the external load can be seen. In addition, the doubling of the external load is represented in almost a doubling of the dynamic component of the EA. The EA response to the external load

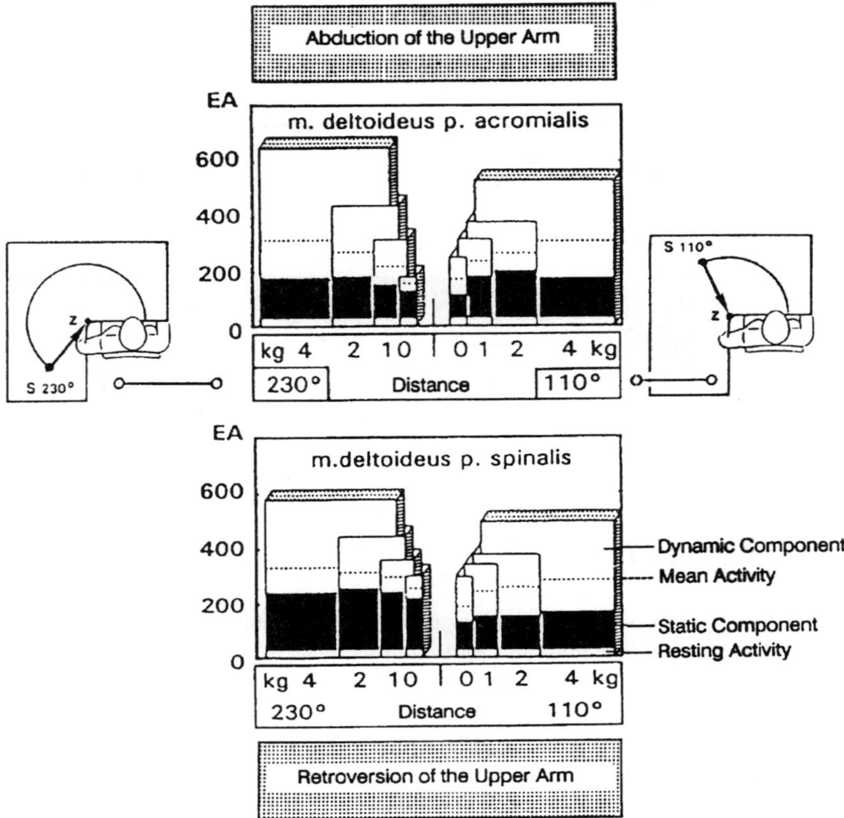

Figure 7.12 Static and dynamic components of EA (black and white areas) from two muscle parts of the deltoid when varying external load (in the directions 110° and 230°).

(restricted almost to the dynamic part) is smaller in muscle groups acting on the shoulder, i.e. the middle part of the trapezius (pars transversa) and in particular the upper part of the trapezius (pars descendens).

Figure 7.14 (page 196) gives insight into the relation between the static and dynamic components of the EA. Here it should become clear that an increasing external load does not necessarily lead to a more unfavorable muscle situation. On the contrary, the physiological situation is positively influenced with a decreasing relation of static to dynamic components of the EA. However, it must be stressed that several muscle parts were statically already very highly loaded when unloaded arm movements at the working speed studied here had to be executed. That means that unfortunate working postures may very often be more decisive on muscular strain than the external loads which have to be lifted during work.

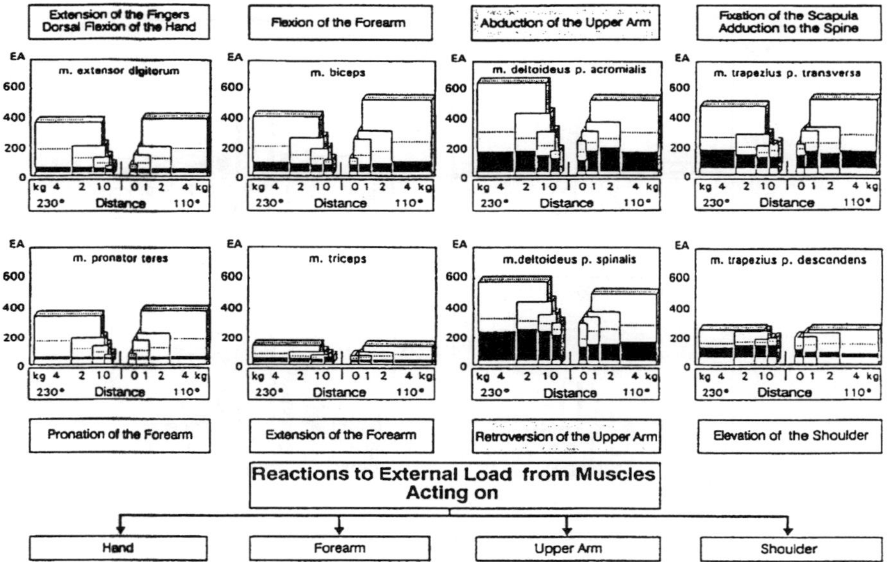

Figure 7.13 Electromyographic responses (static and dynamic components as well as mean EA) from muscles acting on the hand and forearm (left part), the upper arm and shoulder (right part) to a different external load (0, 1, 2, and 4 kg) handled in the directions 110° and 230° (Means from five female Ss.) (Source: Strasser and Müller, 1991).

7.2.4 Standardization of EA Data by Relating Actual Values to Separate Maximum Values from MVC – A Prerequisite for Interindividual and Intermuscular Comparisons and for Calculating Real Strain Values

The measuring of myoelectric signals principally yields only rough arbitrary values in units of micro- or millivolts dependent on the preset gain of the amplifiers, so that both absolute values of the mean EA as well as their components cannot be interpreted as quantitative indicators of muscle strain. Therefore, results from different research groups are mostly comparable only in a qualitative manner. But even with identical working test conditions, as well as identical lead positions and chosen amplifiers and recording set-up, myoelectric data can be confounded by an immense interindividual variation. Figure 7.15 (page 197) shows exemplary time series from two subjects with absolutely identical conditions during a repetitive horizontal manual material handling task.

Manifold causes exist for the large differences in the amplitudes of the two EA curves, e.g. the interindividually varying subcutane fatty tissue which can effectively dampen the biosignals produced by the muscle innervation.

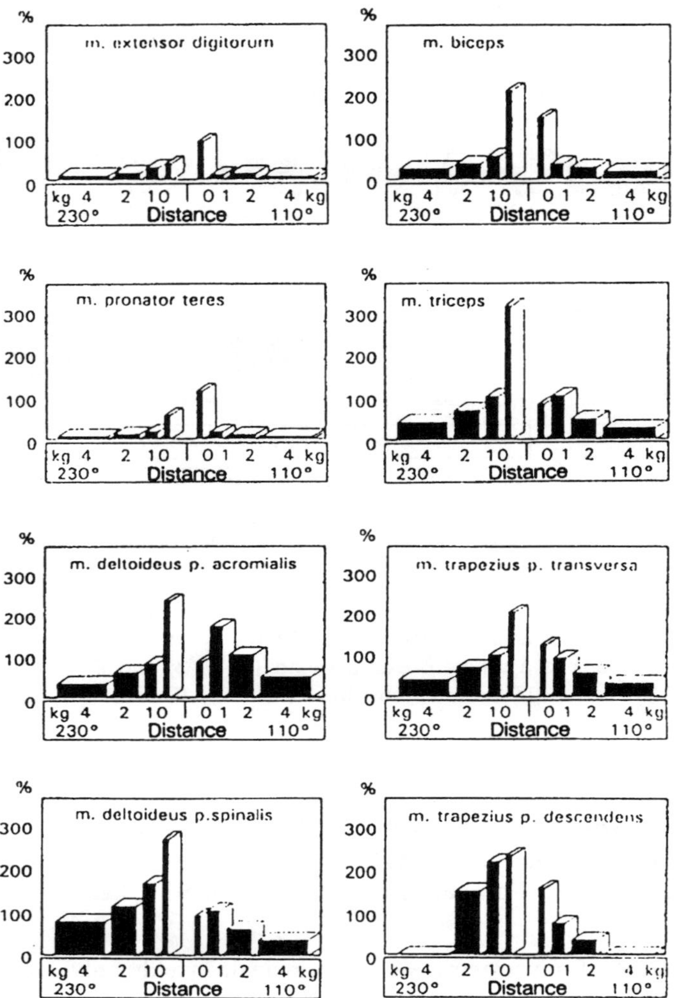

Figure 7.14 Relation between static and dynamic components of EA in percent (calculated by data from Figure 7.13).

Furthermore, the mass of an exerted muscle determines the EA amplitude and, finally, the results can be influenced even by small often unavoidable variations of the lead positions above a muscle during the attachment of the electrodes.

Even when all recommendations made, for example by Zipp (1988, 1989) were obeyed in surface electromyography, intense inter- and intraindividual variations in the amplitudes of the EA cannot be avoided. Therefore, in order to reduce the disturbing effects of individual signal variations and to make the

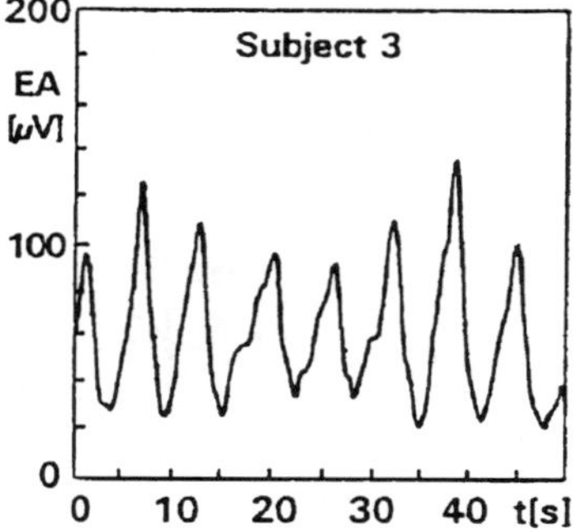

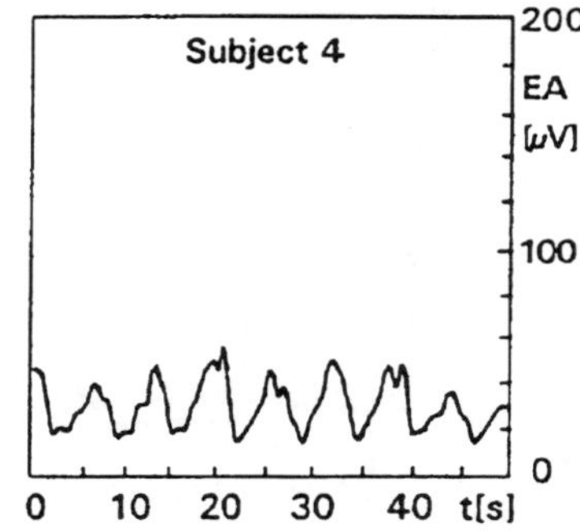

Figure 7.15 EA time series of two subjects resulting from identical working conditions and lead placements.

results of a group of subjects who participate in an investigation comparable, and in order to enable comparison of EA data from different muscle groups, a suitable standardization and relation of EA data to a common reference base is mandatory.

From the standardized EA, it should be possible to draw conclusions on the degree of strain (in the sense of occupying an individual's capacity) which is demanded from the monitored muscles. This becomes absolutely necessary when remaining within or exceeding endurance levels for static or dynamic muscle work is to be examined. In this context, the maximum EA associated with the isometric maximum voluntary contractions (MVCs) was recommended several times in the literature as an appropriate reference base for actual EA data. Yet, EA must not be registered at only one more or less arbitrarily chosen measuring point for MVCs. This, of course, makes a device for measuring maximum strength necessary. Contrary to a pure static workload for unrestricted movements of the upper extremity, a three-dimensional force measurement device is needed.

For applied research into local muscular strain resulting from repetitive movements as shown in the left part of Figure 7.16, MVC measurements should be made in exactly those directions and with arm positions which are congruent with actual repetitive tasks. In this way, the maximum potential of each muscle of a subject should be defined. MVCs necessary for lifting and pulling exertions at several points in the inner and outer reach (see right side of Figure 7.16) should enable the establishment of a reliable and global reference base for standardization of the actual signal amplitudes from EA values obtained during real work.

For this reason, a three-dimensional device for measuring muscle strength during maximum voluntary contractions was integrated into the test set-up with the computer-aided evaluation of multichannel electromyographic registrations. Figure 7.17 represents some results from a separate standardization study

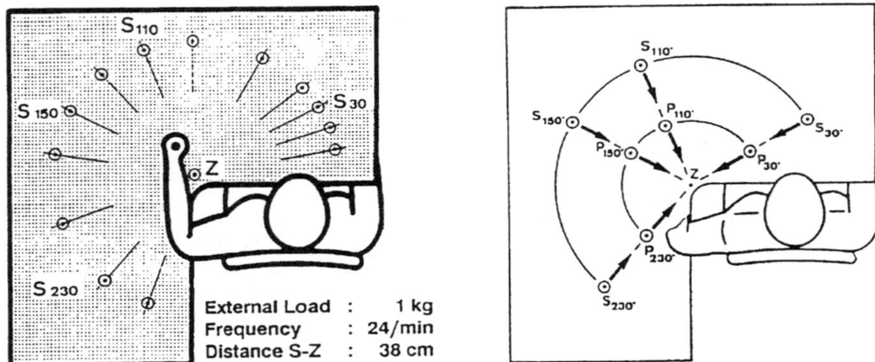

Figure 7.16 Measuring points for maximum voluntary contractions (MVCs) during (pulling and lifting) and associated EA (right part) selected in accordance with the reach and working area for horizontal materials handling (left part).

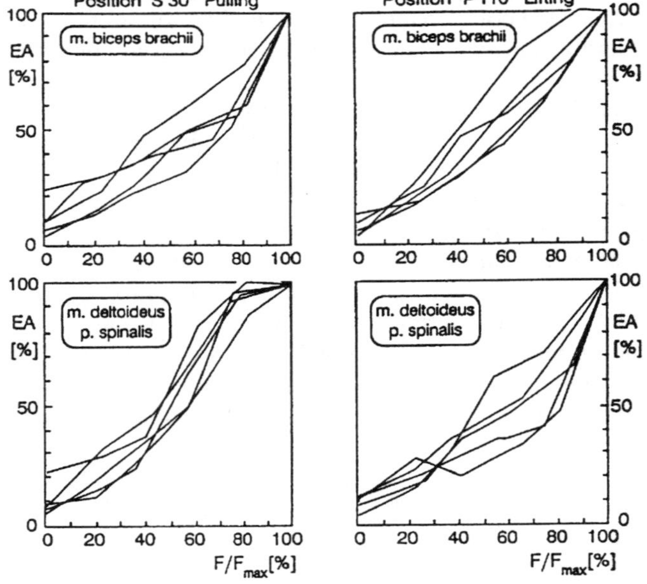

Figure 7.17 Relation between electromyographic activity (EA) and pulling and lifting exertions of two muscles at the measuring points S 30° and P 110° (Characteristic EA curves of five Ss, each.) (Source: Müller, *et al.*, 1989).

and visualizes typical relations between electromyographic activity of the biceps (in the upper two graphs) and the posterior, the spinal part of the deltoid, during pulling and lifting exertions, in two of a total of eight measuring points (on the left and right side, respectively).

In the four graphs in Figure 7.17, each of the characteristic individual EA curves of five subjects shows a more or less linear, a sigmoid, or also an overproportional increase of EA with the degree of contractions in percent of the possible maximum. Because differently shaped curves result from exertions under different biomechanic prerequisites within the same muscle, normalization must be chosen with special circumspection. However, in a pragmatic approach, the linear standardization formula of Figure 7.18 may be a justifiable, tenable compromise.

Therefore, it was suggested relating actual electromyographic activity from one muscle to the maximum EA from the MVC of the same muscle after having subtracted the resting activity. It must be stressed again that the global maximum EA – which does not always coincide with the maximum muscle force output – must be used.

Figure 7.19 visualizes the complete experimental set-up and procedures of computer-aided registering and evaluating of electromyographic data, comprising standardization and splitting-up of EA time series into static and dynamic components as well as a stimulus–response-compatible graphical representation of the results from one muscle part.

When applying these procedures to rough EA values from two muscles measured in microvolts and displayed in circle diagrams on the left side of Figure 7.20, the standardized EA, i.e. muscle strain in percent (shown on the

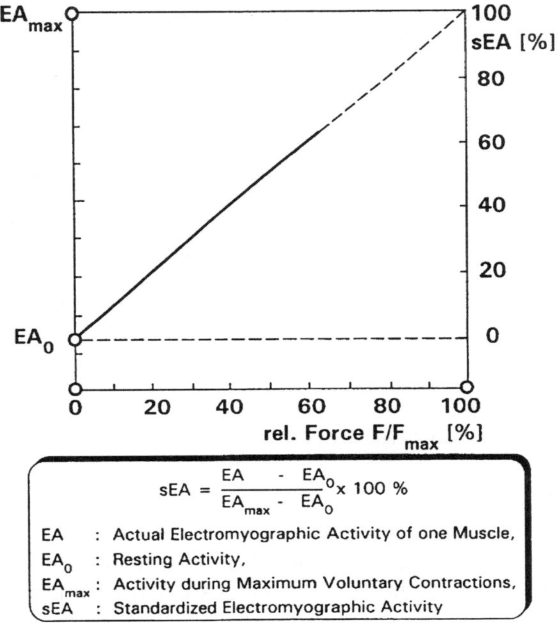

Figure 7.18 Formula for standardization (normalization) of electromyographic activity.

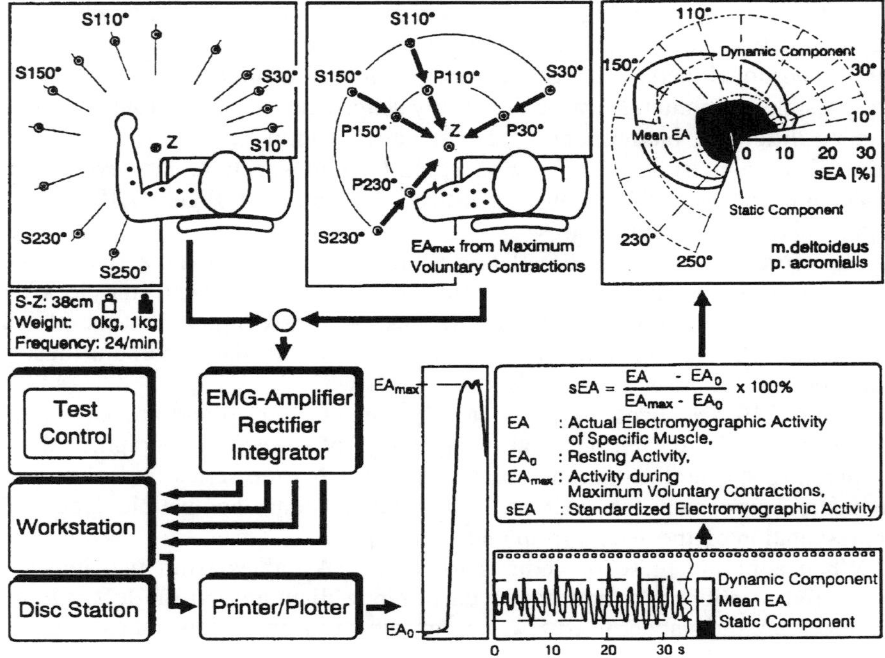

Figure 7.19 Schematic representation of the procedure for assessment and electromyographic evaluation of repetitive horizontal materials handling in different working directions. Test design for work simulation and MVC measurements (upper left part and in the middle), computer-aided evaluation and standardization as well as splitting up of EA data into static and dynamic components (lower part), and visualization of the results compatible with the test design (upper right part).

right side of Figure 7.20), can be determined. When looking at the two profiles formed by absolute data from 11 female subjects (EA in microvolts), it could be concluded that muscle strain of the acromial (middle) part of the deltoid would be markedly lower than that of the spinal part. However, the circle diagrams of the standardized values of EA (sEA in %) reveal an almost identical degree of demanded strain of the middle and spinal part of the deltoid when handling an external load of 1 kg at a frequency of 24 cycles per minute, when arm movements in the same direction, each, had been executed.

In the most unfavorable direction of 150 °, the total strain with more than 30% of the maximum EA is more than twice as high as in a direction of approximately 30 °. The static component of muscular strain – represented by the inner dark area – does not exceed values of approximately 10 and 15%, respectively.

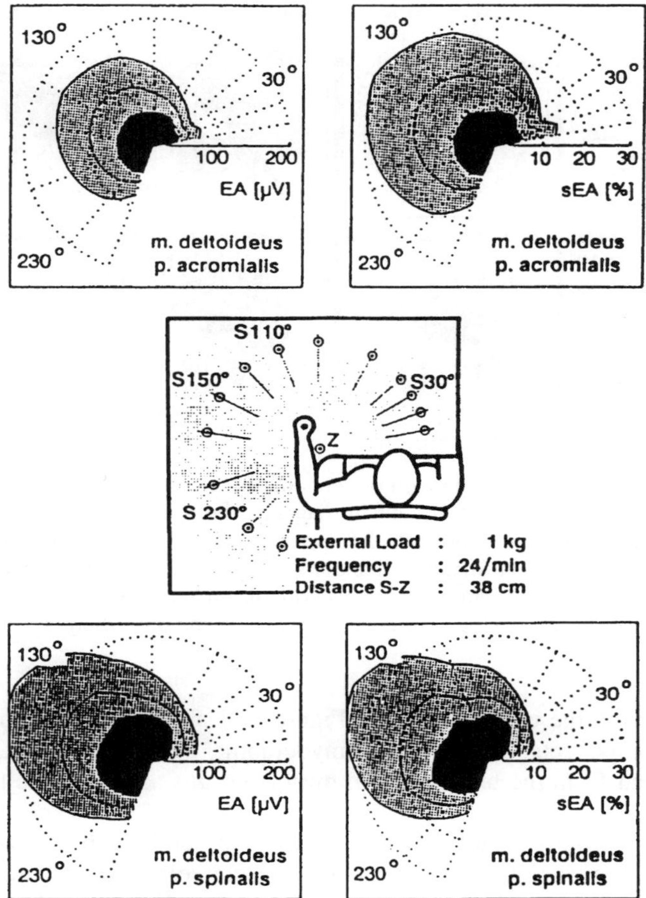

Figure 7.20 Circle diagrams of rough EA values (in μV) and standardized EA (sEA in %) (static and dynamic components of strain) of two muscle parts of the deltoid during horizontal materials handling (means from 11 Ss).

7.3 PROFILES OF MUSCLE STRAIN RESULTING FROM VARIATION OF THE DIRECTION OF MANUAL MOVEMENTS IN THE HORIZONTAL PLANE – MULTICHANNEL ELECTROMYOGRAPHIC MEASUREMENTS VERSUS SUBJECTIVE RATING DATA

Figure 7.21 shows – together with the above-mentioned profiles of muscular strain from the two parts of the deltoid – electromyographic responses of six additional muscle groups to the 13 different working directions in the horizontal plane.

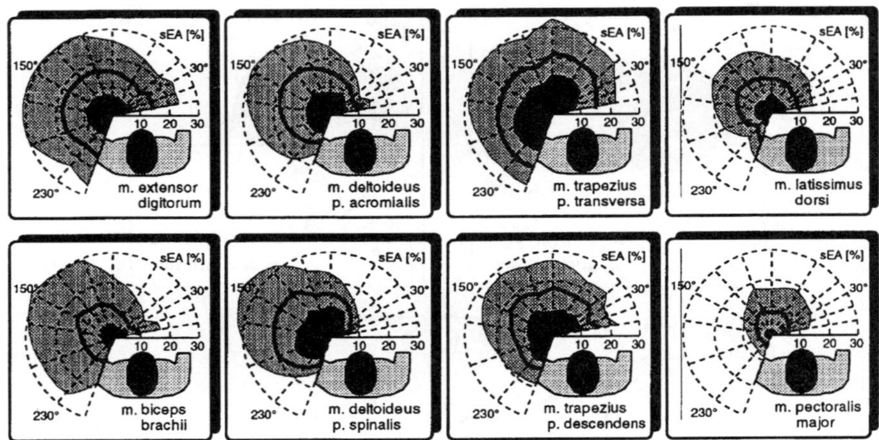

Figure 7.21 Static and dynamic components of the electromyographic activity (black and grey areas, respectively of the circle diagrams, bold line indicating mean EA) of eight muscle groups from the forearm, the upper arm, the shoulder, and the upper parts of the torso during manual materials handling in different directions between 10 and 250° (means from 11 Ss during manipulation of an external load of 1 kg over a distance of 38 cm at a working speed of 24 cycles per minute). (Source: Strasser *et al.*, 1992b).

The graphs in the very left part of Figure 7.21 show standardized static and dynamic components of the electromyographic activity from the extensor digitorum and from the biceps as two muscle groups acting on the fingers and the forearm.

The profile from the biceps reveals – in congruence with the two parts of the deltoid – a distinct dependence of muscle responses according to the direction of work ranging at almost the same level of total strain. Yet, in contrast to the deltoid, relatively low static components of only a few percent of total strain occurred.

Similar results were found for the extensor muscles of the fingers, which was somewhat surprising as this muscle group does not play an active part in work. But this effect is obviously due to the co-contractions of the anatomical antagonist reacting synchronously with a time lag but in line with the flexor muscles of the fingers, which, of course, play an active role in manual materials handling. Furthermore, it may be stressed that static strain remains at a small level for all working directions.

The graphs arranged to the right of those showing the two parts of the deltoid represent the static and dynamic involvement of muscle fibers stabilizing the scapula during arm movements in different working directions. Without going into the detail of the muscular strain which is generally high in these muscle parts, it must be noted that both monitored parts of the triangular cap musculature were not greatly influenced by different working directions. However, the high degree of static strain in almost all directions is remarkable.

Finally, the two graphs in the very right part of Figure 7.21 reveal a generally relatively small degree of strain both in the static and dynamic components via myoelectric activity from muscles acting on the shoulder joint. One reason for this behavior may be the fact that these muscles are not involved in manual work as much as all the others acting on the hand, forearm, upper arm, and shoulder. Static strain is almost neglectable, and the dynamic part reaches up to only approximately 20%, so that indeed most of the workload from an external load of 1 kg seems to be absorbed by the physiological cost of the hand–arm–shoulder muscles. But low values must also be due to the physiological property of multipinnate muscles with wide feathered fibers from which only a small part can be monitored and assessed by electromyographic methods, even in a surface EMG. It is also amazing that the highest physiological cost of the big chest muscle was measured at directions which had previously turned out to be favorable for arm movements. But this is also due to an anatomical property of the pectoralis major, which indeed is more actively involved when arm movements in front of the trunk with medial rotation of the arm are dominating.

The results of this comprehensive and finely structured study, which intended to determine favorable and unfavorable movement directions of the arm within the reach and within an angle area of 10–250 ° in a horizontal plane by means of advanced electromyography, can be summarized in a relatively simple way by data from a global subjective assessment. On the left side of Figure 7.22, means from 11 subjects, which had rated the 13 working directions before the tests on a 13-step scale, are shown. There is no

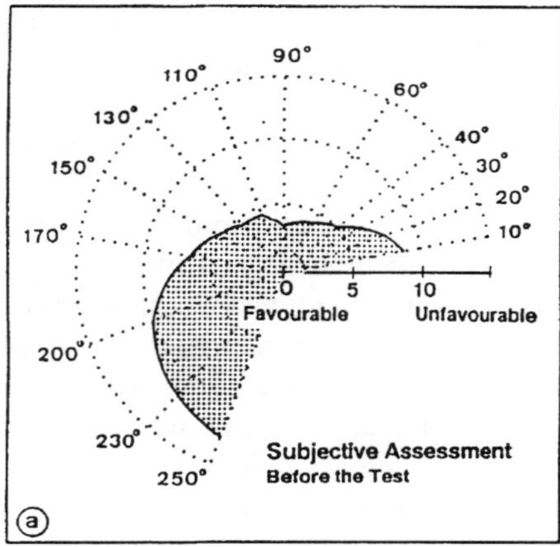

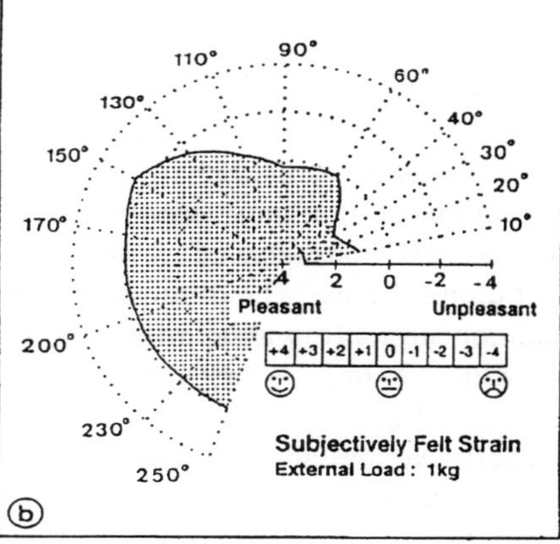

Figure 7.22 Subjective evaluation of the working directions at two rating scales (means from 11 Ss). Assessment of (a) the working directions at a 13-step scale without working experience and (b) after working in each of 13 working directions at a bipolar four-step scale.

physiological confirmation by electromyography for a supposed optimum working direction of approximately 90°, as the subjects themselves expected. But their own work experience must be the cause of subjectively felt strain (see right side of Figure 7.22) – assessed after having finished work in each of the working directions on a bipolar four-step scale – and is revealed to be congruent at least with the strain, shown in circle diagrams, of some 'guide muscles', as, for example the deltoid, which must have a formative influence on the degree of subjectively felt strain.

7.4 LOCAL MUSCLE STRAIN AND INTEGRAL MEASURES OF PHYSIOLOGICAL COST – VALIDATION OF EA DATA BY MEANS OF WORK-RELATED O_2 CONSUMPTION AND WORK PULSES IN A ONE-ARM MATERIALS HANDLING TASK

In view of the complexity of the hand–arm–shoulder musculature with more than several tens of muscles, it cannot be ignored that favorable directions of manual movements, which had repeatedly been found in electromyographic registrations of up to eight muscles, may occur merely by putting the work-load onto some other muscles which are not monitored. Therefore, in order to reduce this residuum of uncertainty with respect to recommendations for favorable directions in movement-based work design, it had to be investigated whether local muscular strain assessed by electromyographic methods – at least in a loose correlation – reacts congruently with oxygen consumption and work pulses.

Therefore, experiments were carried out in the course of which – as shown on the left side of Figure 7.23 – a load of 2 kg had to be lifted repeatedly 24 times per minute during seated work. From three different starting points within reach, S 30°, S 110°, and S 150° measured from the frontal plane of the subject, materials had to be lifted over a distance of 38 cm by arm movements to a fixed point Z near the body. In order to also measure the physiological cost associated with unloaded arm movements, the same tasks had to be performed by handling an almost weightless article, i.e. 0 kg. Each single task lasted 20 min and consisted of 3 min warm-up and 5 min lifting, followed by a 12 min rest pause. The duration of a whole session with all tasks organized in a random sequence was approximately 3 h. A relatively homogeneous group of 11 female subjects with respect to age (25.3 ± 3.0 years), body weight (56.9 ± 7.2 kg), and height (168 ± 4.7 cm) participated in the test sessions. Myoelectric activity from several muscles and oxygen consumption and heart rate as integral measures for stress and strain were measured simultaneously by analyzing the amount and concentration of oxygen in expired air and by counting heart beats via a cardiotachometer. For each subject, mean values of work pulses and oxygen consumption on a 5 min basis due to the different tasks were calculated. Electromyographic activity was registered continuously and evaluated by means of computer-aided methods.

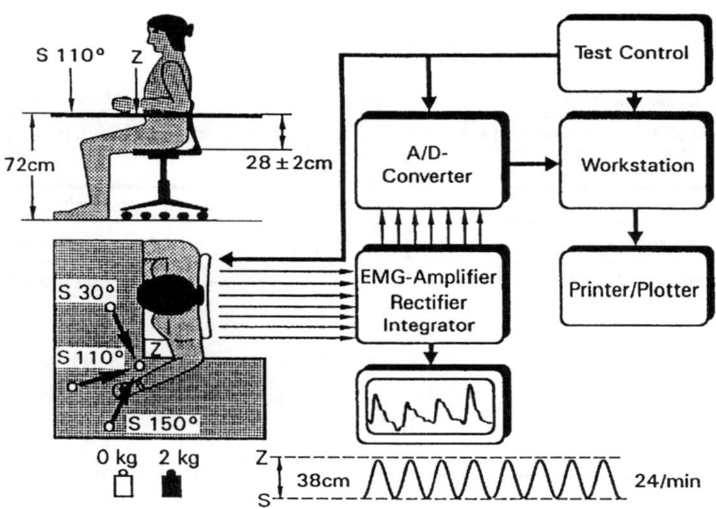

Figure 7.23 Analysis of the influence of repetitive horizontal arm movements on EMG activity of seven muscle groups at three working directions with and without an external load.

When a weight has to be grasped at several points within reach and manipulated to a point near the body at seated work during materials handling in a horizontal plane, the movement elements shown in the lower part of Figure 7.24 must be regarded as essential.

1 A forward movement of the arm, an anteversion, which is mainly brought about by the
   (i) clavicular, i.e. the anterior part of the deltoid and
   (ii) the pectoralis major, i.e. the big chest muscle.
2 A backward movement, a retroversion of the arm, which is affected by the
   (i) spinal part, i.e. the posterior part of the deltoid and
   (ii) some parts of the latissimus dorsi, the long back muscle.
3 A lateral lifting, an abduction of the upper arm, mainly accomplished by the
   (i) acromial, i.e. the middle part of the deltoid.
4 A medial rotation, an inward rotation of the arm towards the body, in which in any case the following muscle groups are involved:
   (i) the anterior part of the deltoid,
   (ii) the pectoralis major, and
   (iii) the latissimus dorsi.
5 The adduction of the arm, caused by the
   (i) pectoralis major, and
   (ii) the latissimus dorsi.

Besides these movement elements mainly acting on the shoulder joint, further

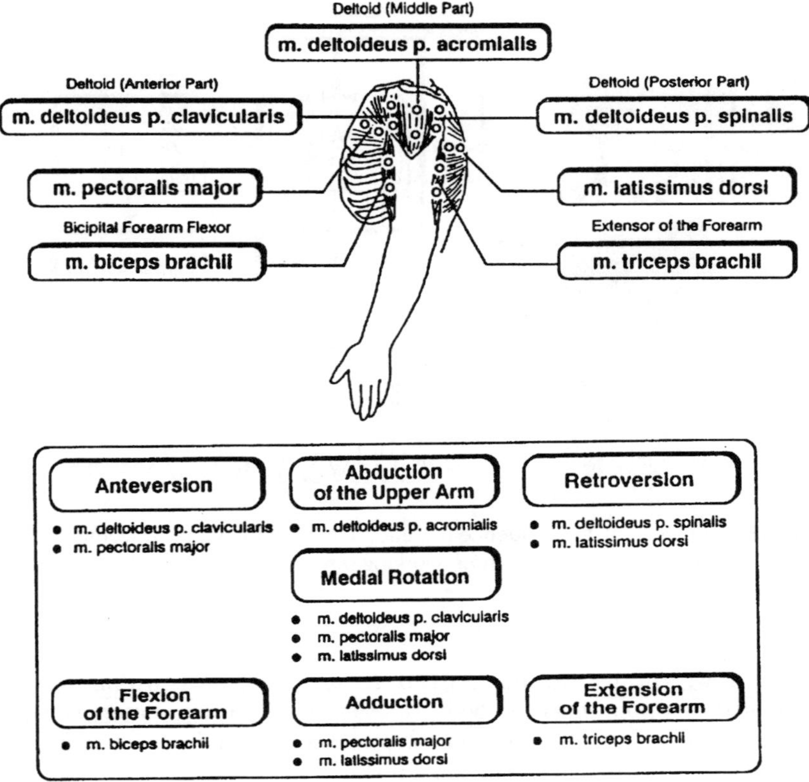

Figure 7.24 Muscles continuously monitored by surface electromyography and some elementary arm movements in which they are involved.

important movement parts during all manual handling processes are:

6 The flexion of the forearm, in which the biceps is involved and
7 The antagonistic stretching, the extension of the forearm, which is executed by the triceps.

Therefore, in order to measure the local physiological cost of repetitive arm movements, the seven muscle parts indicated in the upper part of Figure 7.24 were monitored.

The involvement of the forearm muscles acting on the fingers was not expected to vary essentially due to the different test conditions, and stress imposed on those muscles was not expected to represent a bottle-neck. Therefore, they were not monitored in this study.

Figure 7.25 represents a survey of all the results. The upper row shows maximum torque acting on the shoulder joint, oxygen consumption, and work pulses as global indicators of the work-load. In the lower three rows, local physiological cost assessed by the electromyographic investigations is

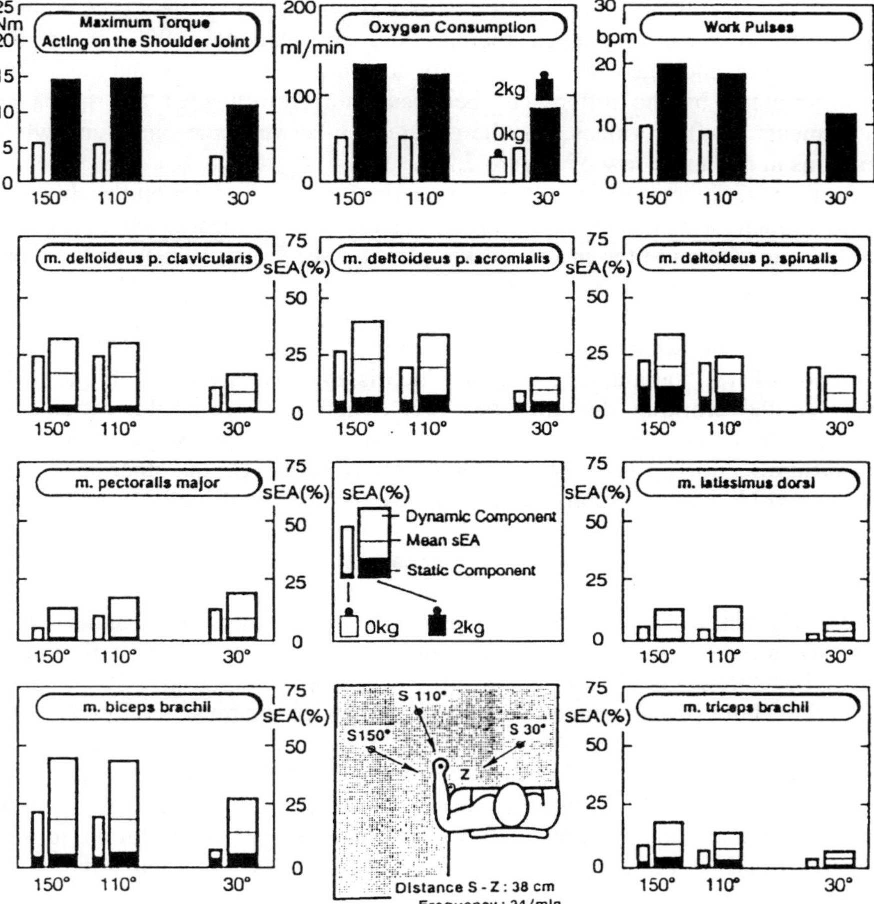

Figure 7.25 Integral measures for stress and strain (maximum torque acting on the shoulder joint, oxygen consumption, and work pulses) and local muscular strain in seven muscle groups (static and dynamic components of standardized electromyographic activity sEA) during materials handling in the horizontal plane (means from 11 female subjects).

visualized in seven graphs compatible to the local arrangement of the different lead positions in Figure 7.24. Static and dynamic components of standardized electromyographic activities from the three parts of the deltoid, the pectoralis major, and the biceps, as well as from the latissimus dorsi and the triceps were shown for the three working directions 30°, 110°, and 150° in each graph. The wide and the small columns represent muscle strain for arm movements with and without an external load.

Oxygen consumption with regard to work (values during work minus values during rest in working posture) is significantly higher at the direction of 150°

in comparison with 30 °. This difference amounts to approximately 60% and the extra physiological cost even increases to approximately 90% when the component is calculated which is directly attributable to the external load. This is represented by the differences between data from loaded and unloaded movements (2–0 kg values, i.e. the differences between the black and white columns in the upper row of Figure 7.25).

Similar relations were found for the maximum torque acting on the shoulder joint (which was calculated from the individual anthropometrics of the subjects in relation to the preset working layout; see Ernst, 1989) and the work pulses. Work pulses were defined as heart rate increases due to manual materials handling above a resting level which was measured in the working posture. The physiological cost for handling external loads of 2 kg in the direction of 150 ° amounts to a level which is approximately 75% higher than that for the working direction of 30 °. When calculating the differences between the black and the white columns, extra physiological cost in the most stressful direction turns out to be approximately 120%. On the other hand, that means working in a favorable direction spares more than half of the cardiac expenses.

The following results become apparent from the graphs for the electromyographic data.

Already without an external load, the local physiological cost indicated by EA values increases when arm movements from a more outward starting point to the body are demanded. The most stressful working direction is 150 °, which – on the whole – is associated with more than twice the muscle effort than has to be exerted in a favorable direction. Static strain predominantly at an important level in the deltoid, but with only marginal proportions in the wide feathered pectoralis and latissimus dorsi, is affected by the working directions much less than the dynamic component. Obviously, an exception to this general trend is represented by the results from the pectoralis major. A reciprocal tendency can be seen with a maximum EA of approximately 20% at 30 °. But this can be explained by anatomical properties of this multipinnate muscle group, which is mostly responsible for the medial rotation of the upper arm (see also Figure 7.21).

In this study, the maximum oxygen consumption necessary for materials handling at 150 ° is in the range of approximately 140 ml min^{-1} and is already close to the value which is generally regarded as the endurance or tolerability level for one-handed work of women. This may be explained as follows. Firstly, it can be assumed that approximately two-thirds of the limit value for continuous effort which is valid for men has to be applied for women. Consequently, in the case of a work performance of 300 W or approximately 900 ml O_2 min^{-1}, respectively, as the value for continuous effort for a whole-body dynamic muscular work of a man, approximately 600 ml O_2 min^{-1} must be estimated for a woman. Secondly, simply assuming that with one-handed work only between one-quarter and one-fifth of the muscles of the whole organism are involved, approximately 200 ml

O_2 min^{-1} (for men) and 130 ml O_2 min^{-1} (for women) have to be regarded as tolerability limits. Thus, from a work-physiological point of view, an optimum movement direction of 30° – with only almost half of the maximum O_2 consumption – is of outstanding importance. Referring to the heart rate measurements, it must be emphasized that the endurance level for this type of work is marked by the maximum of only approximately 20 work pulses (i.e. 20 extra beats per minute attributed to one-handed dynamic work during manual materials handling in a horizontal plane as simulated in this study).

When comparing the results of the integral measures for stress and strain to local muscle strain data, some general conclusions can be drawn. Firstly, assuming the endurance level for the kind of work analyzed is represented by the data of oxygen consumption as explained above, which is also in congruity with classical studies of work physiology (see e.g. Hettinger *et al.*, 1980; Rohdahl, 1989), inferences can also be drawn for a better understanding and meaning of the results from electromyographic studies. Most significant reactions in the dynamic components of EA from the biceps and the different, separately acting muscle groups of the deltoid, mainly involved in work, must be interpreted in the same way as data from corresponding integral measures. That means that a level of strain amounting to at least approximately 30–40% for the maximum EA must be regarded as an upper tolerable endurance limit. This is especially true because these values are also accompanied by important proportions of static strain.

In order to verify the validity of electromyographic methods, correlations between the local and the global characteristics for stress and strain have also been calculated. For details see Strasser and Ernst (1992). Correlation coefficients showed a mutual dependence between global and local indicators. This was not true for all muscles and working conditions, but was the case with most consistent relations for some bottle-neck muscles when handling an external load. For instance, between the mean electromyographic activity of the biceps and especially of the middle part of the deltoid, on the one hand and corresponding oxygen consumption as well as work pulses, on the other hand, highest significance levels were found. As could be expected, the interrelations between the triceps and the integral measures were very low, as this muscle is not involved in the active phase of materials handling which has been analyzed here.

All in all, the study showed that electromyography in congruity with integral measures can be an efficient and reliable method to determine muscle strain, provided that relevant muscles were monitored. Furthermore, an advanced computer-aided multichannel approach proved to be a much more accurate and efficient procedure than global measures in revealing physiological bottlenecks, which are mostly represented in local strain of some muscle groups. These issues can be determined in numbers and figures from electromyographic data.

7.5 ELEMENTS OF A SYSTEM OF PREDETERMINED MUSCLE STRAIN OF THE UPPER EXTREMITY

In work organization, the time fixed for certain operations is an important and often decisive factor. This, however, gives neither an impression of 'how work is done' nor of the effects of work on man himself. The biological effort (e.g. the corresponding strain) cannot be estimated from an equal predetermined time with differing layouts. Therefore, systems of predetermined time (e.g. work factor, WF or methods time measurement, MTM) can be regarded as valuable planning aids for the working process, but from an ergonomics point of view, the traditional movements of working processes should be completed by systems of predetermined strain. Prospective work (place) design involving repetitive movements of the hand–arm system should be made easier by establishing additional guidelines for the consideration of ergonomical tasks. In order to develop elements for such a system of predetermined strain, the corresponding determination of local muscular strain associated with basic movements and sequences of motoric movements of the hand–arm system via multichannel electromyography is absolutely necessary.

How elements of a predetermined system of muscular strain can be determined may be shown by the electromyographic evaluation of the physical load of the left hand–arm–shoulder system during simulated work at different cash register arrangements (for details see Strasser *et al.*, 1991, 1992a).

The manual handling of the goods at cashier work-places in self-service shops necessitates, depending on the spatial work-place design, different manual movements of the left hand–arm–shoulder system, which, over a longer period of time, can cause considerable physical ailments in and complaints from cashiers.

In any case, a cashier in Europe (working with or without a scanner) normally while seated has to move the items over a certain distance either by lifting or pushing.

Depending on the cashier's position, i.e. either parallel to the cash register or counter (as shown in the upper half of Figure 7.26) or sitting in front of the conveyor (as illustrated in the lower part of Figure 7.26), manual movements of the left arm in a direction more or less parallel to the frontal or median plane dominate.

The conventional movement direction of a conveyor in the forward system of a check-out, as seen from the cashier's position, i. e. 'from front to back' and 'from right to left', respectively, (as demonstrated by black arrows with four situations in the left part of Figure 7.26) also necessitates manual movements in the same direction, in other words lifting or pushing 'from front to back' (backward) or 'from right to left.'

In some newly established systems, the conveyor moves in the opposite direction. Depending on the cashier's position, those backward systems (see right side of Figure 7.26) necessitate either forward movements 'from back to

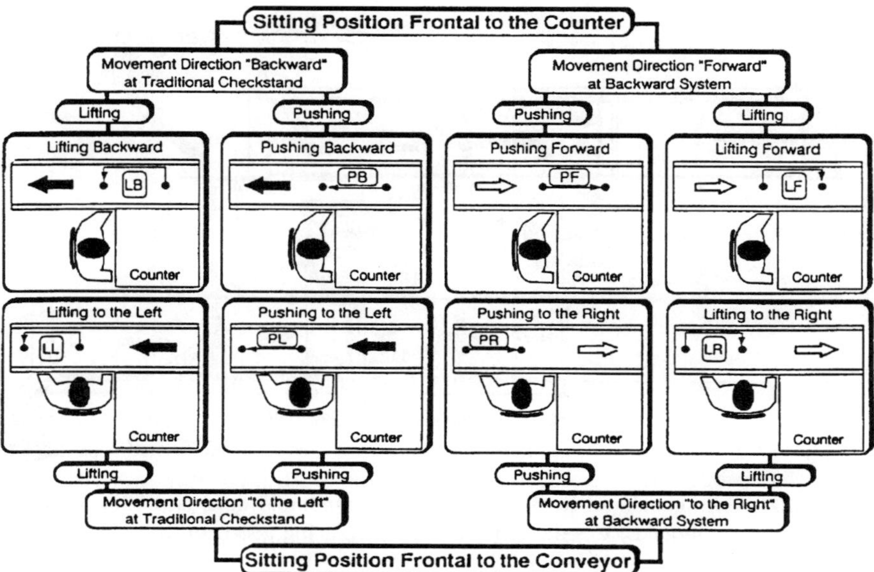

Figure 7.26 Cash register arrangements analyzed regarding muscular load of the left hand–arm–shoulder system of the cashier in self-service shops.

front' or 'from left to right'. In so-called twin check-outs (compare Strasser, 1990) with two cashiers sitting face-to-face (see Figure 7.27), left arm movements from front to back as well as in the reverse direction may become necessary in both parts.

In accordance with the possible working situations, a work-place was set up in the laboratory in order to analyze the effects on important muscle groups of the left hand–arm–shoulder system during different lifting and pushing movements. Figure 7.28 shows the two sitting positions of the cashier 'facing the conveyor' (on the right side) and 'facing the cash register or counter' (on the left side) together with the different dimensions of the work-place for both predetermined movement directions indicating the inner and outer reach, i.e. the small and the large grasping area of the female fifth percentile. Corresponding to the movement direction of the conveyor, the subjects had to execute a movement with the left arm, simulating the transport of an item over a distance of 38 cm on a horizontal working plane. For the predetermined movements 'from right to left' and vice versa, the median plane represented, for example the start or end-point, respectively.

By means of an overproportionally high 2 kg weight, which had to be lifted or pushed at a working speed of 24 movements per minute representing a working process comparable to the registering in self-service shops, the effects of this work on the muscles were recorded by multichannel electromyography during working periods of 2 min for each of the eight constellations.

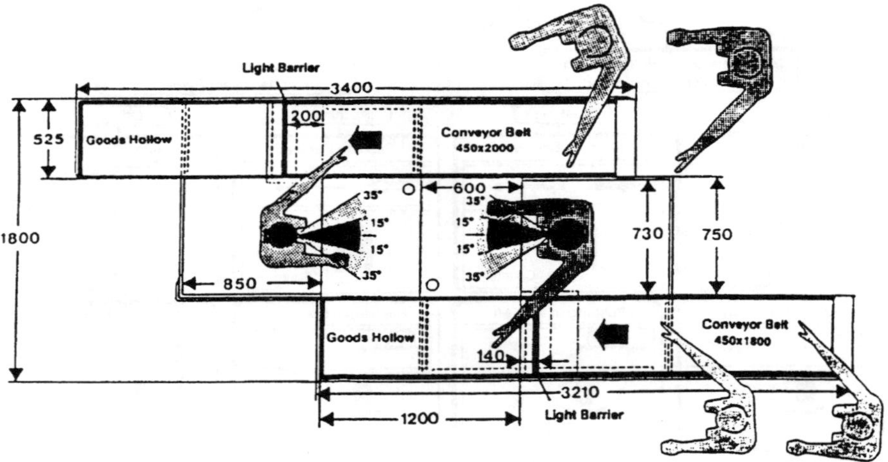

Figure 7.27 Twin check-out workstation with 5th-percentile (left) and 95th-percentile (right) stencils of females, together with customers placing goods on the conveyor belt (from Strasser, 1990).

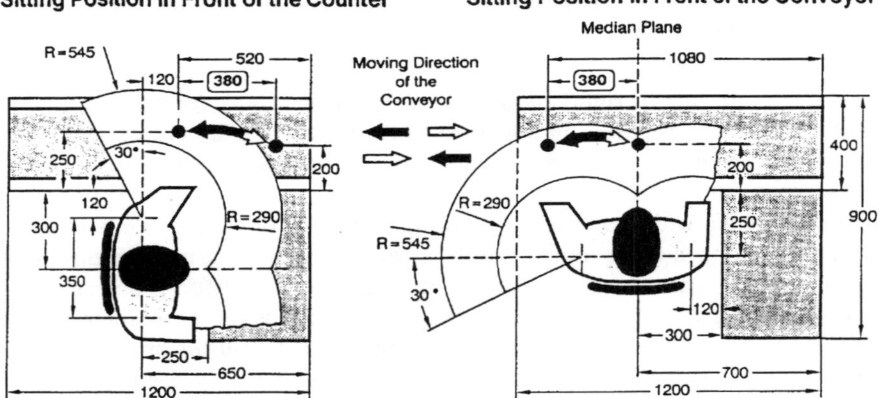

Figure 7.28 Position of the arm movements over a distance of 380 mm, and sitting positions with inner and outer reach of the 5th-percentile woman in a horizontal section plane.

Differences in strain depending on the different movements had been expected especially for the following muscle groups (see also Figure 7.29). In the three parts of the delta muscle, which decisively participate in the abduction (i.e. holding), the anteversion (i.e. forward movement), and the retroversion (i.e. backward movement) of the upper arm, in the musculature bending and stretching the arm, i.e. the biceps brachii together with the brachialis and in the triceps brachii (caput longum), and, finally, in the

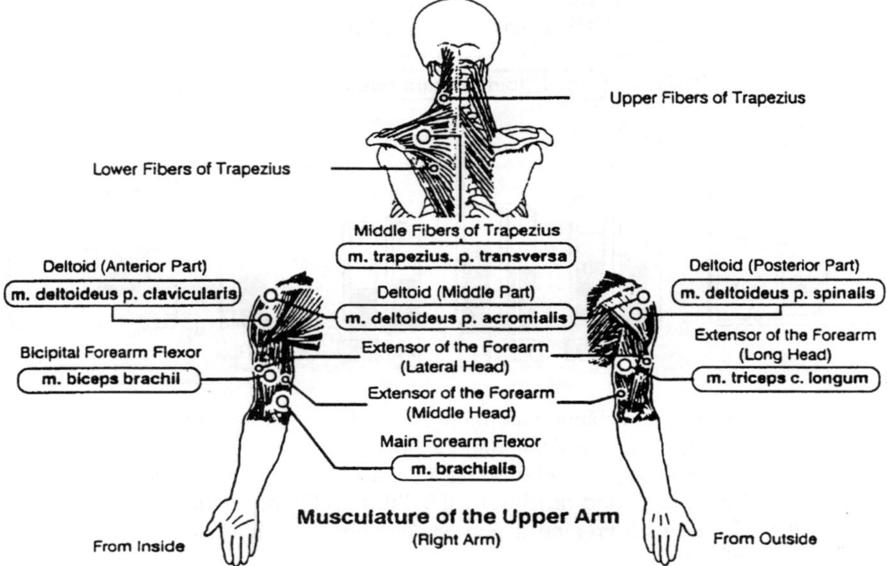

Figure 7.29 Muscle groups of the left upper extremity selected for electromyographic measurements.

triangular cap musculature involved in the stabilization of the scapula and the retraction of the shoulder, respectively. Therefore, simultaneous myoelectric recordings from the muscles whose names are framed in Figure 7.29 was carried out, the method of recording necessitating a limitation to seven channels.

Ten male subjects (aged 22–30 years) performed the different tasks according to a fixed work-rest schedule. Due to repeated measurements for each of the eight test variations and several calibration measurements, the testing period for 1 day (including preparation and follow-up work as well as recovery period) was approximately 4 h for each subject.

Figure 7.30 shows the components of the electromyographic activity for the medium part of the delta muscle measured during the eight different manual movements. The four columns on the left are related to the sitting position 'facing the cash register'. The four columns on the right side represent the results from the tests for the sitting position involving manual movements in the frontal plane.

The following statements can be derived from these results for the musculature abducting and stabilizing the upper arm.

Firstly, manual manipulation by lifting instead of pushing the test item demands a considerably higher strain, whereby the sitting position 'facing the cash register' leads to significantly higher (mean) EA values. Secondly, there are similar great differences between the two sitting positions, the position

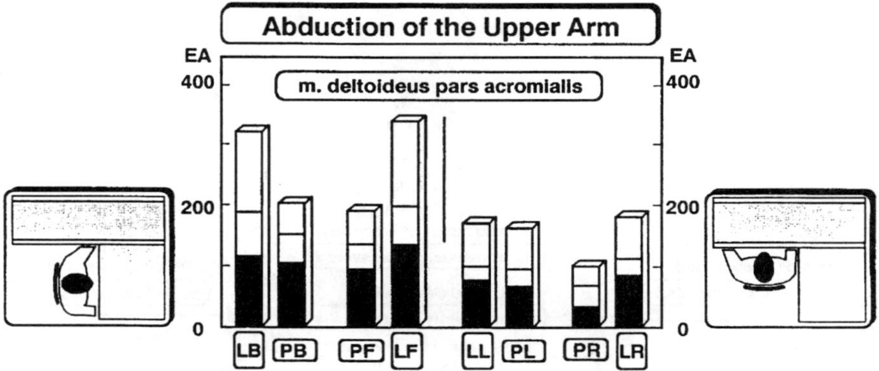

Figure 7.30 Static and dynamic components (black and white areas of the columns) of the EMG activity (EA) from the middle fibers of the deltoid during the manual movement elements lifting (L) and pushing (P) (an external load of 2 kg over the same distance of 38 cm), backward or forward (LB/PB; PF/LF) and to the left or to the right (LL/PL; PR/LR). (Means from nine Ss and working phases with a duration of 2 min and 24 cycles per minute each.)

frontal to the movement direction turning out to be significantly more favorable. Pushing to the right (PR) appears to be the most favorable type of manual work at such a cashier work-place. But before generalizing such an isolated result from one muscle part, it had to be confirmed by solid results from other muscle groups. Although the responses may not be expected to be identical due to physiological reasons (e.g. in the bending and stretching musculature), extreme strain had to be excluded in order to avoid a physiological bottle-neck. Even in the case of generally low values in almost all recordings, the overstraining of one single muscle should be sufficient for a negative assessment.

Figure 7.31 shows a general survey on the EA values of seven muscle responses in the same arrangement as for the middle part of the delta muscle, which is represented here once more in the middle.

The results already mentioned are confirmed to an extensive degree. Nearly all columns associated with lifting are higher than with pushing and sitting positions frontal to groceries' handling are at least more favorable on average. The same is also true for 'pushing to the right', an obviously favorable kind of work involved in the newly established backward systems. In this case, the relatively higher static and dynamic values of the biceps as a powerful flexor and supinator in contrast to the working direction 'pushing to the left' are regarded to be an even desirable strain in the sense of a physiological sequence of movements.

After the subjects had gained experience regarding the work tasks, they had to assess the manual movements via a bipolar four-step scale (see Figure 7.32). The mean values of this subjective evaluation of the eight movement

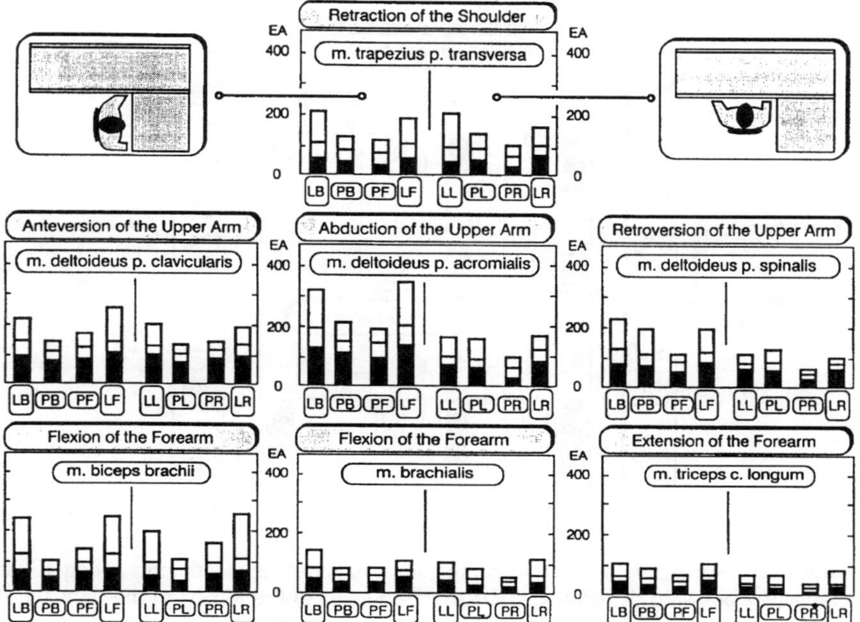

Figure 7.31 Responses of seven muscle groups to the movement elements of Figure 7.26, arranged close to the representation of Figure 7.30, and in the same arrangement as the muscle parts shown in Figure 7.29.

elements, regarded to be favorable or unfavorable, elucidate that the relative differences detected with the aid of electromyography are identical to those of the individual impression (upper part of Figure 7.32). Thus, it turns out that the higher muscular strain caused by all lifting movements is perceived to be more stressful and therefore rather negative. The differences stated by the subjects were very clear in all cases. In principle, all handling tasks with a position frontal to the movement direction were assessed as being significantly more favorable. Furthermore, pushing the items from left to right (PR), as involved in backward systems, was identified to be the absolutely most favorable. There was no negative assessment by any of the ten subjects with regard to this working procedure. This may be due to the fact that they themselves realized that the involvement of the arm bending musculature in this movement direction is always superior to the stretching musculature.

After all electromyographic recordings had been repeated in a second working session the same day (in reverse order), the subjects had to give an additional assessment. After these second recordings, the repeated assessment (as shown in the lower part of Figure 7.32), with generally equal relations between the different movements, remained below those of the first assessment (up to one unit) on the four-step scale. This tendency may be caused by a

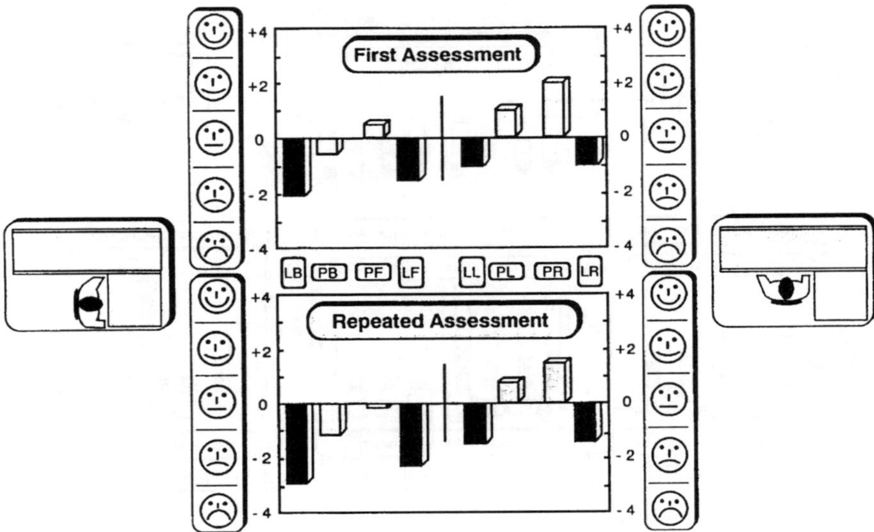

Figure 7.32 Subjective evaluation of the manual movement elements lifting (L) and pushing (P) backward or forward (LB/PB; PF/LF) and to the left or to the right (LL/PL; PR/LR). (Means from 10 Ss.)

general fatigue influencing the results, although the electromyographic data of the second recordings do not support this fact.

With regard to the relevance of the investigations in practice, the following inferences can be drawn: with the aid of the results, which could only be partly presented in this report, it is possible to assess 'in advance' the muscular strain imposed on the left hand–arm–shoulder system by different check-out systems. Therefore, they can be regarded as a first step for a 'system of predetermined muscle strain'. This should be at least of the same importance for both manufacturers and users of check-out systems as, in the assembly range, 'systems of predetermined motion time' (e.g. MTM or WF) are still regarded to be a valuable planning aid for the working process nowadays. But because the allowed time for certain manual work gives neither an impression of 'how work is done' nor of the effects of work on man himself, the biological effort and the corresponding strain of different working methods during identical periods of time cannot be recognized directly; an essential aim of ergonomics should be the development of 'systems for predetermined strain' as a method for preventive and prospective work design. Thus, a systematical evaluation of the results from investigations of strain is an important aid to enable the persons responsible for work design to also consider ergonomic requirements.

Some essential factors influencing the planning and design of assembly work by MTM have been evaluated via electromyography by Müller *et al.* (1988b).

7.6 EVALUATION OF THE ERGONOMIC QUALITY OF HAND TOOLS BY MEANS OF ELECTROMYOGRAPHY – THE EXAMPLE OF SCREWDRIVERS

When tools have to be used during manual tasks and unnecessary physiological cost or possibly negative long-term effects of repetitive manual movements should be avoided, then the equation 'in conformity with man = in conformity with hand' has to be fulfilled. It is meanwhile well known that cumulative trauma disorders (CTD), repetitive strain injuries (RSI), and musculoskeletal complaints can be the issues of Tayloristic work design and unhandy tools (compare among others Putz-Anderson (1988), Osterholz *et al.* (1987), and Fehr, and Krueger (1992)).

The causes of reduced performance and increased absenteeism at work in connection with the above-mentioned disorders can very often be found in unnatural postures of the lower elements of the kinematic chain, which is built by the hand–arm system during coupling with a working tool, in deformations of the curvature of the palm, as well as in constrained postures of the whole body.

Already some years ago, Bullinger and Solf (1978, 1979) made clear that ergonomic approaches in designing hand tools can help to reduce the prevalence of occupational diseases. At least it can be expected that an appropriate ergonomics design in comparison with a traditional layout will reduce biological expenditure while in any case the same output can be reached.

As an example, Figure 7.33 represents two extremely diverging versions of screwdriver grips. The standard grip in the right part with a straight

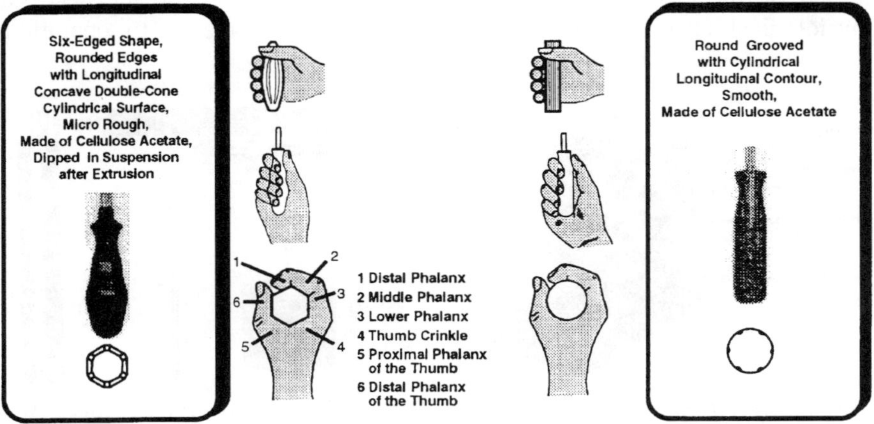

Figure 7.33 Grip contours compatible (left) and incompatible (right) with the hand curvature as a motoric interface of a man–machine system illustrated with the help of screwdriver grips.

longitudinal contour and a round circular cross-section – even with a grooved and rough surface – cannot fit the hand for a power grip or for a pinch coupling. Grips compatible with the natural hand form must have a concave double-cone cylindrical surface and a hexagonal polygon-like cross-section (see left part of Figure 7.33).

Different ergonomic designs must, however, be subject to an objective assessment taking into account the real effects during work. The maximum transferable torque, often used for the evaluation of screwdrivers, can only be one of the criteria to be considered. The electromyographic determination of the physiological cost, necessary for exerting a torque and for screwing itself, represents an important additional direct and 'very close' possibility of gaining insight into muscular strain. In studies published by Strasser (1991) and Strasser *et al.* (1990), the ergonomic qualities of screwdrivers were investigated by applying objective performance measurements, electromyographic methods, and subjective rating. Results concerning electromyography and torque measurements shall be briefly reported in the following.

The left part of Figure 7.34 shows a device for measuring the maximum exertable torque and for predetermining a submaximum static-isometrical demand for torque. In the right part, a device for a dynamic screwing test can be seen, in which a weight had to be screwed upwards over a shaft to a fixed point within a predetermined period of time.

The operation of this device is shown in the lower right part of Figure 7.35, with a subject grasping the screwdriver in a pinch coupling. The job height of this screwing device can be individually adjusted (to elbow height) in such a way that the longitudinal axis is aligned with the lower arm. The photo on the left side shows bipolar electrodes attached on top of the flexor digitorum as the muscle for closing the hand (EA 1) and the biceps as supinator (EA 2). With the aid of the electromyographic activity (EA 1 and EA 2), the necessary physiological effort could thus be determined simultaneous to operational

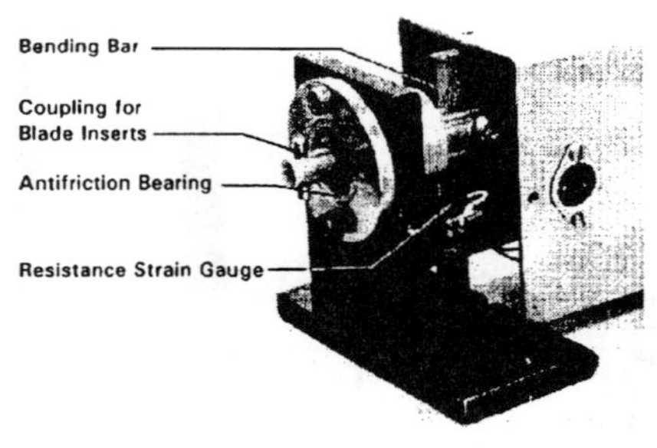

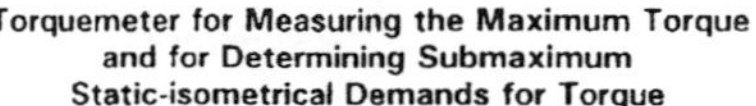
Torquemeter for Measuring the Maximum Torque and for Determining Submaximum Static-isometrical Demands for Torque

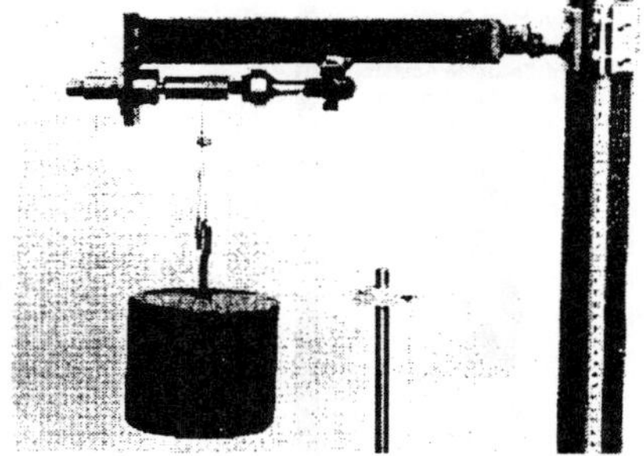
Device for Determining the Dynamic Screwing Work

Figure 7.34 Devices for measuring maximum torque (left) and for simulating dynamic screwing work (right).

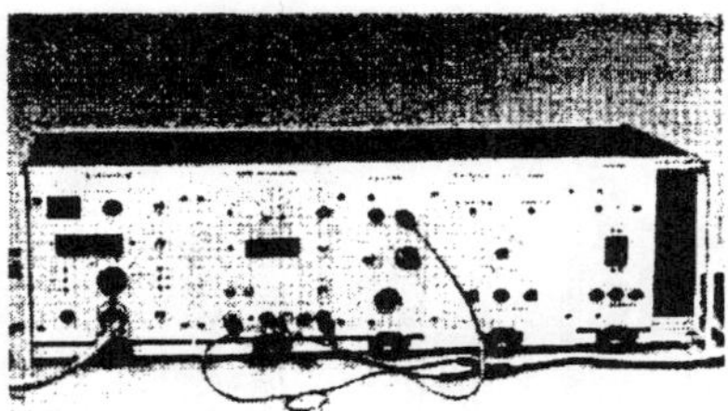

Device for the Evaluation of Electromyographic Data and Display of Power Measurement

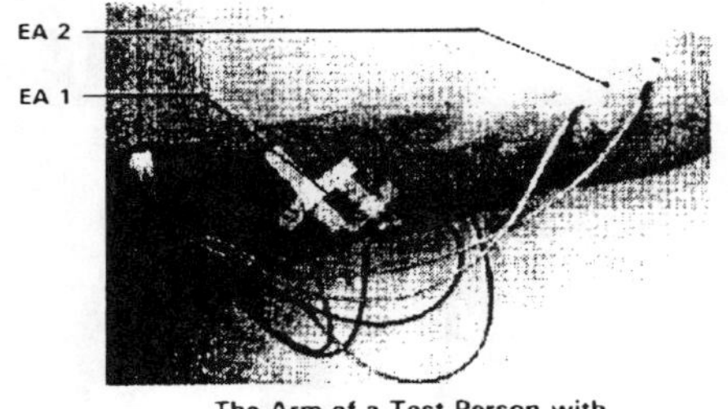

The Arm of a Test Person with Fixed Surface Electrodes

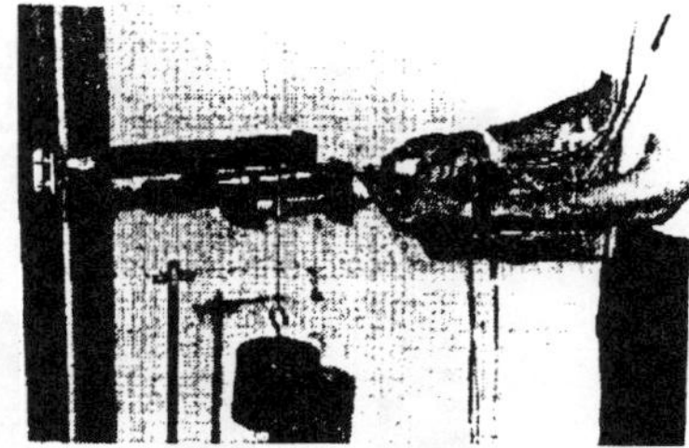

Device for Measuring Dynamic Screwing Work in Action

Figure 7.35 Equipment for measuring EMG signals (EA) and torque (a relatively simple version with integrators), electrode placements above the flexor digitorum (EA1) and the biceps brachii (EA2), and the device for simulating screwing tasks in action.

parameters. For this purpose, a very simple device shown in the upper part of Figure 7.35 for measuring force and evaluating electromyographic data had been employed.

From the wide selection of commercially available screwdrivers, seven typical models – as shown in the lower part of Figure 7.36 – were selected for a systematical investigation. Besides five types claiming to be of ergonomic design, there was also a new design solution (G 4) and finally G 6, a typical representative of conventional mass production (see also Figure 7.33).

The upper part of Figure 7.36 contains (in the black columns) the mean values of the exertable maximum torque. On both sides of the black columns, standardized electromyographic activity of the two muscle groups is represented, which has been measured during working with the seven different handles.

The torques differed considerably, whereby the values correlated in a certain manner with the grip volume as a considerably influencing factor. The design grip G 4, for instance, brought approximately only half of the torque which could be achieved with grips G 5 and G 2.

The electromyographic activity of the musculature closing the hand (EA 1) and the biceps (EA 2), which is nearly equally high for all grips, can be regarded as reliable evidence for the always equally high voluntary muscular contraction of the test persons. Therefore, the measurement of the EA values

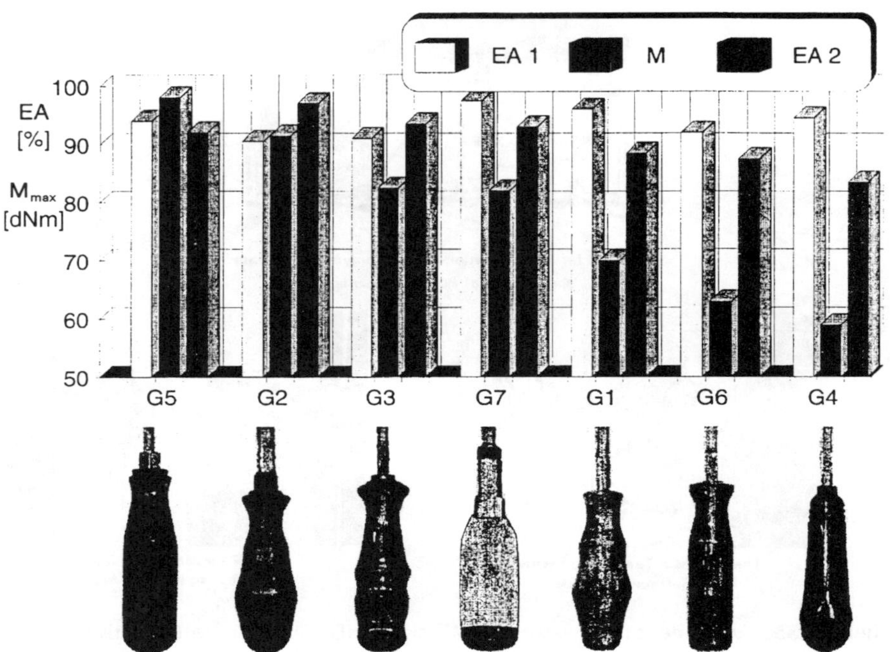

Figure 7.36 Maximum torque M_{max} to be exerted (in deci-Newton-meter) with standardized electromyographic activity of the musculature closing the hand (EA 1) and outward rotating of the forearm (EA 2) (in % of individual electromyographic maximum values). Mean values of all screwdriver grips (G1–G7) from five Ss.

represents a supervising criteria when measuring maximum force or torque. However, equal biological energy led to considerable differences of the operational performance resulting from more or less favorable grasping and coupling conditions, as can be seen in this picture.

If the maximum possible operational performance is not required, but a submaximum equal output is demanded, the total variability due to the different models must turn out in physiological data. This more practical application should make clear how more or less ergonomically designed tools require more or less 'effort'.

Figure 7.37 demonstrates the muscular strain in the case of two submaximum isometric levels of demand when mobilizing 20% of the maximum torque during 20 s or 40% of the maximum torque M_{max} during a halved period of only 10 s in accordance with the principle of equal work. The column diagram arrangements show (for the seven different grips) a development of the physiological cost, which is mostly inverse to the maximum torque.

It should be noticed in Figures 7.36 and 7.37 that grip G 1 – though almost identical with G 2 with respect to its shape and outfit – failed both in

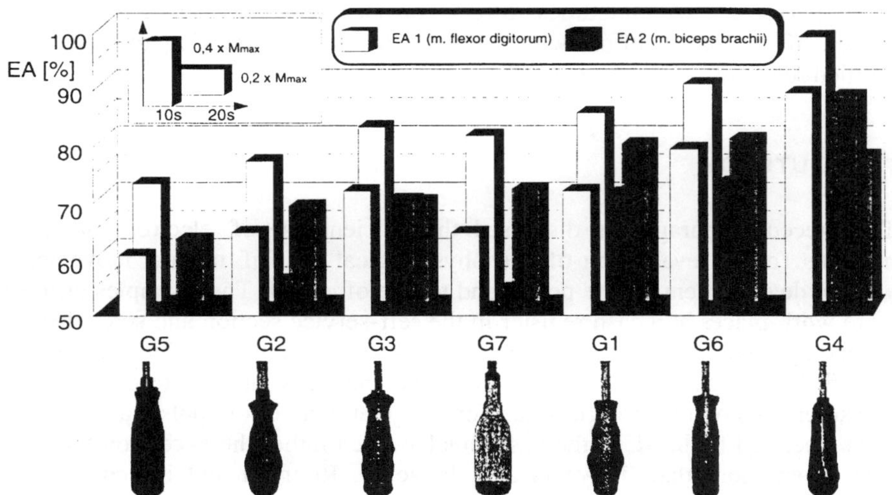

Figure 7.37 Relative height of the static strain regarding the musculature for closing the hand (EA 1) and outward rotating of the forearm (EA 2) (standardized electromyographic activity in % of individual electromyographic maximum values) during equal submaximum (static) screwing performance (mean values from five Ss).

operational and physiological data. Its wax-like surface and the therefore reduced frictional resistance led to a distinctly less favorable result.

When regarding the results of the study presented here only in part, and bearing in mind the tool design method developed by Bullinger and Solf (1979) which is meanwhile well known and has proven useful in Germany, the importance of the ergonomical idea once more becomes evident. The manufacturer of screwdriver G 1 apparently did not realize the adequate importance of even one of the criteria, that is the right friction resistance which results from the material and surface. And by only missing this single objective, all further efforts for obtaining an ergonomically optimum design as well as the hand-adequate dimensions were considerably diminished so that model G 1, for example, is no longer competitive with model G 2.

In conclusion, ergonomic approaches necessitate an investigation of all concrete influencing factors of a working system with regard to their effects on the whole, the same also being true for the macro range of work design. As sectoral optimization efforts often remain measures of doubtful value or sometimes only represent the glossing over of a poorly designed working system, an incompletely investigated ergonomic design becomes perhaps only a kind of cosmetic repair. This might be true to an even greater degree for quite a few design solutions in this range. Finally, it should be pointed out once more that the measurement of the electromyographic activity of the grasping and supination musculature which is mainly involved in screwing work is a simple method for evaluating the necessary muscular strain, which, despite equal performance, sometimes differs considerably. With this method, the degree of biological effort with different designs of working tools can be verified relatively easily by exact data.

7.7 OUTLOOK

The preceding paragraphs discussed the efficient use of electromyographic methods for the evaluation of the physiological cost of manual movements and the development of the power and torque of a tool. The examples utilized were work-places at a cash register in the self-service section and screwdrivers as perhaps the most widely used tool. Those electromyographic methods – specified for each individual case – can and should be utilized to obtain exact measurements of the motoric strain for the great number of tools that still have to be operated by hand. In the past, this has been rather the exception than the rule. Even more than 10 years ago, however, Rohmert and his co-workers (Zipp *et al.*, 1981), among others, despite very limited technical possibilities of electromyography, came up with important ideas for the development of improved ergonomic keyboards. In the meantime, the author and his co-workers have succeeded in improving computer-based multichannel electromyographic methods at least to a point where a sufficiently sensitive and reliable as well as affordable measurement of local physiological cost becomes possible for various applications on the job (Keller *et al.*, 1991; Böhlemann *et al.*, 1993, 1994; Strasser *et al.*, 1993, 1994; Wang and Strasser, 1993; Kluth *et al.*, 1994).

In the Anglo-American literature, electromyographic methods (however, often limited to short-time measurings at certain points) combined with subjective and performance survey methods for the layout of tools have – after the pioneering feat of Tichauer (1978) – almost become state-of-the-art (see among others Kumar *et al.*, 1987; Marras, 1990; Degani *et al.*, 1993; Eklund and Freivalds, 1993; Kilbom *et al.*, 1993; Lewis and Narayan, 1993; Kendall *et al.*, 1994). It is our hope that ergonomically sensible and satisfying solutions will be found more often for the layout of a work-place and the design of work tools in a pluralistic market.

Functional interfaces for man – that is, such possibilities that are favorable for the execution of tasks and are compatible with the anatomic and muscle–physiological fundamental laws of the hand–arm system – must not be sacrificed to aesthetic design demands or technical restrictions of a usually solely cost-based production. Although a nice appearance and low production costs are desirable and can usually be marketed fairly well, they certainly are not sufficient criteria for an efficient and satisfying product in the long run (see also Sperling *et al.*, 1993).

When, for instance, the production of tools – with regard to suitable shapes and fitting sizes (for small, medium large, and large hands) and with regard to gripping materials and abrasive-resistant surfaces – is not oriented to the fingers, the hand's arch, and the anatomic joints of the hand–arm system's kinematic chain, then the work with those tools inevitably leads to unnecessary physiological cost. It is not rare that at least the often-required sensitive proprioceptive feedback for the exact leading of a tool gets lost in muscle problems and pain from forming blisters. Finally, physiological cost should not be completely omitted when the usually dominant economic–technical goals and their monetary cost are considered.

References

BÖHLEMANN, J., KLUTH, K. & STRASSER, H. (1993) A system for the analysis of stress and strain in repetitive manual work. In *The Ergonomics of Manual Work* (Ed. W.S. MARRAS, W. KARWOWSKI, J.L. SMITH & L. PACHOLSKI), pp. 348–8. Taylor & Francis, London/Washington, DC).

BÖHLEMANN, J., KLUTH, K., KOTZBAUER, K. & STRASSER, H. (1994) Ergonomic assessment of handle design by means of electromyography and subjective rating. *Appl. Ergonom.*, **25**(6), 346–54.

BULLINGER, H.-J. & SOLF, J.J. (1978) Produkt-Ergonomie hilft Berufskrankheiten vermeiden. *REFA-Nachrichten*, **31** (1), 17–21.

BULLINGER, H.-J. & SOLF, J.J. (1979) *Ergonomische Arbeitsmittelgestaltung II – Handgeführte Werkzeuge – Fallstudien*. Wirtschaftsverlag NW, Bremerhaven.

DEGANI, A., ASFOUR, S.S., WALY, S.M. & KOSHY, J.G. (1993) A comparative study of two shovel designs. *Appl. Ergonom.*, **24** (5), 306–12.

EINARS, W. (1979) Elektromyographie als arbeitswissenschaftliche Methode zur Beurteilung von lokaler Belastung, Beanspruchung und Ermüdung bei dynamischer Muskelarbeit. PhD thesis, Institut für Arbeitsphysiologie der TU München.

EKLUND, J. & FREIVALDS, A. (Ed.) (1993) Special issue on hand tools for the 1990s. Introduction – Handtools for the 1990s. *Appl. Ergonom.*, **24** (3), 146–7.

ERNST, J. (1989) *Elektromyographische und biomechanische Analysen zur Optimierung von horizontalen Umsetzrichtungen*. VDI-Fortschrittberichte, Reihe 17, Biotechnik Nr. 56, VDI-Verlag, Düsseldorf.

FEHR, K. & KRUEGER, H. (Eds) (1992) *Occupational Musculoskeletal Disorders: Occurrence, Prevention and Therapy*. Eular Publishers, Basel.

HETTINGER, Th., KAMINSKY, G. & SCHMALE, H. (1980) *Ergonomie am Arbeitsplatz – Daten zur menschengerechten Gestaltung der Arbeit*. Kiehl Verlag, Ludwigshafen/Rhein.

KELLER, E., BECKER, E. & STRASSER, H. (1991) Objektivierung des Anlernverhaltens einer Einhand-Akkord-Tastatur für Texteingabe. *Zeitschrift für Arbeitswissenschaft* **45** (17 NF) 1, 1–10.

KENDALL, C.B., SCHOENMARKLIN, R.W. & HARRIS, G.F. (1994) A method for the evaluation of an ergonomic hand tool. In *Advances in Industrial Ergonomics and Safety VI* (Ed. F. AGHAZADEH), pp. 539–45. Taylor & Francis, London.

KILBOM, A., MÄKÄRÄINEN, M., SPERLING, L., KADEFORS, R. & LIEDBERG, L. (1993) Tool design, user characteristics and performance: a case study on plate-shears. *Appl. Ergonom.*, **24** (3), 221–30.

KLUTH, K., BÖHLEMANN, J. & STRASSER, H. (1994) A system for a strain-orient analysis of the layout of assembly workplaces. *Ergonomics*, **37**(9), 1441–8.

KUMAR, S., CHENG, C.K. & MAGEE, D.J. (1987) Comparison of two rake handles. In *Trends in Ergonomics/Human Factors IV* (Ed. S.S. ASFOUR), pp. 631–8. Elsevier Science Publishers B.V., North-Holland.

LAURIG, W. (1970) *Elektromyographie als arbeitswissenschaftliche Untersuchungsmethode zur Beurteilung von statischer Muskelarbeit.* Beuth-Verlag, Berlin/Köln/Frankfurt.

LAURIG, W. (1974) *Beurteilung einseitig dynamischer Muskelarbeit.* Beuth-Verlag, Berlin/Köln/Frankfurt.

LEWIS, W.G. & NARAYAN, C.V. (1993) Design and sizing of ergonomic handles for hand tools. *Appl. Ergonom.*, **24** (5), 351–6.

MARRAS, W.S. (1990) Guidelines industrial electromyography (EMG). *Int. J. Indust. Ergonom.*, **6**, 89–93.

MÜLLER, K.-W., ERNST, J. & STRASSER, H. (1988a) Eine Methode zur Fraktionierung der elektromyographischen Aktivität bei Beanspruchungsanalysen von repetitiven Tätigkeiten. *Zeitschrift für Arbeitswissenschaft.* **42** (14 NF) 3, 147–53.

MÜLLER, K.-W., ERNST, J. & STRASSER, H. (1988b) Lokale Muskelbeanspruchung und bewegungstechnische Arbeitsgestaltung bei horizontalen Umsetzungstätigkeiten. *Zeitschrift für Arbeitswissenschaft 42* (14NF) 4, 226–31.

MÜLLER, K.-W., ERNST, J. & STRASSER, H. (1989) Ein Normierungsverfahren der elektromyographischen Aktivität zur Beurteilung einseitig dynamischer Muskelbeanspruchung. *Zeitschrift für Arbeitswissenschaft* **43** (15 NF) 3, 129–35.

OSTERHOLZ, U., KARMAUS, W., HULLMANN, B. & RITZ, B. (Ed.) (1987) Work-related musculo-skeletal disorders. *Proceedings of an International Symposium.* Schriftenreihe der Bundesanstalt für Arbeitsschutz, Tb 48, Wirtschaftsverlag NW, Bremerhaven.

PUTZ-ANDERSON, V. (Ed.) (1988) *Cumulative Trauma Disorders. A Manual for Musculoskeletal Diseases of the Upper Limbs.* Taylor & Francis, London/New York/Philadelphia.

RAU, G. (1977) Anwendung der Elektromyographie bei der Beurteilung körperlicher Momentan- und Langzeitbeanspruchung. *Zeitschrift für Arbeitswissenschaft* **31** (3 NF), 112–20.

ROHDAHL, K. (1989) *The Physiology of Work.* Taylor & Francis, London/New York/Philadelphia.

SPERLING, L., DAHLMAN, S., WIKSTRÖM, L., KILBOM, A. & KADEFORS, R. (1993) A cube model for the classification of work with hand tools and the formulation of functional requirements. *Appl. Ergonom.*, **24** (3) 212–20.

STRASSER, H. (1990) Evaluation of a supermarket twin-checkout involving forward and backward operation. *Appl. Ergonom.*, **21** (1), 7–14.

STRASSER, H. (1991) Different grips of screwdrivers evaluated by means of measuring maximum torque, subjective rating and by registering electromyographic data during static and dynamic test work. In *Advances in Industrial Ergonomics and Safety III* (Ed. W. KARWOWSKI & J.W. YATES), pp. 413–20. Taylor & Francis, London/New York/Philadelphia.

STRASSER, H. & ERNST, J. (1992) Physiological cost of horizontal materials handling while seated. *Int. J. Indust. Ergonom.*, **9**, 303–13.

STRASSER, H. & MÜLLER, K.-W. (1991) Zur Muskelbeanspruchung des Hand–Arm–Schulter–Systems bei repetitiven horizontalen Armbewegungen in abhüngigkeit von der umsetzlast. In *Verhandlungen der Deutschen Gesellschaft für Arbeitsmedizin e.V.* (Ed. F. SCHUCKMANN & S. SCHOPPER-JOCHUM), pp. 151–8. Gentner Verlag, Stuttgart.

STRASSER, H., MÜLLER, K.-W., ERNST, J. & KELLER, E. (1989) Local muscular strain dependent on the direction of horizontal arm movements. *Ergonomics*, **32** (7), 899–910.

STRASSER, H., LAUBER, M. & KOCH, W. (1990) Produkt-ergonomische Beurteilungsmethoden für handbetätigte Arbeitsmittel – Leistungsdaten und Beanspruchung des Hand-Arm-Systems beim Test verschiedener Schraubendrehergriffe. *Zeitschrift für Arbeitswissenschaft* **44** (16 NF) 4, 205–13.

STRASSER, H., GROSS, E. & KELLER, E. (1991) Electromyographic evaluation of the physical load of the left hand–arm–shoulder system during simulated work at eight different cash register arrangements. In *Advances in Industrial Ergonomics and Safety III* (Ed. W. KARWOWSKI & J.W. YATES), pp. 457–63. Taylor & Francis, London/New York/Philadelphia.

STRASSER, H., BÖHLEMANN, J. & KELLER, E. (1992a) Elektromyographische und subjektive Ermittlung der Muskelbeanspruchung bei arbeitstypischen Bewegungen an Kassenarbeitsplätzen zur Entwicklung von Bausteinen eines Systems vorbestimmter Beanspruchung. *Zeitschrift für Arbeitswissenschaft* **46** (18 NF) 2, 70–6.

STRASSER, H., ERNST, J. & MÜLLER, K.-W. (1992b) *Günstige Bewegungen für die ergonomische Arbeitsgestaltung – Elektromyographische Untersuchungen des Hand–Arm–Systems.* Schriftenreihe 'Arbeitsmedizin-Arbeitsschutz-Prophylaxe und Ergonomie', Band 11, 136 Seiten, 70 Abbildungen, 10 Tabellen, 110 Literaturstellen, C. Haefner Verlag, Heidelberg.

STRASSER, H., ERNST, J. & MÜLLER, K.-W. (1993) Electromyographic investigations of favourable movements of the hand–arm system as a basis for ergonomic workdesign. In *The Ergonomics of Manual Work* (Ed. W.S. MARRAS, W. KARWOWSKI, J.L. SMITH & L. PACHOLSKI, pp. 175–8. Taylor & Francis, London/Washington DC.

STRASSER, H., WANG, B. & HOFFMANN, A. (1994) Electromyographic and subjective assessment of mason's trowels equipped with different handles. In *Advances in Industrial Ergonomics and Safety VI* (Ed. F. AGHAZADEH), pp. 553–60. Taylor & Francis, London/New York/Philadelphia).

TICHAUER, E.R. (1978) *The Biomechanical Basis of Ergonomics.* John Wiley & Sons, New York.

WANG, B. & STRASSER, H. (1993) Left- and right-handed screwdriver torque strength and physiological cost of muscles involved in arm pronation and supination. In *The Ergonomics of Manual Work* (Ed. W.S. MARRAS, W. KARWOWSKI, J.L. SMITH & L. PACHOLSKI), pp. 223–6. Taylor & Francis, London/Washington DC.

ZIPP, P. (1988) *Optimierung der Oberflächenableitung bioelektrischer Signale.* Fortschritt-Berichte, Reihe 17, Biotechnik Nr. 45, VDI-Verlag, Düsseldorf.

ZIPP, P. (1989) Leitregeln für die Oberflächen-Myographie: Ausgewählte Beispiele: In

Motorikforschung aktuell – Die Elektromyographie in der Motorikforschung (Ed. R. DAUGS, K.-H. LEIST & H.-V. ULMER), pp. 68–73. DVS-Protokolle Nr. 35, Clausthal-Zellerfeld.

ZIPP, P., HAIDER, E., HALPERN, N., MAINZER, J. & ROHMERT, W. (1981) Untersuchungen zur ergonomischen Gestaltung von Tastaturen. *Zentralblatt für Arbeitsmedizin, Arbeitsschutz, Prophylaxe und Ergonomie*, **31** (8) 326–30.

CHAPTER EIGHT

EMG of neck and shoulder muscles: the relationship between muscle activity and muscle pain in occupational settings

R.H. WESTGAARD, T. JANSEN AND C. JENSEN

The Norwegian Institute of Technology, Trondheim, National Institute of Occupational Health, Oslo

8.1 INTRODUCTION

The application of electromyography (EMG) in ergonomics is motivated by the problem of musculoskeletal complaints in the work-place. It is commonly assumed that such complaints are mediated through biomechanical forces acting in the body, since awkward postures and repetitive movement are identified as risk factors in epidemiological studies (Winkel and Westgaard, 1992). However, biomechanical forces may also result in improved health (Åstrand, 1988) and a key issue is then to distinguish forces that are health promoting from those that cause impaired health. The chain of events postulated to result in health effects is summarized in Figure 8.1. Work demands (external exposure, i.e. physical demands quantified independently of the worker) exert their effects through muscle activity patterns generated to meet these demands (internal exposure). The muscle activity causes short-term responses, e.g. increased circulation, fatigue, and a variety of metabolic responses, that on a longer time scale may result in health effects. Factors independent of muscle activity influence the outcome and thus modulate the assumed relationship. Later in this paper evidence to suggest that muscle pain may arise independently of muscle activity is discussed, a mechanism not considered in Figure 8.1. In this model, a surface EMG recording may be used to estimate muscle activity (internal exposure) and

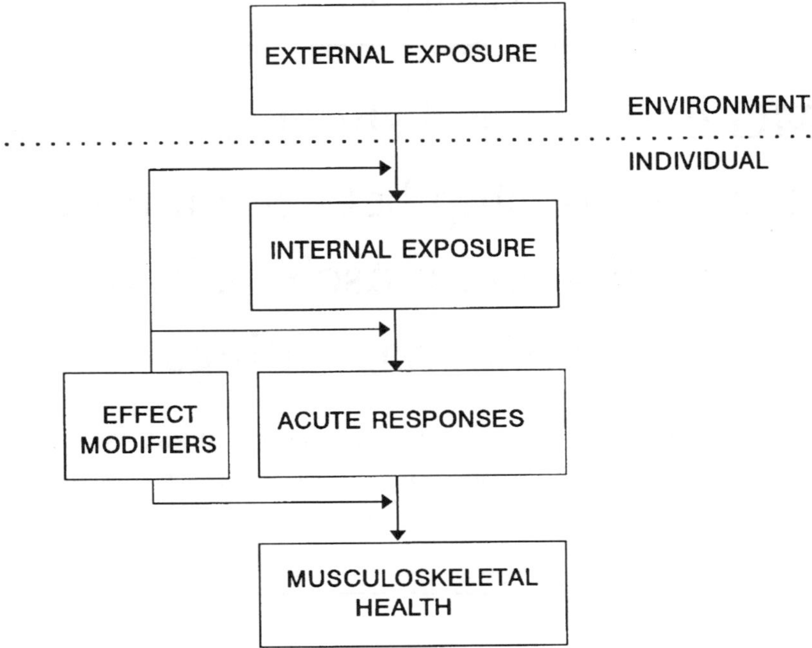

Figure 8.1 Model to indicate relationship between physical work-load, specified by work demands independently of the worker (external exposure) and long-term health effects. Intermediate stages in this relationship are biomechanical forces generated (internal exposure) and short-term physiological responses.

thereby, hopefully, provide an estimate of the risk of a future development of a health problem. This scheme depends on the existence of a physical exposure–health effect relationship and an important research objective within ergonomics is to provide information about this relationship, if existing.

Research to this aim, using the development of musculoskeletal complaints in the trapezius as a model, has been carried out in our laboratories over the last 15 years. This chapter presents a summary of our results to date. In this work other research questions have surfaced, such as possible pathophysiological mechanisms in the association between psychosocial factors and muscle pain, and evidence relating to this question is presented.

8.2 MATERIAL AND METHODS

8.2.1 Material

This paper reports selected results from three cross-sectional, one longitudinal, and one case–control studies, aiming to look at the relationship between activity in the trapezius muscle and pain in the shoulder and neck region. One cross-sectional and the case–control study included two separate groups of workers, performing light manual and office work, the data presented here as separate studies. The studies were all independent of each other, e.g. the two groups performing chocolate manufacturing (study D, E) were employed in different companies. Table 8.1 shows the main work task, study design, time period of collecting the material, number of subjects with EMG recordings, age, employment time, the variables used to quantify vocational EMG activity, and additional tests with recording of the EMG responses. All subjects included in the table were female.

There is a clear trend of reducing age and employment time from light to heavy work tasks. In the heavy assembly work (study B) the labour turnover rate per year was more than 60% for those working full time and more than 30% for the part-time workers. A concern when conducting that study was to ensure that enough time had elapsed to allow all subjects fair time to develop muscle pain symptoms. Analysis of sick leave (e.g. Wærsted and Westgaard, 1991), and later the development of pain symptoms in relation to employment time (Veiersted and Westgaard, 1993), suggested that the employment period should be at least 2 years in the case of light work and at least 1 year in case of heavy or repetitive industrial tasks. This was difficult to fulfill in the case of the heavy assembly work due to the large labour turnover and some subjects with an employment time between 0.5 and 1 year and pain symptoms in the shoulder and neck region were included in study B.

In addition to those studies listed in Table 8.1, a study of electromechanical assembly work has been carried out (Westgaard and Aarås, 1984, 1985; Aarås and Westgaard, 1987). In this study the median muscle activity level varied in a slow, systematic manner during the working day due to variation in the working height, in contrast to the other studies where the median level did not change. This made it difficult to estimate the critical work load variables and that study is therefore not included in Table 8.1.

8.2.2 Effect Measures

In studies of the relationship between muscle activity and muscle pain, the measurement of health effect must relate to the same body structure as the exposure measurement, preferably the same muscle. The pain should also reasonably be considered an effect of muscle activity, i.e. patients with arthritis or other systemic disease must be excluded. It can be difficult to ascertain that pain symptoms are originating in the muscle used for exposure measurement. Referred pain is a problem; the pain usually has a wider location than the limited pick-up area of a bipolar surface electrode and it may originate in structures other than muscle, e.g. joints or connective tissue, thereby making the relationship to the muscle activity measurements more uncertain. The shoulder is a complicated biomechanical structure and causes a number of

Table 8.1 Material

Study	Work task	Study design	Study period	Number of subjects	Age (mean, range)	Employment time (mean, range)	EMG variables	Additional tests
A	Sewing machine operation	Cross-sectional	1983–1984	30	32 (20,54)	6 (3,14)	Static, median activity	
B	Assembly work (high-static loads)	Cross-sectional	1980–1985	28	27 (17,46)	4 (0.5,10)		
C	Secretarial/office work	Cross-sectional	1989–1990	32	42 (23,64)	11 (2,19)	Static, median activity EMG gap variables	Muscle coordination tests Mental activation test
D	Light manual work	Cross-sectional	1989–1990	39	35 (21,65)	9 (2,35)		
E	Light manual work	Longitudinal	1987–1990	30	22 (17,46)	– (0,2)	Static, median activity EMG gap variables	EMG activity at rest
F	Secretarial/office work	Case–control	1990–1993	48	38 (27,50)	14 (2,32)	Static, median activity EMG gap variables	Muscle coordination tests mental activation test
G	Light manual work	Case–control	1990–1993	30	30 (22,53)	5 (1,18)		EMG activity at rest

problems in this respect. It is, at the same time, a structure of considerable interest as the frequency of musculoskeletal complaints in industry has shown a particular prominent increase for this body region in the last decade. In our own research we have focused on the trapezius muscle. It is accessible to surface EMG measurements and is known as a source of myalgic symptoms with a presumed muscle activity-related etiology. However, pain symptoms in the shoulder may arise from structures other than the trapezius, in which case trapezius muscle activity at best is a poor indicator of the critical exposure variables.

Figure 8.2 illustrates this point in case of study A. A larger group of sewing machine operators (210 workers) and office workers (35 workers) in the company underwent a clinical examination with palpation, passive and active arm movement, force exertion against resistance and a static holding test. In Figure 8.2 those (113 workers) reporting pain symptoms in the shoulder region, including the trapezius, in the previous 12 months with a score of at least 4 on a scale from 0 to 6 are compared with those reporting no pain (45 workers) in the shoulder and neck. The symptom scale takes into account information on both the intensity and occurrence of the complaint (Westgaard and Jansen, 1992a). The figure shows the fraction of workers with a positive response to various clinical tests, including pain provoked upon maximal force exertion against resistance (muscle load tests), palpation of the muscle belly or tendons (palpation tests), pain upon passive movement of the arm in flexion and abduction with fixation of the scapula (shoulder movement tests), and pain or discomfort developing after a static holding test with the arms held for 1 min in 90° abduction in the scapular plane (fatigue test). In the case of the palpation tests tenderness was recorded unless the palpation of trigger points (tr.p.) is indicated.

A high fraction of those reporting shoulder pain the previous 12 months had clinical signs localized to the belly or tendons of the trapezius muscle. However, many such signs were also located to the belly or tendons of the neck and rotator cuff muscles (supraspinatus, infraspinatus, and subscapularis). Those with pain were also more inclined to report pain in active head movement and showed restricted rotation of the head. There was a similar pattern of clinical signs both for the sewing machine operators and office workers, but such signs were less frequent and predominantly localized to the right side for the office workers (T. Jansen and R.H. Westgaard, unpublished). Several workers without symptoms showed the same clinical signs. The two factors with highest eigenvalue in a factor analysis of this material related to clinical signs in the trapezius muscle and in the rotator cuff muscles. The two tests that best distinguished those with and without symptoms were trigger points in the trapezius and pain developing during the fatigue test. These two tests gave a correct classification of 75% of the subjects reporting or not reporting pain symptoms (discriminant analysis). In the longitudinal study clinical signs relating to the rotator cuff muscles developed additionally to those relating to the trapezius muscle (Veiersted and Westgaard, 1993).

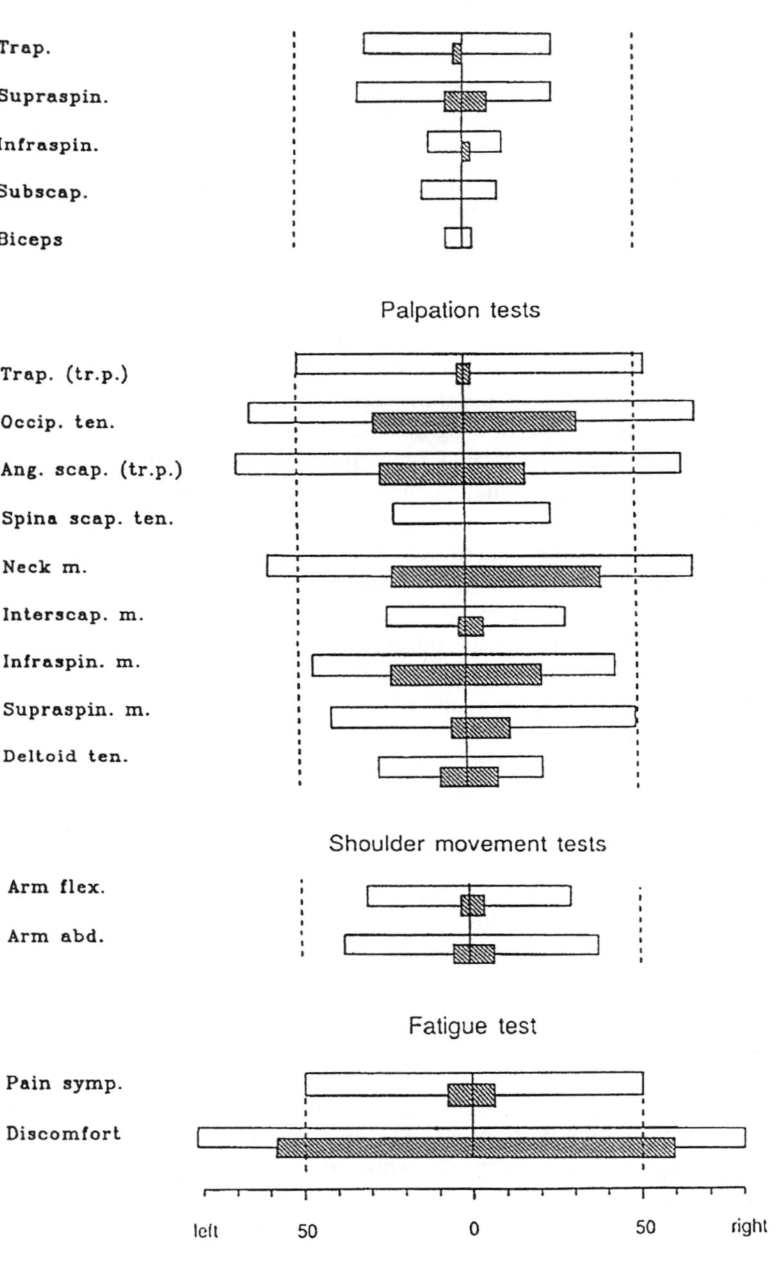

Figure 8.2 Percent workers, either reporting significant pain symptoms in the shoulder and neck region (open columns) or without such symptoms (hatched columns), responding to clinical tests. Results for the right and left side are given separately. See text for details.

The location of the trapezius muscle was considered in the pain reports, ensuring that pain symptoms in the shoulder included the trapezius region. A clinical examination of the shoulder was carried out in all projects. However, in the cross-sectional and case–control studies some subjects reported symptoms in the previous 12 months before the clinical examination, but were symptom free at examination, making it difficult to verify clinically the location of the complaint. Those reporting shoulder pain at the clinical examination always indicated a location of symptoms that included the trapezius muscle. However, there is a possibility that the trapezius muscle was not affected for a small number of subjects with a positive shoulder pain score.

8.2.3 EMG Measurement of Trapezius Activity

EMG methods are considered in detail in other chapters of this book and our procedure for obtaining EMG measurements from the trapezius is described elsewhere (Westgaard, 1988; Jensen *et al.*, 1993b). In short, conventional bipolar recording of surface EMG signals was performed for a period lasting from 30 min to 2 h. In each case the recording represented a typical work pattern or a combination of work patterns when this was indicated. In this section some specific problems relating to surface EMG recording from the trapezius are summarized.

The standard bipolar electrode position, midway between the acromion and the spine of the C7 vertebra (Zipp, 1982), was used in studies A–D. This position normally spans the end-plate region, where a marked depression of the EMG amplitude is observed (Figure 8.3a). The position of the minimum amplitude moves with the arm posture (Figure 8.3b; see Figure 8.4 for illustration of the different arm positions), presumably due to movement of the skin relative to the underlying muscle. Thus, the calibration of vocational recordings with the electrode in this position and based on the EMG amplitude at maximal voluntary contraction in shoulder elevation, becomes unreliable in situations where the arm posture changes during the recording. In studies F and G the bipolar electrode was placed in a more lateral position, avoiding this problem.

In the early studies with strenuous work tasks (studies A and B) the EMG signal was calibrated to indicate force by simultaneous recording of EMG activity and force in shoulder elevation. However, muscle activity rather than muscle force appeared the most relevant variable at the low activity levels observed in light manual and sedentary work. Also, the EMG–force calibration reflected the combined effects of several synergistic muscles and is a questionable indicator of trapezius force output (Johnson *et al.*, 1994). A calibration in terms of EMG activity at maximal force generation ('proportional EMG') was therefore preferred and the early studies were recalibrated by this method. This corresponds to a linear EMG–force calibration curve. The highest EMG response (EMG_{max}) either in shoulder

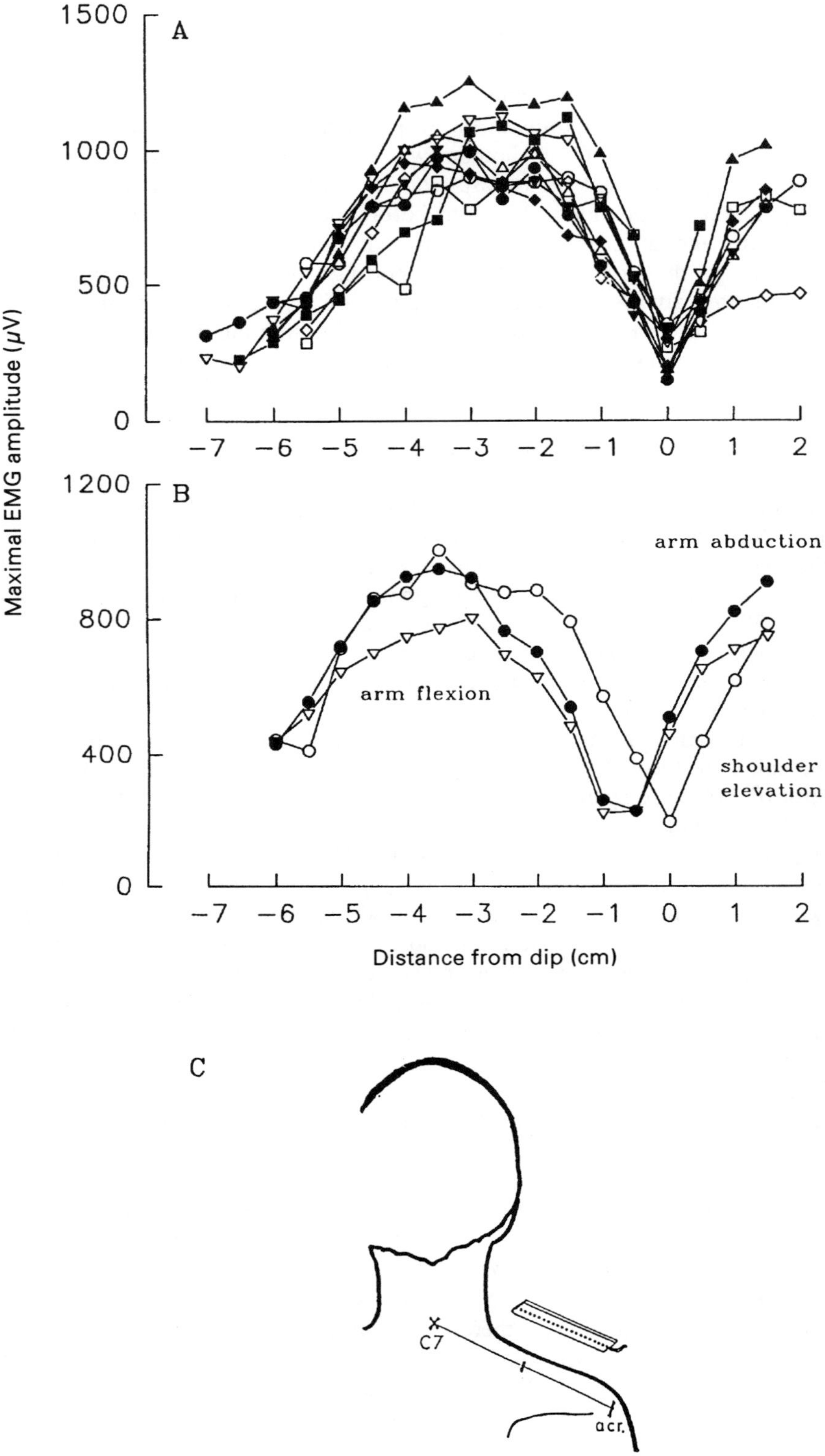

A
B
C
Maximal EMG amplitude (μV)
1500
1000
500
0
1200
800
400
0
−7
−6
−5
−4
−3
−2
−1
0
1
2
Distance from dip (cm)
arm abduction
arm flexion
shoulder
elevation
C7
acr.

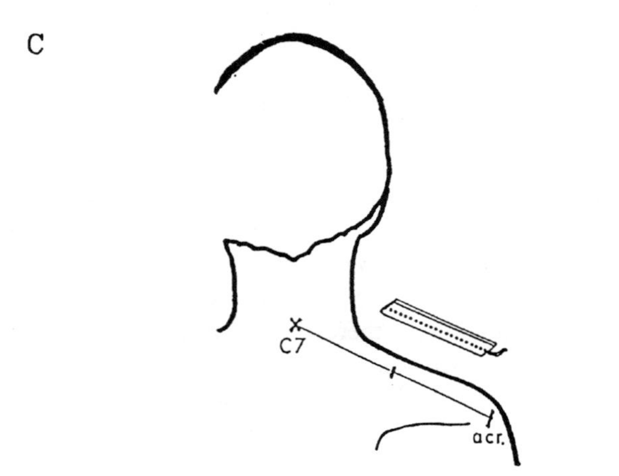

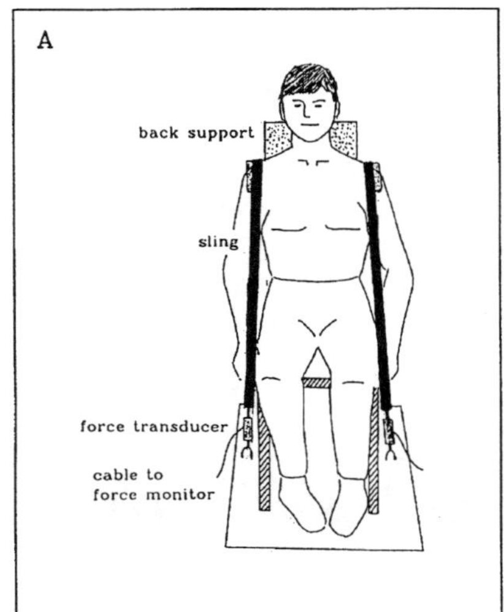

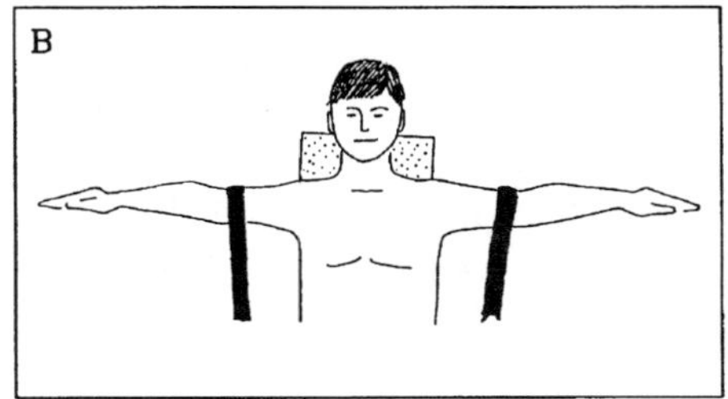

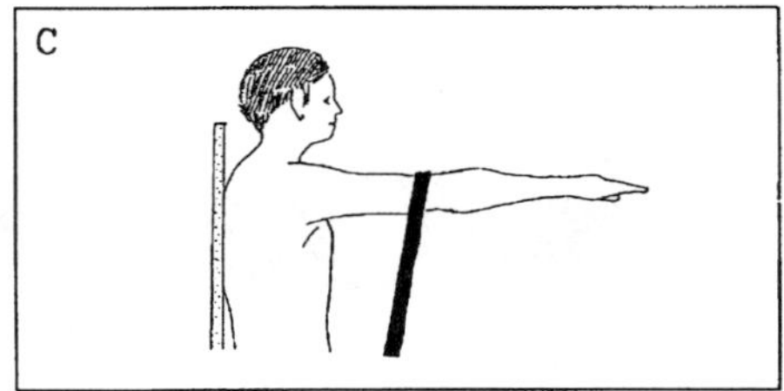

Figure 8.4 Force measurement set-up and arm postures during (a) shoulder elevation, (b) arm abduction and (c) arm flexion.

elevation or in 90° arm abduction, each movement performed twice, was used as basis for the EMG calibration (Figure 8.4). The EMG recordings were performed bilaterally and the highest of the responses used in the further analysis. The reliability of the estimate of EMG_{max} was investigated in a separate study (Jensen *et al.*, 1995), showing that 95% of the differences between two estimates of EMG_{max} performed on each subject were smaller than approximately 20% of the mean value of the two estimates (each estimate was based on the described calibration procedure repeated after a 2 h interval). In other words, the normalized EMG activity level can be expected to vary within a range corresponding to ±20% due to variation in the EMG_{max}. In our recordings of EMG activity during the performance of a work task the described calibration procedure was performed both before and after the work recording and the highest of all responses used in the analysis. This can be presumed to further reduce the variation in EMG_{max}, but the residual variation has not been determined.

Figure 8.3 *opposite* (a) Individual EMG amplitude profiles recorded during maximal voluntary contractions in shoulder elevation and with the arms in a vertical position. The position along the length of the right upper trapezius muscle is indicated as the distance from the dip position to C7 (positive values) or to the acromion (negative values). (b) EMG amplitude profiles recorded during maximal voluntary contractions with the arms in a vertical position (shoulder elevation), at 90° abduction (arm abduction) or 90° flexion (arm flexion) for one subject. (c) The position of the 16 channel array electrode on the lead-line between the seventh cervical vertebra (C7) and the acromion.

In some circumstances a single recording with the electrodes on a line from C7 to the acromion may not give an adequate description of activity in the whole upper trapezius muscle. Figure 8.5 illustrates an experimental situation with three bipolar recordings, one on the line between the acromion and C7 (middle panel), one positioned 2 cm towards the clavicula (top panel), and one positioned 2 cm towards the scapula (lower panel). The subject was asked

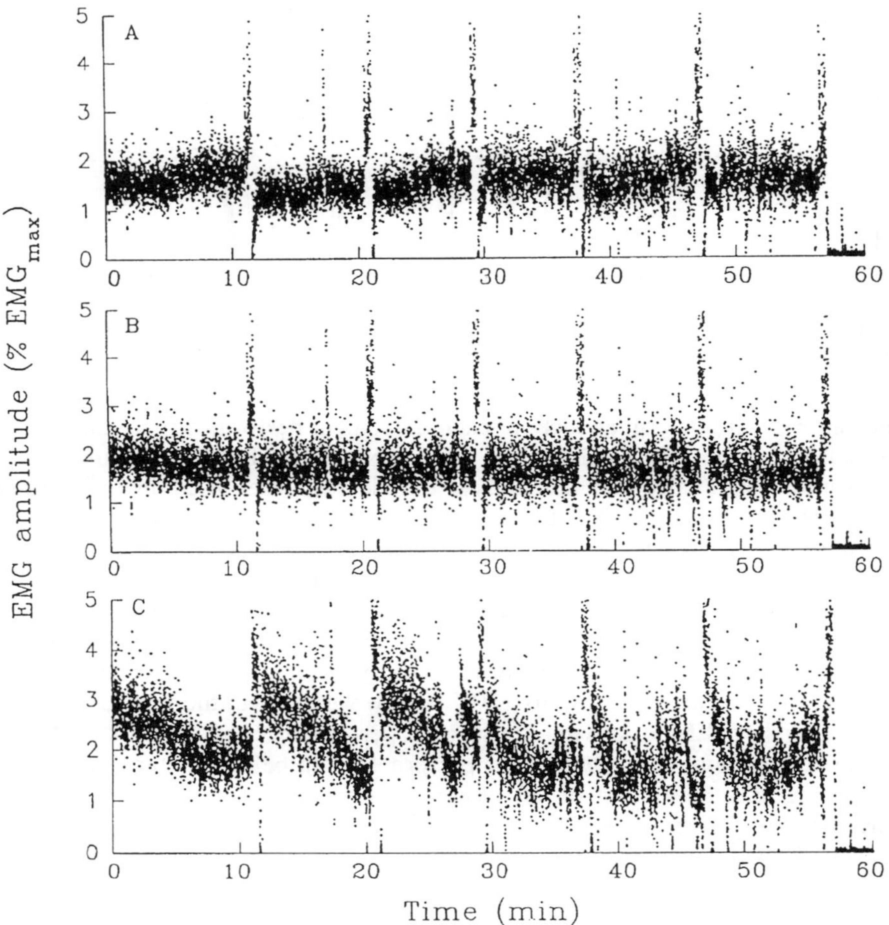

Figure 8.5 EMG activity patterns of three simultaneous, bipolar recordings from the upper right trapezius muscle, with the middle electrode pair positioned 2 cm lateral to the midpoint between C7 and the acromion and with 2 cm center distance between the electrodes (middle panel). The other two sets of bipolar electrodes were positioned parallel to the middle electrode, either 2 cm towards the clavicula (top panel) or 2 cm towards the scapula (lower panel). The subject was maintaining a contraction of 2% EMG_{max} for 1 h by help of biofeedback from the middle electrode.

to maintain a constant force set at 2% EMG_{max} through visual feedback from the middle EMG recording. While that task was well achieved, other parts of the trapezius showed marked changes in EMG level with time and was reset every 10 min when subjective data were collected. The same phenomenon can be observed in situations where EMG activity is generated during attention-related tasks. Such effects point to independent neural control of several motor neurone pools innervating the upper trapezius. Figure 8.5 is reproduced from a study of regional upper trapezius EMG activity where a few subjects performed the test that produced the EMG recordings plotted in the figure. Sixteen other subjects performed three different tasks of less than 5 min duration each and the median EMG activity recorded at each electrode position was calculated for each task and subject. For the attention-related task the mean EMG activity level of these 16 subjects was 1.4% EMG_{max} at the electrode position on the line between C7 and the acromion. The mean difference between the EMG activity levels recorded at this position and the position 2 cm toward the clavicula or 2 cm toward the scapula were 0.4% EMG_{max} and 0.7% EMG_{max}, respectively. When the subjects were instructed to keep their arms abducted 90 °, the mean EMG activity level at the position on the midline was 10.6%EMG_{max}, whereas the EMG activity levels at the other two positions differed on average by 2.6% EMG_{max} and 2.2% EMG_{max}, respectively, from the former value. Thus, the total EMG activity level of the whole upper trapezius is not identical to the EMG activity level recorded at one electrode position, but may vary within a range indicated by the examples just shown. The relative difference is most pronounced at activity levels of a few percent EMG_{max}, whereas a single recording usually provides an adequate estimate of trapezius activity in vocational recordings with higher activity levels.

The data reduction of the vocational EMG recordings aimed to extract variables that can be used as risk indicators for muscle pain. The recordings were quantified in terms of static, median, and peak EMG levels according to the method of Jonsson (1978, 1982). These variables are based on the averaged EMG response over a whole recording. Periods with low EMG activity, indicating short pauses in the muscle activity pattern (Veiersted *et al.*, 1990), were additionally quantified in the more recent projects. These were defined as periods of at least 0.2 s duration with full-wave-rectified and averaged EMG activity below 0.5% EMG_{max} ('EMG gaps', Figure 8.6). Other variables developed on the basis of this analysis were 'long gaps' (duration ≥0.6 s) and 'gap time' (total time with the EMG level lower than 0.5% EMG_{max}, quantified as seconds per minute recording). This choice of variables related both to theoretical considerations and to the sensitivity of the recording equipment. The gap time of 0.2 s is long enough to ensure that cancellation of the surface EMG signal due to stocastic processes is unlikely. The level chosen to represent 'resting muscle', 0.5% EMG_{max}, is approximately three times higher than the noise level of our recording equipment (<1.5 μV; the typical maximal EMG amplitude is 1 mV in the 'plateu region', see Figure 8.3a) to

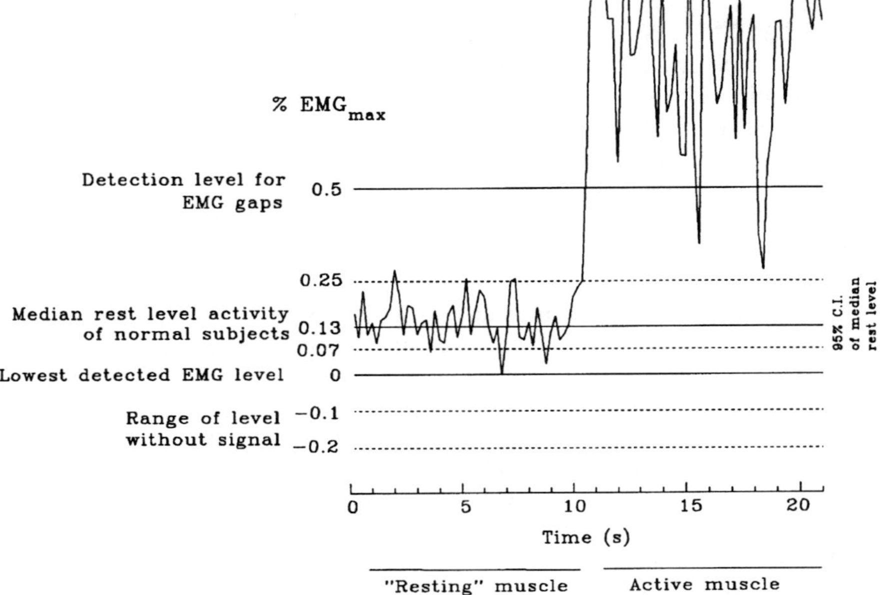

Figure 8.6 EMG gap analysis. The lowest detected EMG level (full-wave-rectified and averaged EMG signal over 0.2 s) is defined as the system noise level, usually 1–1.5 μV and corresponding well with the signal level when measuring over a 1 kohm resistance. Median muscle activity during rest is usually of the same magnitude, as indicated by the section of recording marked 'resting level'. Periods of EMG activity (≥0.2 s) below 0.5% EMG_{max} are defined as EMG gaps. Two such periods are seen in the section of the recording marked 'active muscle'. 'Long gaps' (≥0.6 s) and total gap time were additionally quantified.

ensure that the quantification of EMG gaps is not confounded by random noise. The lowest frequency of sustained firing of motor units in the human trapezius is 6–7 Hz (Wærsted *et al.*, 1996), so that at least one EMG potential of a continuously firing motor unit would be observed within a period of 0.2 s. Thus, an EMG gap may serve as an indicator of interruption of activity in continuously firing motor units.

The quality of the EMG equipment was not adequate for quantification of the EMG gaps in the early studies, however, these studies were likely to yield very low numbers in view of the high static loads recorded. In the later studies variables quantifying the EMG responses during psychomotoric tests and during rest were included in an attempt to measure the general activation level of the subjects. These variables were quantified as the mean full-wave-rectified and averaged EMG response during the test period.

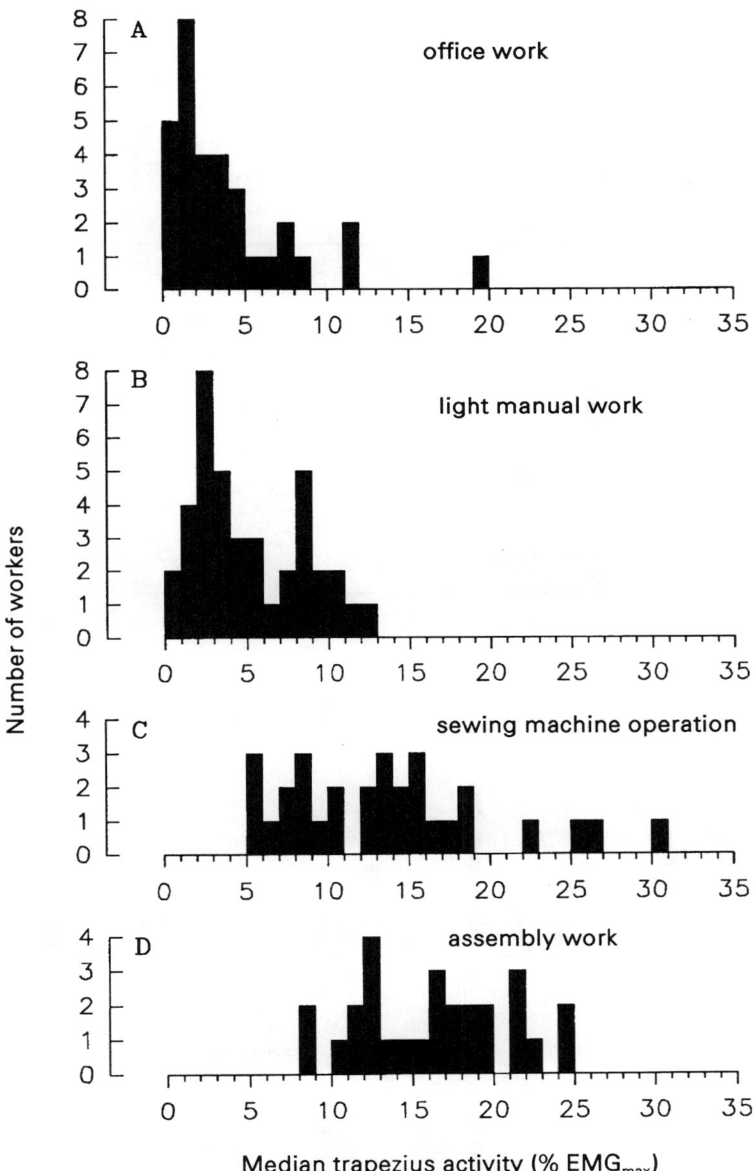

Figure 8.7 Distribution of median trapezius EMG activity (highest of right and left trapezius for each subject) of workers in four different work situations.

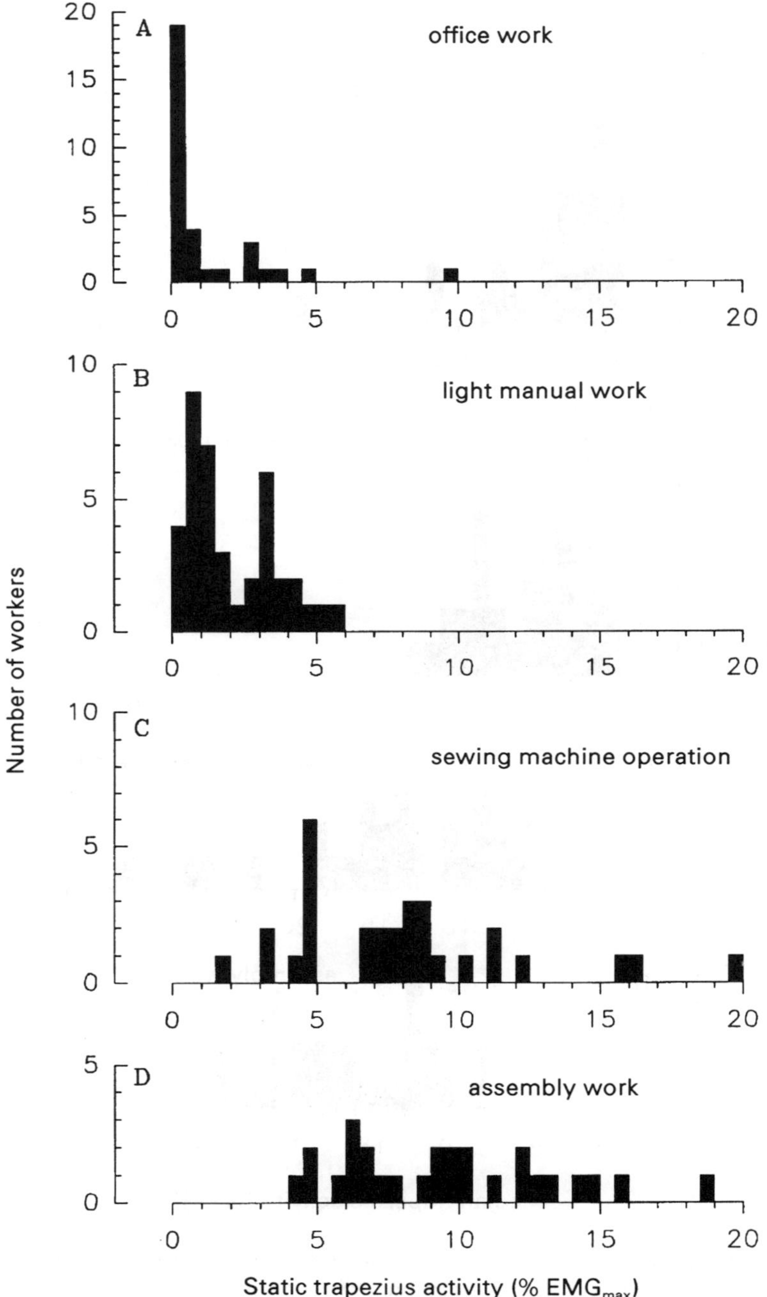

Figure 8.8 Distribution of static trapezius EMG activity (highest of right and left trapezius for each subject) of workers in four different work situations.

8.3 RESULTS

8.3.1 The Association Between Muscle Activity and Muscle Pain: Cross-sectional Studies

The median muscle activity levels of the upper trapezius for workers in studies A–D are shown in Figure 8.7. A trend of higher activity levels from office work and light manual work to sewing machine operation and heavy assembly work is evident. The manual, highly repetitive work of chocolate manufacturing was performed at a similar median activity level to the non-paced work tasks of the office workers. However, the large variation in the median activity level within each group is perhaps more striking. A similar variation within each group was observed for the static activity level (Figure 8.8). For this variable it proved possible to differentiate office and chocolate manufacturing workers on account of the very low static load levels for the office workers. Within each group the variation in muscle activity level could not be ascribed differences in work tasks as the work tasks chosen for the recordings were similar for all workers in a group and typical of that work situation.

The observed variation in activity level may provide a basis for understanding individual differences in pain patterns of workers within the same group, i.e. those who develop pain symptoms may have the highest muscle activity level. This was generally not the case. Figure 8.9 shows scatter plots of symptom score versus median activity level. There was evidence of a correlation between median activity level and symptom score only for the sewing machine operators, but then accounting for only 14% of the variance in symptom data. In the case of office work and chocolate manufacturing some workers with very low median activity level, <5% EMG_{max}, reported high symptom scores. Scatter plots of static activity level versus symptom score showed similar patterns.

The association between median activity level and symptom score across work groups was statistically highly significant, but again only accounting for 15% of the variance (Figure 8.10a). The striking feature in the figure is the wide scatter of data points around the regression line. The association between median trapezius activity and symptom score, when considered on a group basis, was however clear (Figure 8.10b). The equivalent plots with static activity showed the same pattern.

Several factors may contribute to explain the variation around the regression line in Figure 8.10a. Potential methodological problems include the following.

1 Exposure level is only one of three variables needed to quantify mechanical exposure. The mechanical exposure quantity, which is the appropriate variable to correlate with health symptoms in a mechanical exposure–health effect relationship, conceptually consists of a combination of the variables: level, repetitiveness, and duration of the exposure (Winkel and Westgaard, 1992).

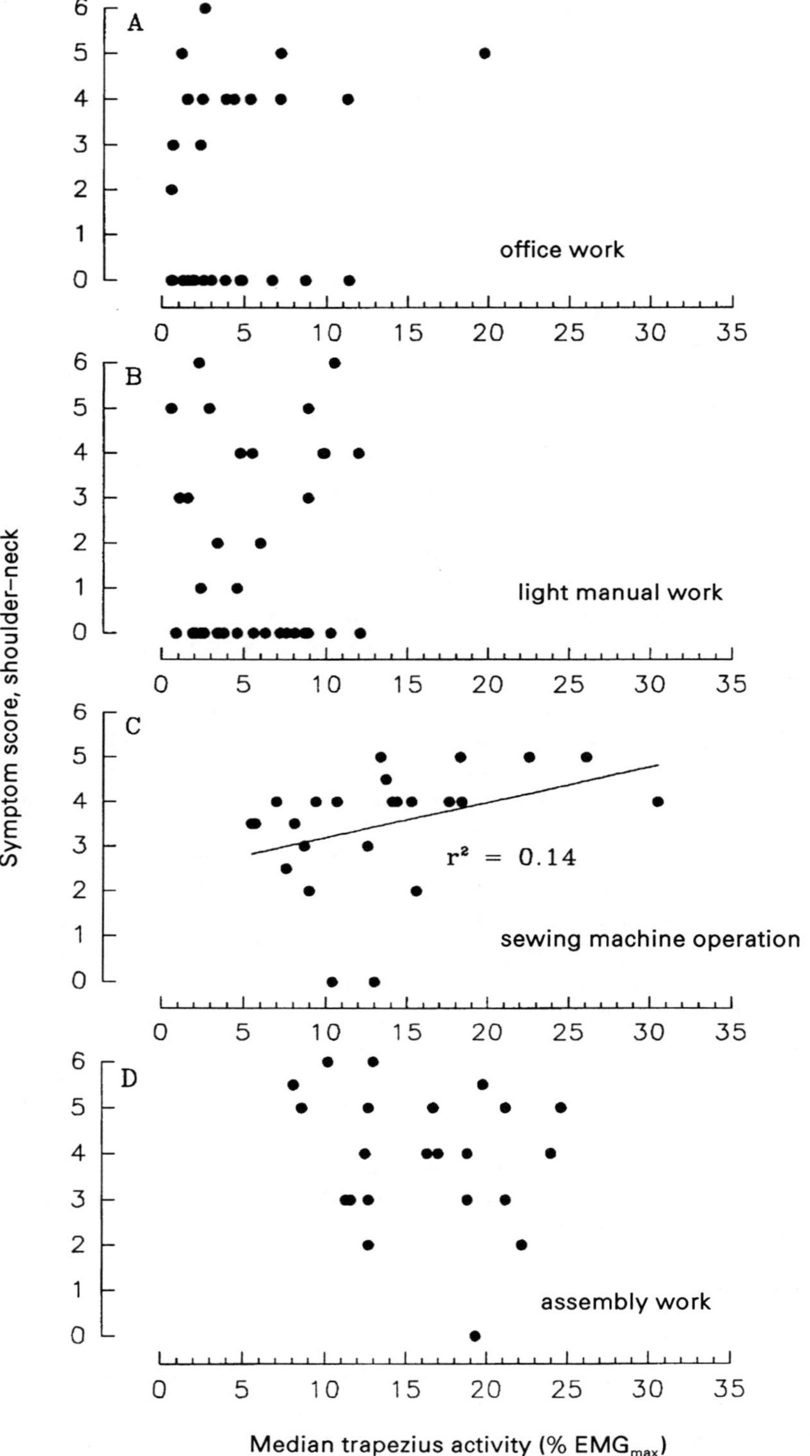

Figure 8.9 The relationship between symptom score in the shoulder and neck region and median trapezius EMG activity for the four groups of workers illustrated in Figures 8.7 and 8.8.

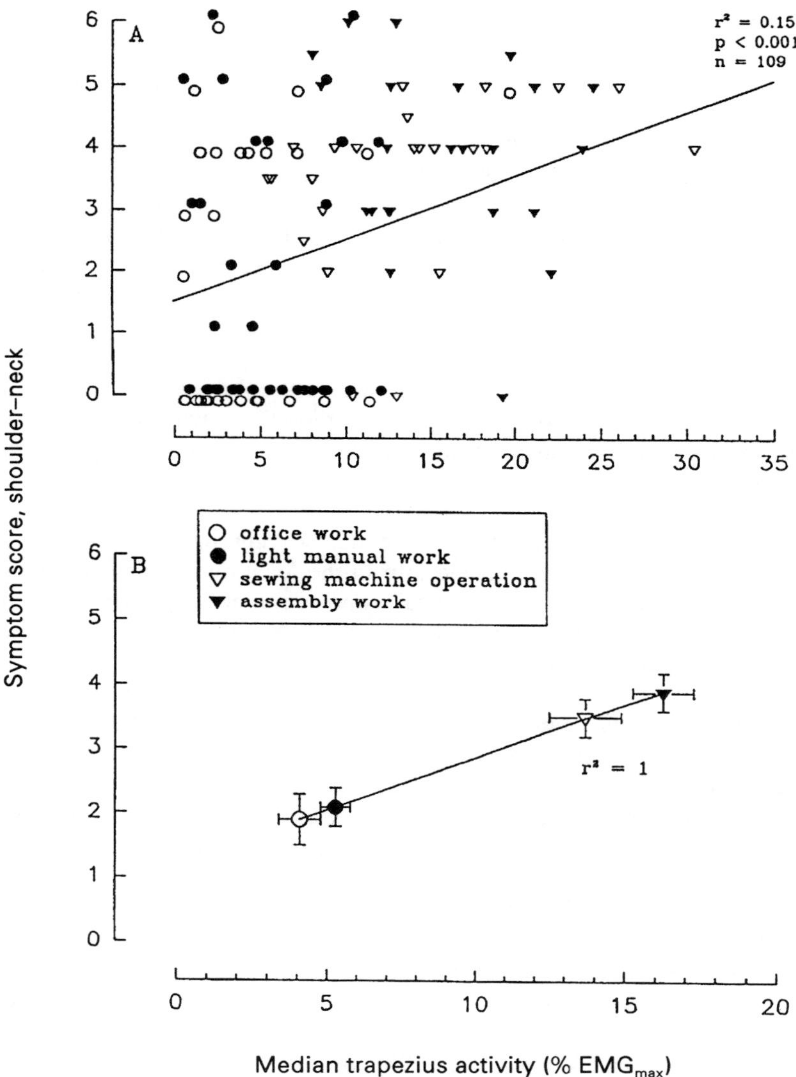

Figure 8.10 (a) Same data as in Figure 8.9, but plotted in one panel. (b) Same as (a), but plotted as group mean values with SEM.

However, all manual work tasks were short-cycled and highly repetitive, indicating a similar level of repetitiveness. Regarding duration, all workers performing office work and chocolate manufacturing were employed full time. In the case of the sewing machine operators and the heavy assembly work, some workers had a 5 h working day. The part-time workers did however show a similar risk of developing musculoskeletal complaints (Wærsted and Westgaard, 1991; Westgaard and Jansen, 1992b). Exposure level may thereby serve as a reasonable indicator of the mechanical exposure quantity in our material.

2 Different historical exposure should be taken into account. In our material this was done only by excluding subjects with a short employment time. No correlation with employment time was observed. In fact, some workers with high muscle activity and low symptom level had been employed for more than 10 years. The difference in historical exposure may influence the results either way as an increased risk of impaired health with increasing exposure time can be counteracted by 'healthy worker' effects.

3 There may be problems in estimating the daily exposure level on basis of a 30 min to 2 h EMG recording, due to different combinations of work tasks over the working day or week. In all studies there was some variation in work tasks and physical work-load. However, the intraindividual variation in muscle activity level between work tasks was much less than the interindividual variation observed.

4 There is uncertainty associated with the scoring of pain symptoms, in addition to the mentioned uncertainty in the location of the symptoms and thereby the validity of the critical exposure recording. The validity of surface EMG recordings performed at one electrode position for an estimate of whole upper trapezius activity is also a concern.

While these problems cannot be discounted, we feel there is little doubt that the data in Figure 8.10 indicate a general trend of increasing pain symptoms with increasing muscle activity at work.

In recent studies we have tried to eliminate some of the uncertainty regarding the assessment of the physical exposure. In the longitudinal study with repeated measurement of trapezius activity every 10 weeks (study E), care was taken to make the measurements representative of the overall internal exposure of the trapezius in their job situation, and the same work tasks were assessed on repeated measurements (Veiersted *et al.*, 1993). The location of pain symptoms to include the upper trapezius was ensured by frequent pain reports and clinical examination of the shoulder and neck region (Veiersted and Westgaard, 1993). In this study there was a difference in the level of muscle activity between prospective patients and workers remaining healthy on first measurement, performed at the time of job recruitment. This difference disappeared on subsequent measurements, performed at 10 weeks intervals.

In the case–control study with office and repetitive manual work each case–control pair was matched for gender, placement in the work organization, and present and historical exposure. The cases of manual workers with symptoms in the trapezius region differed significantly from the symptom-free controls with regard to vocational muscle activity, while there was no sign of such differences for the office workers (Vasseljen and Westgaard, 1995).

A tentative conclusion on the basis of all studies is that there is evidence of an underlying mechanical exposure–health effect relationship, but EMG variables quantifying the average activity level in vocational recordings provide only a coarse risk indicator of muscle pain. In particular, such measurements are not very helpful in distinguishing those at risk within the same work group. The difference in EMG variables observed when, e.g. sedentary work is compared with heavy industrial work, is clearly associated with increased pain symptoms. However, differences in mechanical exposure of that magnitude are readily quantified by simple observational methods.

A similar conclusion seems to hold for the EMG gap variables, quantifying periods with very low EMG activity. It has proved difficult to distinguish workers with and without musculoskeletal symptoms in cross-sectional studies on the basis of EMG gap analyses (Jensen *et al.*, 1993a); however, in the longitudinal study a consistent difference was observed with the prospective patients showing fewer EMG gaps. The case–control study gave results consistent with the longitudinal study for the manual group, but there were no differences in the EMG gap variables for the office group. Thus, the EMG gap variables are not a general predictor of muscle pain symptoms, although interesting associations between gap variables and pain symptoms are observed for some of the work groups. The gap analysis may prove relevant only at low vocational loads, as load levels below 0.5% EMG_{max} are seldom when performing heavy industrial tasks.

8.3.2 Work-place Interventions to Reduce the Muscle Activity Level

The large variation in data indicating an exposure–effect relationship can be interpreted to suggest that a moderate reduction in exposure level is not an effective intervention against the development of musculoskeletal complaints. Furthermore, work-place interventions often result in relatively small changes in the muscle activity variables. In the case of the sewing machine operators of study A, neither the mechanical exposure nor the sick leave due to musculoskeletal complaints in the shoulder and neck region changed as a result of a work-place intervention (Westgaard *et al.*, 1986). The intervention in the study of heavy assembly work (study B) had as its main feature a reduction in working height from 30 to 15 cm above elbow height. The consequent reduction in median and static muscle activity is shown in the scatter plots of Figure 8.11. The EMG activity in the new work-places was recorded after several months to more than a year's experience with the modified work-places. For most workers only a moderate reduction in muscle activity was observed (points near the line of identity in Figure 8.11), despite work-place modifications that were perceived as significant by the ergonomists and were favourably received by the workers. The overall effect was a reduction in the median level from 17.1 to 12.3% EMG_{max} and in static level from 10.1 to 6.8% EMG_{max}, a considerably smaller effect than expected from the substantial reduction in working height.

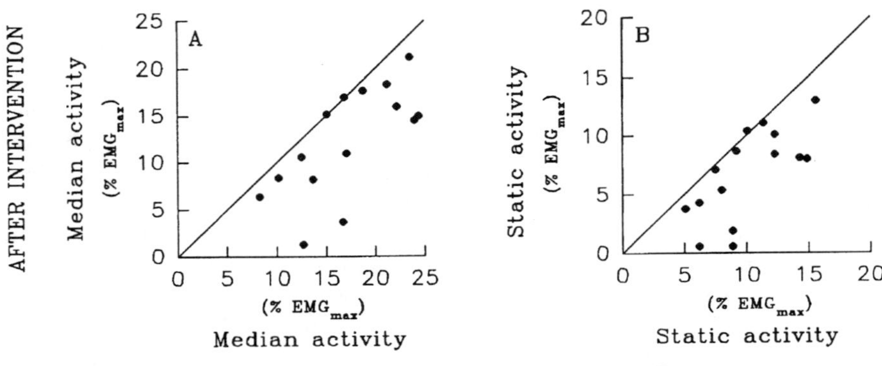

Figure 8.11 (a) Median and (b) static trapezius muscle activity levels before and after an ergonomic work-place intervention in study B, reducing the working height from 30 to 15 cm above elbow height.

In contrast, laboratory studies detect marked differences in short-term subjective assessment of effort and fatigue and in physiological responses with moderate changes in exposure level or pattern (e.g. Sjøgaard *et al.*, 1988). Ergonomics intervention studies with recording of muscle activity and musculoskeletal symptoms pre-intervention and after a follow-up period also indicate beneficial health effects of only a moderate reduction in the level of muscle activity. In an early study the redesign of the work-place clearly reduced the muscle load and a reduction in sick leave both for new and existing workers followed (Westgaard and Aarås, 1985; Aarås and Westgaard, 1987). There was a significant reduction in sick leave per man-labour year for a group of 15 workers exposed both to the old and new work situation. The average reduction in static muscle activity was 1.4% EMG_{max} (95% CI, 0.3–4.1% EMG_{max}) for a smaller group of eight subjects (Aarås and Westgaard, 1987).

Aarås (1993) reported a similar result for female data dialogue workers. The average symptom level in the shoulder region was reduced from 3.4 in the previous 12 months before the work-place intervention to 2.2 in the following 12 months, using a 10 cm VAS scale. The anchor points were 'no pain at all' and 'worst pain imaginable'. The number of days with pain did not change. The static muscle activity in the trapezius was reduced by 1.5% EMG_{max} (95% CI, 0.5–2.6% EMG_{max}) from the old to the new work situation. EMG gaps and gap time, based on a threshold value of 1% EMG_{max}, showed similar changes and were also strongly correlated with static muscle activity level. In both these studies there was little or no change in the repetitiveness and duration of the physical exposure, as work tasks and working hours did not change.

These results suggest that each worker can be quite sensitive to changes in the physical work-load, and benefit from a moderate reduction in exposure level. The results together with the data in Figures 8.7 and 8.10 may also suggest that a critical level of muscle activity ('threshold') exists for each individual, where the risk of developing musculoskeletal complaints increases considerably when it is exceeded. This level would vary between individuals and may vary with time for each individual, depending on health status, on-the-job training and other variables. This is a somewhat speculative interpretation of data, and other risk factors, not indicated by variables quantifying the averaged EMG responses, must also be considered. Examples of such variables are given in the following sections.

8.3.3 Muscle Pain Symptoms not Correlated to Vocational Muscle Activity

Stress symptoms and psychosocial problems at work are increasingly recognized as risk factors for musculoskeletal complaints (Bongers *et al.*, 1993). In the cross-sectional studies the subjects were asked to indicate psychological or psychosocial disturbances, either as a general problem relating to the individual worker (studies A and B) or as a specific assessment of psychosocial working conditions (studies C and D). In studies A and B the data were collected by interviewing the workers regarding such problems and later constructing a score on the basis of the notes. In studies C and D the workers responded to a structured interview regarding psychosocial problems at work. The results were dichotomized so that workers giving indications of psychological or psychosocial problems in general or at work, or indicating a high level of general tension (here considered a stress symptom) were classified as stress positive, while the others were given a negative score.

Figure 8.12 shows that those with a positive stress score recorded a higher symptom level in the shoulder and neck relative to those with a negative score

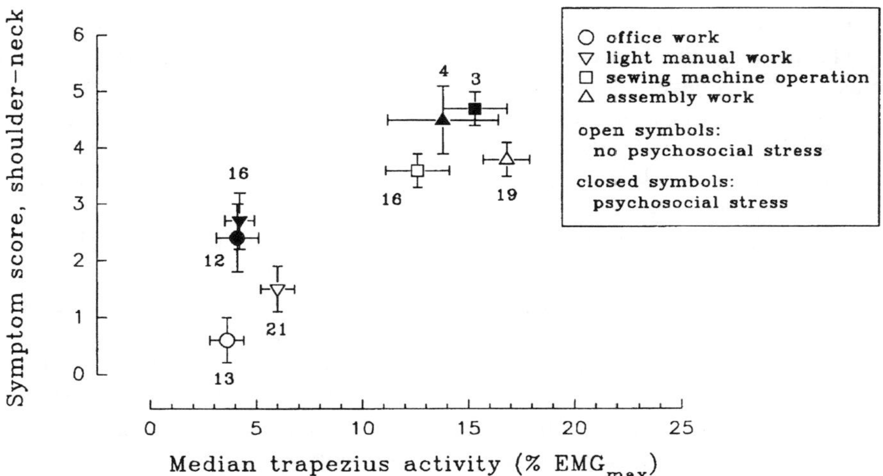

Figure 8.12 Same data as in Figure 8.10, but stratified according to positive or negative score on the stress variable. Numbers indicate the number of subjects in each group.

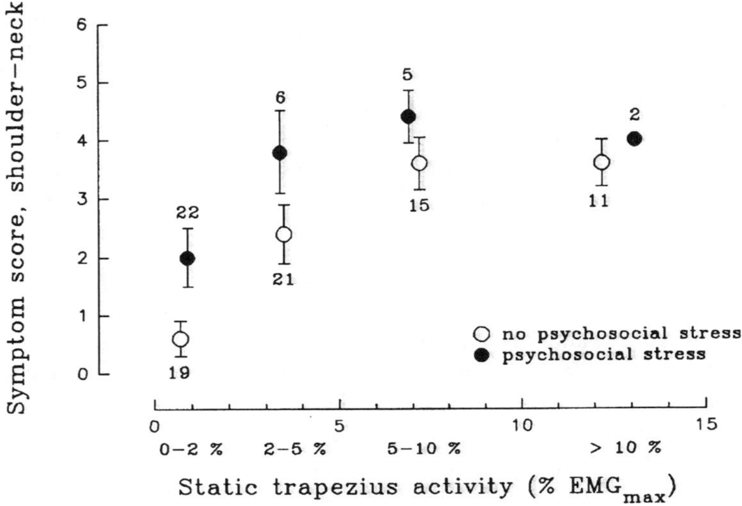

Figure 8.13 Symptom score (mean, SEM) based on the same data as in Figure 8.10, but grouped according to static muscle activity level and stratified on stress score. Numbers indicate the number of subjects in each group.

for an unchanged median muscle activity level in all work groups (see also Jensen *et al.*, 1993a). The 'explained variance' of the regression (based on the Pearson correlation coefficient) improved from 0.15 (Figure 8.10a) to 0.30 when those with a positive stress score were excluded from the material. An improved correlation was also observed for static muscle activity versus symptom score when workers with a positive stress score were excluded.

The data of the cross-sectional studies were grouped according to static muscle activity level across all work groups, stratified on stress score. Figure 8.13 shows the mean symptom score for workers with static muscle activity levels of 0–2, 2–5, 5–10, and >10% EMG_{max}. A non-linear exposure–effect relationship appeared. Workers with low static muscle activity and a negative stress score had very low symptom levels. There was an increase in symptom score with increasing static muscle activity, leveling out at static levels above 5% EMG_{max}. Workers with a positive stress score reported consistently higher symptom scores, and showed no further increase above 2% EMG_{max}. The largest difference between subjects with positive and negative stress scores was observed at low static muscle activity. The data obtained at low static activity levels were mainly based on subjects performing office work and light manual work. For these two groups there was no differential between-group effect in the association between symptom score and static activity level, neither for positive nor negative stress scores (Figure 8.14).

The association between median muscle activity and symptom score was less clear. In particular, the results at low activity levels were less clear when

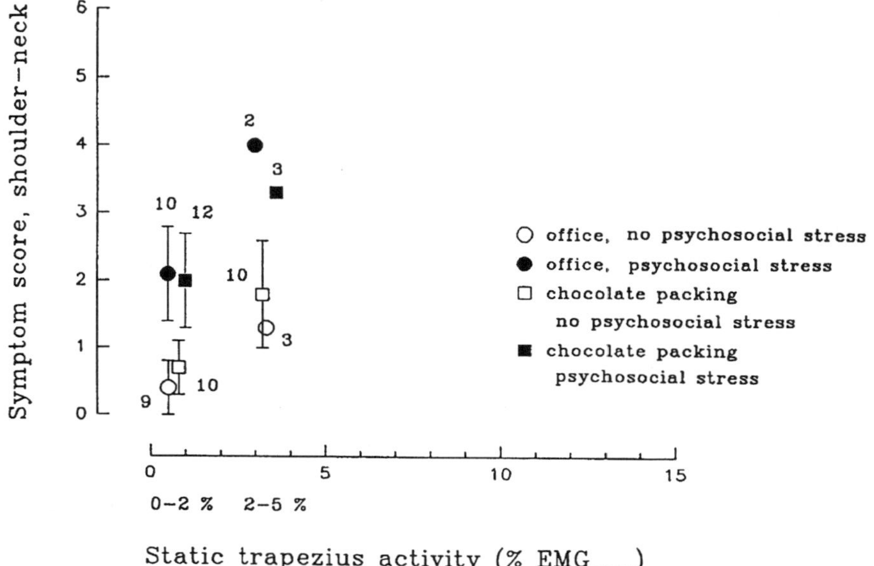

Figure 8.14 Symptom score (mean, SEM) of office and light manual workers (studies C and D), grouped according to static muscle activity level and stratified on stress score. Numbers indicate the number of subjects in each group.

median rather than static muscle activity was used, consistent with the observation that office work and light manual work were distinguished by static, but not by median muscle activity. This can be an indication that the static muscle activity level is a better risk indicator than median muscle activity, possibly because static muscle activity to some extent reflects both the level and repetitiveness of the mechanical exposure.

Other data supplement these results. In studies C and D there was no difference in the EMG gap analysis between those with positive and negative stress scores (Jensen *et al.*, 1993a). In the longitudinal study the subjective indication of stressful working conditions or conflicts with colleagues appeared a significant risk factor for patient status in the survival analysis (Veiersted and Westgaard, 1994), but relatively few workers had a positive stress score and there was no evidence of a correlation with the EMG responses (Veiersted, 1994).

In the case–control study a high level of subjectively reported general tension was best able to differentiate cases and controls both for manual and office workers (Vasseljen *et al.*, 1995). Subjective tension correlated with variables both relating to individual factors (e.g. neuroticism) and to psychosocial stress at work in the case of the office workers who were in a high-stress situation, either awaiting lay-offs or having a high work-load with considerable overtime. Such correlations were generally absent in case of the manual workers who had better psychosocial conditions at work. There was no correlation between vocational EMG variables and subjective tension for cases and controls of the office group. For cases of manual workers there was a negative correlation between EMG variables and subjective tension, which may suggest that a condition of myofacial pain in the shoulder can develop either due to excessive muscle activation or due to stress symptoms occurring independently of muscle activity (Vasseljen and Westgaard, 1996).

Thus, stress or stress symptoms as a risk factor for musculoskeletal complaints do not appear to be mediated through increased vocational muscle activity, as measured by surface EMG. There may still be an association with activity in a subset of low-threshold motor units (see next sections for a discussion of this aspect), but the possibility that muscle pain may develop independently of muscle activity is intriguing. Recently it was reported that the EMG activity is much higher in trigger points of the upper trapezius than in adjacent areas for patients with tension headache and fibromyalgia (Hubbard and Berkoff, 1993). This was thought to represent muscle spindle activity and it was postulated that the EMG activity in trigger points was generated by intrafusal muscle fibers. Sympathetic activation of muscle spindles, detectable as an increase in muscle tension has been demonstrated (Passatore *et al.*, 1985). It may therefore be worth exploring the hypothesis that muscle pain can be triggered by prolonged contraction of intrafusal muscle fibers, activated both through normal servo-type activation by γ motor neurones, e.g. as a response to mechanical demands and through increased sympathetic nerve activity, e.g. as a reaction to stress. However, studies on fibromyalgia patients

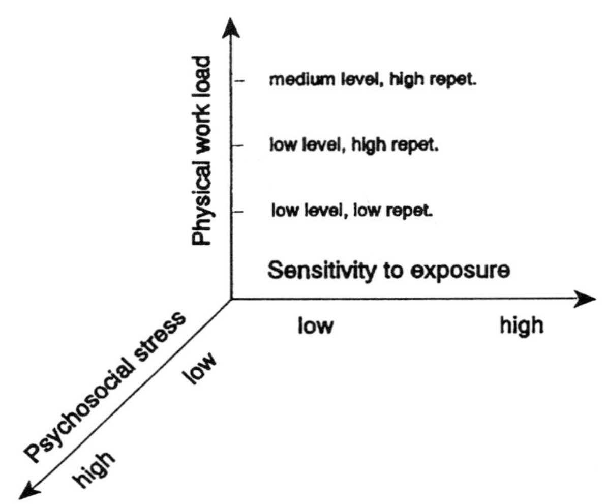

Figure 8.15 A three-dimensional model of risk factors for the development of musculoskeletal pain symptoms. See text for details.

have not shown any increase in sympathetic activity at rest or during a mental stress test (Elam *et al.*, 1992).

Another variable consistently appearing as a risk factor in our studies is previous pain symptoms, presumably indicating individual sensitivity to work-load exposure (Westgaard *et al.*, 1993a). We have no evidence to indicate that this risk factor is associated with muscle activity. These results made us propose the model in Figure 8.15 as a basis for a risk assessment of the development of musculoskeletal complaints in the shoulder and neck region at work. The risk is increasing along all three axes, with physical work-load being the only variable assessed by observation or recording of biomechanical exposure in the work-place. In this assessment level (amplitude), repetitiveness and duration should be independently quantified. Static muscle activity and/or EMG gap parameters may, however, serve as reasonable risk indicators for physical work-load in this scheme. It should be noted that our material is not optimized to provide contrasting data with regard to the temporal pattern of the biomechanical exposure, and the model in Figure 8.15 may be modified regarding risk assessment of the physical work-load when such material becomes available.

8.3.4 Possible Physiological Mechanisms in the Development of Pain Symptoms at Low Muscle Activity Levels

Apparent work-related muscle pain syndromes that may or may not relate to psychosocial problems or stress symptoms, can develop at very low muscle activity levels. The pathophysiological mechanisms underlying this phenomenon have proved elusive, although hypotheses have been suggested (Edwards, 1988). Thus, such pain symptoms are in a similar situation to a number of other pain syndromes where muscle activity is considered a possible etiological factor. These include tension headache, temporomandibular pain, pain in other muscles of the neck (e.g. sternocleidomastoid), and possibly fibromyalgia (Mersky, 1986). Although muscle activity has been suggested as an etiological factor in all these syndromes, it has proved difficult to show such an association. In all these syndromes, the painful condition may develop without exposure to heavy loads.

Several lines of evidence point to prolonged activation of low-threshold muscle fibers contributing to the pathophysiological process for work-related trapezius myalgia: patients with trapezius myalgia show hypertrophy of slow type I fibers, i.e. fibers recruited early in the activation of the muscle (Lindman *et al.*, 1991). There also appear to be many type I fibers with damaged mithocondria ('ragged-red fibers') in regional myalgia and fibromyalgia, indicative of the exhaustion of metabolic processes in some low-threshold fibers (Larsson *et al.*, 1988, 1990). However, such phenomena are observed in workers without muscle pain, both exposed and not exposed to similar muscle activity patterns (Larsson *et al.*, 1992). Thus, the observed associations do not represent a causal relationship between ragged-red fibers and pain, but may suggest that prolonged activation of low-threshold fibers contributes to the pathophysiological process in some muscle pain syndromes. This hypothesis makes obvious sense in relation to the Henneman size principle, stating that there is a relatively fixed order of recruitment of motor units at increasing levels of muscle force (Henneman *et al.*, 1965; see also Chapter 6 by Hägg and Kadefors).

We have been particularily interested in the phenomenon of psychologically mediated muscle activity ('psychogenic tension'; Westgaard and Bjørklund, 1987). The phenomenon was first observed in the 1930s when Jacobsson (1930) observed muscle activity in response to imagined movements. Psychologically mediated stressors such as difficulty and frustration also elicit a muscular response (Davis, 1938). We have observed an increase in muscle activity relating to task complexity and motivational factors, relevant for many work situations (Wærsted *et al.*, 1991, 1994). The trapezius EMG responses in these experiments were at a level similar to that observed in many work situations with light work tasks, i.e. 1–4% EMG_{max}. Recording of single motor units showed that an invariant EMG pattern as observed by surface electrodes is mediated by prolonged activation of a small number of low-threshold motor units (Wærsted *et al.*, 1993, 1996). Thus, if muscle pain syndromes developing at low activity levels are related to prolonged activation of low-threshold motor units, psychogenic muscle activity is an attractive candidate to initiate the pathophysiological processes.

Such activity patterns are difficult to assess on the basis of averaged EMG variables. The EMG gap analysis has a better potential in that respect and was developed with this in mind. Psychogenic tension often results in muscle activity higher than 0.5% EMG_{max} in the surface EMG recording, which would then preclude the recording of EMG gaps. The gap analysis has, however, several shortcomings. The possibility that psychogenic tension is generated intermittently or only in response to special situations must be considered, causing a sampling problem. The level of 0.5% EMG_{max} may also be too high to detect critical incidences of such activity. Alternatively, higher activity levels may be generated in parts of the trapezius muscle not observed by a single bipolar surface electrode. Thus, a negative result of the EMG gap analysis cannot be taken as definite evidence that prolonged low-level muscle activity is not important in the development of muscle pain syndromes and the gap analysis should be supplemented with other methods to detect such activity.

8.3.5 Psychophysiological Tests as Indicators of Risk of Occupational Muscle Pain

In trying to meet some of the above concerns an alternative strategy has been explored: EMG recordings in combination with psychophysiological tests. The rationale for this approach is that an inherent tendency of muscle activation as a response to stress can be observed in tests designed to elicit this response. Alternatively, a tendency to excess muscle activity may show up in situations where there is no biomechanical need to contract the muscles of interest, i.e. a condition of rest. This strategy is not new, but has a long tradition in patient-based studies of syndromes with an assumed muscle activation-related etiology. Studies of patient groups with headache, temporomandibular pain, and low back pain, using this approach, were reviewed by Flor and Turk (1989). Our experiments were originally performed by using a two-choice reaction-time test, designed to promote attention-related muscle activity (Westgaard and Bjørklund, 1987; Wærsted *et al.*, 1991, 1994). We also designed a test to describe muscle coordination patterns during forced arm movement (Westgaard and Bjørklund, 1987).

In study D with office workers, but not in study C with manual workers there was a tendency for those with 'high' muscle activity (>1% EMG_{max}) during the reaction time test to also have a higher symptom score (Westgaard *et al.*, 1993b). This tendency was strengthened if the EMG response of the passive trapezius in the muscle coordination test was also considered. The office workers in study D reporting psychosocial problems showed a higher response in the reaction time test. These findings were mostly not repeated in the subsequent case–control study also based on office workers and manual workers with low physical exposure (studies F and G): the reaction-time test did not distinguish cases and controls in the manual group where other EMG variables were sensitive, nor in the office group. The failure of this test for the manual workers may be understood in terms of the low response levels during the test. Very few workers showed muscle activity level higher than 1% EMG_{max}. However, a majority of those with a high test response in that study also showed high symptom scores.

A problem with such tests is that the context, social background, motivational aspects, etc., influence the response, making it difficult to use the results in cross-sectional comparisons. Students undergoing the same test in a separate laboratory study responded with on average twice the EMG level of the workers in the case–control study. The EMG response may also be sensitive to interactions between the subject and the task. A specific test may not trigger a response with subjects that may however be quite sensitive to other stimuli. Thus, the most obvious application of these tests is experimental designs to study within-subject effects of different stressors.

Alternative measures that give a reliable indication of a tendency to low-level muscle tension at work or in general are therefore needed. We have looked at muscle activity in conditions of rest, including forced rest during the vocational recording (machine stops, Veiersted, 1994) and in a laboratory setting (Vasseljen and Westgaard, 1995). The latter included EMG recordings in an upright standing posture, part of the calibration procedure for angular transducers (Aarås *et al.*, 1988), and during periods of rest in between instructed activities, either seated or standing. In the first instance the subjects were not aware of this secondary purpose of the test ('uninstructed rest'), while they were aware of our interest in resting muscle activity before the test with instructed activities.

These experimental situations presumably still contain an uncontrolled element of stress that can vary from person to person. The same criticism regarding cross-sectional comparisons of the EMG response in these tests may therefore be raised. However, recording muscle activity during work breaks has an ecological validity in that the recording relates to the vocational situation. The laboratory rest is more artificial, but there are no performance requirements. Awareness of the purpose of the test may matter, as the EMG response was lower in the test with a known purpose than in the test where resting posture was adopted without such knowledge.

The problem of ecological validity has been approached by designing tests where EMG activity is measured when the subject is imagining an individually chosen stressful image (Flor *et al.*, 1992). There are problems also in this approach, given that the EMG variables are to serve as risk indicators for vocational muscle pain and the best approach at present is probably to collect empirical data regarding several such tests.

Our limited experience with this approach has shown that prospective patients in the longitudinal study had higher muscle activity than controls during forced pauses in the production due to machine stops (Veiersted, 1994). In the case–control study of manual workers the EMG activity during uninstructed rest was the variable that best distinguished cases and controls, giving a 73% correct classification in a discriminant analysis. There were highly significant correlations between this variable and vocational EMG variables. However, no other EMG variable could improve on the classification, indicating that the differentiation between cases and controls obtained by other EMG variables were contained in the uninstructed rest variable (Vasseljen and Westgaard, 1995).

In contrast, cases and controls in the office group, who were distinguished on the basis of subjective tension and a number of psychosocial variables, could not be differentiated by the uninstructed rest variable, nor were they differentiated by any other EMG variable. This discrepancy in the results between the two studies is further evidence of a lack of association between stress-related muscle pain symptoms and muscle activity.

These data suggest that EMG activity recorded during rest periods, particularly when the subject is not aware of the purpose, can be a powerful differentiating variable between subjects suffering or later developing muscle pain symptoms and those who do not develop such symptoms. In the two studies with a positive result in this respect (studies E and G), an association between muscle activity and muscle pain was also indicated by other EMG variables.

8.4 CONCLUSIONS

The studies summarized in this paper provide evidence to suggest that at least three sets of variables can serve as risk indicators of trapezius pain syndromes.

1 Median or static muscle activity variables as defined by Jonsson (1978) to indicate the general level of muscle activity at work. The relationship between trapezius muscle activity and musculoskeletal complaints is particularly in evidence for those not reporting stress symptoms or stressful working conditions. Static muscle activity is probably the better risk indicator, as this variable is to some extent influenced both by the level and repetitiveness of the physical exposure. This conclusion is also supported by many studies from other laboratories. The lack of sensitivity of these variables in cross-sectional risk assessments within a group of workers with homogeneous work tasks should be noted.

2 Variables that may indicate continuous activity in low-threshold motor units (EMG gaps) are probably more closely associated with musculoskeletal complaints than variables that estimate averaged muscle activity in work situations with a low physical work-load. Furthermore, variables that indicate a general tendency of generating low-level muscle activity may provide more sensitive discrimination between those developing and not developing musculoskeletal complaints than variables quantifying vocational muscle activity.

3 Variables that indicate stressful working conditions or stress symptoms, including subjectively reported general tension, constitute a powerful risk factor for musculoskeletal complaints. No physiological variable has so far been shown to correlate consistently with this risk factor. In particular, muscle activity during work, during psychophysiological tests or at rest have not shown any consistent association with subjectively reported, stress-related variables. Evidence of consistent associations may still come forward, given the appropriate recording circumstances and technical recording conditions, and it is noted that psychogenic tension can be elicited by manipulation of stress-related variables. However, the evidence so far is that this risk factor is not always mediated through increased muscle activity.

References

AARÅS, A. (1993) Relationship between trapezius load and the incidence of musculoskeletal illness in the neck and shoulder. In *Advances in Industrial Ergonomics and Safety V*, (Ed. R. NIELSEN & K. JØRGENSEN), pp. 121–4. Taylor and Francis, London.

AARÅS, A. & WESTGAARD (1987) Further studies of postural load and musculoskeletal injuries of workers at an electro-mechanical assembly plant. *Appl. Ergonom.*, **18**, 211–19.

AARÅS, A., WESTGAARD, R.H. & STRANDEN, E. (1988) Postural angles as an indicator of postural load and muscular injury in occupational work situations, *Ergonomics*, **31**, 915–33.

ÅSTRAND, P.-O. (1988) From exercise physiology to preventive medicine. *Ann. Clin. Res.*, **20**, 10–17.

BONGERS, P.M., DE WINTER, C.R., KOMPIER, M.A.J. & HILDEBRANDT, V.H. (1993) Psychosocial factors at work and musculoskeletal disease. *Scand. J. Work, Environ. Health*, **19**, 297–312.

DAVIS, R.C. (1938) The relation of muscle action potentials to difficulty and frustration. *J. Exp. Psychol.*, **23**, 141–58.

EDWARDS, R.H.T. (1988) Hypotheses of peripheral and central mechanisms underlying occupational muscle pain and injury. *Eur. J. Appl. Physiol.*, **57**, 275–81.

ELAM, M., JOHANSSON, G. & WALIN, B.G. (1992) Do patients with primary fibromyalgia have an altered muscle sympathetic nerve activity? *Pain*, **48**, 371–5.

FLOR, H. & TURK, D.C. (1989) Psychophysiology of chronic pain: do chronic pain patients exhibit symptom-specific psychophysiological responses? *Psychol. Bull.*, **105**, 215–59.

FLOR, H., BIRBAUMER, N., SCHUGENS, M.M. & LUTZENBERGER, W. (1992) Symptom-specific psychophysiological responses in chronic pain patients. *Psychophysiology*, **29**, 452–60.

HENNEMAN, E., SOMJEN, G. & CARPENTER, D.O. (1965) Excitability and inhibitability of motoneurons of different sizes. *J. Neurophysiol.*, **28**, 599–620.

HUBBARD, D.R. & BERKOFF, G.M. (1993) Myofacial trigger points show spontaneous needle EMG activity. *Spine*, **18**, 1803–7.

JACOBSON, E. (1930) Electrical measurements of neuromuscular states during mental activities I. Imagination of movement involving skeletal muscle. *Am. J. Physiol.*, **91**, 567–608.

JENSEN, C., NILSEN, K., HANSEN, K. & WESTGAARD, R.H. (1993a) Trapezius muscle load as a risk indicator for occupational shoulder–neck complaints. *Int. Arch. Occupat. Environ. Health*, **64**, 415–23.

JENSEN, C., VASSELJEN, O. & WESTGAARD, R.H. (1993b) The influence of electrode position on bipolar surface electromyogram recordings of the upper trapezius muscle. *Eur. J. Appl. Physiol.*, **67**, 266–73.

JENSEN, C., VASSELJEN, O. & WESTGAARD, R.H. (1995) Estimating maximal EMG amplitude for the trapezius muscle: on the optimization of experimental procedure and electrode placement for improved reliability and increased signal amplitude. *J. Electromyogr. Kinesiol.*, in press.

JOHNSON, G., BOGDUK, N., NOWITZKE, A. & HOUSE, D. (1994) Anatomy and actions of the trapezius muscle. *Clin. Biomech.*, **9**, 44–50.

JONSSON, B. (1978) Kinesiology – with special reference to electromyographic kinesiology. In *Contemporary Clinical Neurophysiology (EEG Supplement 34)* (Ed. W.A. COBB & H. VAN DUIJN, pp. 417–28. Elsevier, Amsterdam.

JONSSON, B. (1982) Measurement and evaluation of local muscular strain in the shoulder during constrained work. *J. Human Ergol.*, **11**, 73–88.

LARSSON, B., LIBELIUS, R. & OHLSSON, K. (1992) Trapezius muscle changes unrelated to static work load. *Acta Orthoped. Scand.*, **63**, 203–6.

LARSSON, S.-E., BENGTSSON, A., BODEGÅRD, L., HENRIKSSON, K.G. & LARSSON, J. (1988) Muscle changes in work-related chronic myalgia. *Acta Orthoped. Scand.*, **59**, 552–6.

LARSSON, S.-E., BODEGÅRD, L., HENRIKSSON, K. G. & ÖBERG, P. Å. (1990) Chronic trapezius myalgia – morphology and blood flow studied in 17 patients. *Acta Orthoped. Scand.*, **61**, 394–8.

LINDMAN, R., HAGBERG, M., ÅNGQVIST, K.-A., SÖDERLUND, K., HULTMAN, E. & THORNELL, L.-E. (1991) Changes in muscle morphology in chronic trapezius myalgia. *Scand. J. Work, Environ. Health*, **17**, 347–55.

MERSKY, H. (Ed.) (1986) Classification of chronic pain: description of chronic pain syndromes and definition of pain terms. *Pain*, (Suppl. 3), S1–S226.

PASSATORE, M., GRASSI, C. & FILIPPI, G.M. (1985) Sympathetically-induced development of tension in jaw muscles: the possible contraction of intrafusal muscle fibres. *Pflügers Arch.*, **405**, 297–304.

SJØGAARD, G., SAVARD, G. & JOEL, C. (1988) Muscle blood flow during isometric activity and its relation to muscle fatigue. *Eur. J. Appl. Physiol.*, **57**, 327–35.

VASSELJEN, O. & WESTGAARD, R.H. (1995) A case–control study of trapezius muscle activity in office and manual workers with shoulder and neck pain and symptom-free controls. *Int. Arch. Occupat. Environ. Health*, **67**, 11–18.

VASSELJEN, O., WESTGAARD, R.H. & LARSEN, S. (1995) A case–control study of psychological and psychosocial risk factors for shoulder and neck pain at the work place. *Int. Arch. Occupat. Environ. Health*, **66**, 375–82.

VASSELJEN, O. & WESTGAARD, R.H. (1996) Can stress-related shoulder and neck pain develop independently of muscle activity? *Pain* (in press).

VEIERSTED, K.B. (1994) Sustained muscle tension as a risk factor for trapezius myalgia. *Int. J. Indust. Ergonom.*, **14**, 333–9.

VEIERSTED, K.B. & WESTGAARD, R.H. (1993) Development of trapezius myalgia among female workers performing light manual work. *Scand. J. Work, Environ. Health*, **19**, 277–83.

VEIERSTED, K.B. & WESTGAARD, R.H. (1994) Subjectively assessed occupational and individual parameters as risk factors for trapezius myalgia. *Int. J. Indust. Ergonom.*, **13**, 235–45.

VEIERSTED, K.B., WESTGAARD, R.H. & ANDERSEN, P. (1990) Pattern of muscle activity during stereotyped work and its relation to muscle pain. *Int. Arch. Occupat. Environ. Health*, **62**, 31–41.

VEIERSTED, K.B., WESTGAARD, R.H. & ANDERSEN, P. (1993) Electromyographic evaluation of muscular load pattern as a predictor of trapezius myalgia. *Scand. J. Work, Environ. Health*, **19**, 284–90.

WÆRSTED, M., EKEN, T. & WESTGAARD, R.H. (1993) Psychogenic motor unit activity: a possible muscle injury mechanism studied in a healthy subject. *J. Musculoskeletal Pain*, **1**, 185–90.

WÆRSTED, M., EKEN, T. & WESTGAARD, R.H. (1996) Activity of single motor units in attention-demanding tasks: Firing pattern in the human trapezius muscle. *Eur. J. Appl. Physiol.* (in press).

WÆRSTED, M. & WESTGAARD, R.H. (1991) Working hours as a risk factor in the development of musculoskeletal complaints. *Ergonomics*, **34**, 265–76.

WÆRSTED, M., BJØRKLUND, R. & WESTGAARD, R.H. (1991) Shoulder muscle tension induced by two VDU-based tasks of different complexity. *Ergonomics*, **34**, 137–50.

WÆRSTED, M., BJØRKLUND, R.A. & WESTGAARD, R.H. (1994) The effect of motivation on shoulder–muscle tension in attention-demanding tasks. *Ergonomics*, **37**, 363–76.

WESTGAARD, R.H. (1988) Measurement and evaluation of postural load in occupational work situations. *Eur. J. Appl. Physiol.*, **57**, 291–304.

WESTGAARD, R.H. & AARÅS, A. (1984) Postural muscle strain as a causal factor in the development of musculoskeletal illnesses. *Appl. Ergonom.*, **15**, 162–74.

WESTGAARD, R.H. & AARÅS, A. (1985) The effect of improved workplace design on the development of work-related musculoskeletal illnesses. *Appl. Ergonom.*, **16**, 91–7.

WESTGAARD, R.H. & BJØRKLUND, R. (1987) Generation of muscle tension additional to postural muscle load. *Ergonomics*, **30**, 911–23.

WESTGAARD, R.H. & JANSEN, T. (1992a) Individual and work related factors associated with symptoms of musculoskeletal complaints. I A quantitative registration system. *Bri. J. Indust. Med.*, **49**, 147–53.

WESTGAARD, R.H. & JANSEN, T. (1992b) Individual and work related factors associated with symptoms of musculoskeletal complaints. II Different risk factors among sewing machine operators. *Bri. J. Indust. Med.*, **49**, 154–62.

WESTGAARD, R.H., WÆRSTED, M., JANSEN, T. & AARÅS, A. (1986) Muscle load and illness associated with constrained body postures. In *The Ergonomics of Working Postures*, (Ed. N. CORLETT, J. WILSON & I. MANENICA), pp. 5–18. Taylor and Francis, London.

WESTGAARD, R.H., JENSEN, C. & HANSEN, K. (1993a) Individual and work-related risk factors associated with symptoms of musculoskeletal complaints. *Int. Archiv. Occupat. Environ. Health*, **64**, 405–13.

WESTGAARD, R.H., JENSEN, C. & NILSEN, K. (1993b) Muscle coordination and choice-reaction time tests as indicators of occupational muscle load and shoulder–neck complaints, *Eur. J. Appl. Physiol.*, **67**, 106–14.

WINKEL, J. & WESTGAARD, R. (1992) Occupational and individual risk factors for shoulder–neck complaints: part II – The scientific basis (literature review) for the guide. *Int. J. Indust. Ergonom.*, **10**, 85–104.

ZIPP, P. (1982) Recommendations for the standardization of lead positions in surface electromyography. *Eur. J. Appl. Physiol.*, **50**, 41–54.

CHAPTER NINE

Surface electromyographic assessment of low back pain

SERGE H. ROY AND CARLO J. DE LUCA
NeuroMuscular Research Center, Boston University, Boston MA

9.1 INTRODUCTION

Among the many possible uses of surface electromyographic (EMG) signals, the application to assessing paraspinal muscle function has been of growing interest to researchers and professionals in the field of ergonomics. The interest in this application of EMG measurement is undoubtedly related to the unique opportunity it provides in understanding and evaluating the muscular component to lower back pain (LBP) syndromes. Surface EMG techniques provide a window to the neuromuscular system by non-invasive means. Recent technological developments have overcome some of the previous limitations of multichannel EMG signal data acquisition and processing which have prevented this method from achieving widespread clinical or occupational use (Gilmore and De Luca, 1987; Merletti *et al.*, 1990; Roy, 1992). The primary challenge at this time is the development and validation of protocols to characterize normal back muscle functioning and identify specific impairments associated with LBP during various tasks. This chapter will summarize the progress being made in achieving these objectives.

The most common diagnosis reported for LBP is acute or chronic musculoskeletal injury (Bigos and Battie, 1987; Andersson *et al.*, 1989; Teufel and Traue, 1989). Although it is not possible at this time to identify definitively paraspinal muscles as the etiological site of LBP, muscular and other soft-tissue injuries are suspected when no other structural or neural abnormalities can be identified on the basis of radiographs or bone scans (White and Gordon, 1982; Andersson *et al.*, 1989). Regardless of whether the soft-tissue injury is located in the muscle or other components of the spinal complex, such as the intervertebral disc, facet joint, or ligament, normal muscle functioning is likely to be impaired secondary to pain or mechanical disorders (De Luca, 1993b). The importance of these structures to normal back functioning is underscored by the fact that the most common conservative treatment approaches currently recommended for LBP are targeted at reversing musculoskeletal dysfunction through exercise (Darling, 1993; D'Orazio, 1993). Prevention is an important adjunct to this approach and includes muscle sparing techniques, redesign of the work-site or modification of the job task (Isernhagen, 1993). The discipline of ergonomics and occupational safety and health have contributed greatly to these latter important aspects of LBP management (Isernhagen, 1993).

No other single musculoskeletal disorder has had such a profound effect on work productivity, medical costs and compensation claims as LBP (White and Gordon, 1982; Deyo, 1987). The high prevalence of LBP is second only to the common cold as a cause of work absenteeism (Deyo, 1987). Its cost to this society in monetary terms is currently estimated to be anywhere from $30 to $70 billion annually (Deyo and Tsui Wu, 1987; Deyo *et al.*, 1991). While the prevalence of LBP has not changed over the past 20 years, the costs have increased exponentially and show no signs of abatement (Deyo, 1987). The spiraling costs have created a crisis for the employer who has had to assume a greater responsibility in meeting the expenses incurred when an employee suffers a work-related back injury. Modifications to the work-site based on the implementation of ergonomic principles have proven to be cost-effective and have therefore assumed greater importance in decreasing the incidence of LBP (Bigos and Battie, 1987; Isernhagen, 1993). In addition, the chronicity of LBP in some patients has been reduced by new treatment approaches that have moved away from bed-rest and inactivity to recommendations favoring early mobilization and physical function (Mayer *et al.*, 1987). This functionally directed rehabilitation has emerged as the new basis for the therapeutic management of LBP. The approach relies more heavily upon objective musculoskeletal assessment procedures and prevention than pain measurement and is proving to be more effective than previous treatment approaches based primarily on pain management (Mayer *et al.*, 1986, 1987). As a result of these developments, more objective and reliable musculoskeletal evaluation techniques are needed to provide the basis for initiating and monitoring treatment progression. It is hoped that the techniques used to assess musculoskeletal function in the clinical research environment may also be incorporated into the work-site evaluation, particularly as these techniques become easier to use and therefore more adaptable outside of the laboratory.

This chapter provides a historical perspective on the use of surface EMG procedures to identify and characterize disturbances to normal paraspinal muscle function related to LBP. For the most part, it is intended as a general review rather than a critical review. The studies were selected for their relevance to the ergonomist or occupational health professional interested in new procedures to reduce the risks of LBP associated with the demands of the work-place. We begin with an overview of paraspinal muscle impairment and its role as either a primary or secondary component to LBP. EMG studies that describe paraspinal muscle function in patients with LBP will be discussed; first, for amplitude parameters of the EMG signal, and then for EMG spectral parameters. The topics are further organized according to the posture and task being studied and whether static or dynamic tasks were specified. Studies in which EMG spectral parameters were utilized to measure fatigue during sustained or repetitive tasks will also be reviewed. Work currently in progress to develop clinical assessment procedures and instrumentation for quantifying LBP impairment on the basis of these and other surface EMG parameters will be presented separately. The chapter will conclude with suggestions for future research.

9.2 MUSCLE IMPAIRMENT AND LOW BACK PAIN

Although the evidence linking paraspinal muscle impairment to LBP is compelling and of obvious relevance to the field of ergonomics, the precise nature of the linkage between muscle and LBP is not known. Our current understanding of paraspinal muscle impairment can best be understood on the basis of considering it as either a primary or a secondary disorder. This classification is based on whether the impairment is a direct or an indirect result of back injury.

9.2.1 Primary Muscle Disorders

Skeletal muscle disorders may occur as the result of direct muscle injury causing contusions, lacerations, compartment syndromes, or ischemia. Primary muscle disorders associated with LBP, however, are more commonly the result of muscle strain injuries rather than direct trauma. Muscle strain injury can be defined as 'indirect injury to the myotendinous unit caused by too much tension, stretching, or a combination of the two' (Garret *et al.*, 1989b). It is often cited as the most frequent type of LBP injury, particularly among the working population (Holbrook *et al.*, 1984). The likelihood of back muscle strain appears to increase after the age of 25 years, as does the risk of LBP (White and Gordon, 1982; Garret *et al.*, 1989b). Even though many of the clinical characteristics of acute lumbosacral strain are similar to the characteristics of muscle strain in the extremities, muscle is more often considered the locus of pain and disability in the extremities than in the back (O'Donoghue, 1984). This disparity may be the result of the historical emphasis placed on lumbosacral disc herniation as the source for almost all serious LBP conditions.

Of the few experimental studies conducted to investigate the cause of strain injuries, almost all were conducted on muscles located in the extremities rather than the back or trunk (Garret *et al.*, 1989b). Their findings indicated that injuries usually occur as a response to excessive load or stretch and are most common during eccentric contractions in muscles that span two or more joints (Brewer, 1960). Paraspinal muscles, particularly the erector spinae and multifidus muscles, cross two or more vertebral levels and are usually pre-loaded eccentrically during lifting, an activity commonly associated with musculoskeletal LBP injury (Andersson *et al.*, 1976a,b, 1979; Andersson and Schultz, 1979). Biomechanical and epidemiologic data also show that exposure of the lumbar spine to large loads is associated with a higher prevalence of LBP (Frymoyer *et al.*, 1983). Of the many work-place factors that have been considered as causes for LBP (for a review, see Bigos and Battie, 1987) exposure to heavy physical work (Magora, 1970; Chaffin and Park, 1973; Frymoyer, 1989) and vibration (Wilder *et al.*, 1982; Frymoyer *et al.*,) are often mentioned as possible sources of strain injury.

In addition to these well-accepted factors contributing to musculoskeletal injury, other more speculative mechanisms have been proposed. Some have theorized that the anatomical characteristics of motor units may contribute to muscle strain injuries. Muscle regions containing a relatively high concentration of motor units may produce strong localized muscle tension which could result in strains and other soft-tissue damage (Tidball, 1984). Secondly, if each muscle fiber is connected to a common tendon, rather than an individual tendon, the resulting strain would be more severe since it would be proportional to the sum total of the muscles being activated (Andersson *et al.*, 1989). The presence of muscle fatigue, defined as the reduced ability of a muscle to maintain a desired force (Edwards, 1981), may also contribute to the likelihood of muscle strain injury. Paraspinal muscle fatigue can decrease the muscular support to the spine and result in increased mechanical stress to its functional components (Seidel *et al.*, 1987). External loads are transmitted more readily to the soft tissue of the spine when the paraspinal musculature loses its ability to generate tension as a result of fatigue (Nicolaisen and Jorgensen, 1985; Garret *et al.*, 1989b). Ergonomic studies have demonstrated that muscle fatigue can impair motor coordination and control which in turn may lead to muscle strain injury (Bigos and Battie, 1987). A few experimental studies have indicated that appropriate muscle activation may protect muscle and limit strain injury by limiting the transference of loads to soft tissue (Garret *et al.*, 1987, 1989a,b). The possible benefits of pre-conditioning and 'warm-up' in preventing muscle strain injury is inconclusive at this time (Ekstrand and Gillquist, 1983; Wiktorsson-Mollter *et al.*, 1983; Safran *et al.*, 1988); however, it has been shown in an animal model that the force needed to rupture a muscle increases significantly in pre-conditioned muscles (via maximum isometric contraction exercise) when compared to a control group (Safran *et al.*, 1988).

9.2.2 Secondary Muscle Disorders

The onset of pain invariably initiates neuromuscular and behavioral responses that, for the most part, likely represent efforts to prevent or reduce further pain or injury by either 'splinting' the spinal segment(s), as for instance by muscle spasm, or by repositioning the back or altering muscle tension to relieve mechanical impingement on nerves or other sensitive tissue. Sustained or recurring episodes of these musculoskeletal compensations may result in rapid structural and functional adaptations of the muscular tissue. Generalized physical inactivity related to prolonged bed-rest or pain avoidance behavior can precipitate a deconditioning of back muscles. Deconditioning may lead to specific physiological adaptations, such as muscle fiber atrophy and changes to the relative fiber type proportions of a muscle (Kraus and Rabb, 1961; Booth, 1987; Andersson *et al.*, 1989).

Specific hypotheses have been proposed to explain the relationship between secondary sources of paraspinal muscle impairment and LBP (Nouwen and Bush, 1984; Lund *et al.*, 1991; Cassissi *et al.*, 1993). The notion that a pain–spasm–pain cycle underlies at least some LBP disorders is credited to Travell *et al.* (1942). A more recent version of this hypothesis, referred to as the reflex-spasm model, proposed a muscle spasm reflex mechanism to immobilize injured tissue (Collins *et al.*, 1982). The reflex-spasm model predicts that the body responds involuntarily to pain or injury by the production of a muscle spasm which immobilizes or protects the painful area to allow for recovery. Muscle spasm may even aggravate the sensation of pain by restricting circulation and promoting the accumulation of muscle metabolites which are irritants to nerve endings. Armstrong (1984) proposed that the accumulation of these acidic waste products either stimulates the nerve endings directly or through pressure from increased osmotic tension. Ischemic muscle conditions, considered to be an additional LBP factor, may also result from the prolonged tension of muscle spasm (Armstrong, 1984). Since it may be assumed that high metabolite levels are present in a muscle during ischemia and increased tension, pain may be produced following exercise of that muscle. Price *et al.* (1948) proposed that this mechanism is an important factor in explaining the relationship between muscle activity and aggravation of LBP. Muscle deficiency resulting from inactivity or disuse may also contribute to back pain. Kraus and Rabb (1961) proposed that when a muscle is used beyond its limit and is either constricted, unable to yield to fast movements, or unable to overcome resistance, pain will ensue. They further argued that the muscle constriction or 'tightness' not only exposes the muscles to pain and spasm but will produce the typical 'jelling' pain that one feels when arising in the morning or after sitting for a prolonged period of time.

Another hypothesis relates muscle spasm to psychological stress and is referred to as the 'stress-causality' model (Jacobsen, 1944). The model predicts that abnormal tension due to psychological stress may cause the back muscles to go into spasm. As a result, a spasm–pain–spasm cycle occurs as in the reflex-spasm model. There is also evidence that prolonged tension shortens muscles and deprives them of elasticity (Kraus and Rabb, 1961). Once a muscle has reached a sufficient level of tension and has weakened from lack of activity, it may become increasingly susceptible to muscle strain injury.

9.3 EMG STUDIES OF PARASPINAL MUSCLE FUNCTION DURING STATIC POSTURE

Paraspinal EMG activity has been studied extensively in patients with LBP, either at 'rest' or while exerting isometric trunk extension torques during static positions such as standing, sitting, or prone-lying (De Vries, 1968a,b; Jayasinghe *et al.*, 1978; Jorgensen and Nicolaisen, 1987; Garret *et al.*, 1989b; Biedermann, 1990; Biedermann *et al.*, 1991; Cassissi *et al.*, 1993; De Luca 1993b). Muscular models of LBP dysfunction that include spasm or hyperactivity of muscle as a key component, predict that during resting activity or static postures, the EMG amplitude should be higher in the paraspinal muscles of LBP patients compared to pain-free control subjects (Nouwen and Bush, 1984; Ahern *et al.*, 1988; Arena *et al.*, 1989). In a review that evaluated the likelihood of this possibility, Nouwen and Bush (1984) concluded that the evidence for higher EMG activity in chronic LBP was minimal, particularly when only well-matched control studies were considered. For those studies in which patients and controls were matched according to sex, age, and other factors (Collins *et al.*, 1982; Ahern *et al.*, 1988), no significant differences between groups were found. In another review of the EMG literature, higher resting levels of integrated EMG activity were reported in lumbar paraspinal muscles of patients with chronic LBP compared to control subjects for tests conducted during prone-lying at rest (Grabel, 1973; Miller, 1985) and during standing for a 10 min period (Hoyt *et al.*, 1981; Miller, 1985). In this latter study, it was also reported that no significant increases in integrated EMG were observed for subjects tested during a semi-Fowler's position (a reclined sitting position), or in upright sitting. Other studies conducted from a sitting posture indicated that patients with chronic LBP had consistently lower integrated EMG activity than normal controls (Cassissi *et al.*, 1993). The lower EMG activity when combined with lower peak torque capability has been referred to as a 'muscle deficiency model' of chronic LBP (Cassissi *et al.*, 1993). In another study in which various activities were compared while subjects were asked to produce a minimal contraction of their lower back muscles (Kravitz *et al.*, 1981; Miller, 1985), only one activity out of the 11 activities tested resulted in a significantly different EMG activity between LBP patients and controls.

It should be recognized, however, that many of these studies have been criticized for their failure to normalize the integrated EMG data, their use of questionably high band-pass filtering techniques to attenuate signal noise artefact, their failure to report variability of the data between subjects, and the possibility that the increases in EMG activity were fatigue related (Miller, 1985). Comparing the EMG signal to a standardized or reference contraction is considered essential to the validity of intersubject comparisons (De Luca, 1979; Perry and Bekey, 1981; Basmajian and De Luca, 1985; Miller, 1985). In a well-controlled and normalized EMG study comparing three experimental tasks (quiet sitting, standing, and sitting during a repetitive unilateral upper extremity task), no significant differences in integrated EMG activity were observed between chronic LBP patients and non-pain controls (Miller, 1985). In one of the few publications in which the evidence for a pain–spasm–pain cycle was reviewed (Roland, 1986), methodological limitations were also identified. These limitations included the observation that patient populations were poorly described, the role of muscle spasm in the LBP patients remained obscure, and technical difficulties were often present regarding the positioning of the subjects and the selective recording of EMG signals from the muscles of interest. In spite of these limitations, however, the authors concluded that there was clear experimental evidence consistent with a pain–spasm–pain cycle.

Although muscle spasm is most commonly diagnosed in patients with acute LBP, most of the studies in the literature on muscle spasm involve chronic LBP patients. In general, these studies are equally divided between those that report increased EMG activity associated with muscle spasm (Kravitz *et al.*, 1981; Soderberg and Barr, 1983; Sherman, 1985) and those that do not (Collins *et al.*, 1982; Nouwen, and Bush, 1984; Cohen *et al.*, 1986; Miller *et al.*, 1987). Localized muscle abnormalities associated with long-loop inhibitory reflexes, have been offered as a possible mechanism for muscle spasm (De Andrade *et al.*, 1965; Garret *et al.*, 1989b; Hides *et al.*, 1994). Paraspinal muscles as well as some of the connective tissues in the region of the spine contain sensory nerve endings sensitive to changes in length, tension, position and movement which may be initiated by the presence of pain or other factors (Garret *et al.*, 1989b). In addition, some sensory endings respond to change in position by inhibiting muscle activity while others, possibly related to muscle spasm, increase muscle activity. In some studies, increased myoelectric activity has been observed at muscle sites corresponding to palpable abnormalities (Denslow and Clough, 1941; Elliot, 1944; Arroyo, 1966; England and Delbert, 1972; Fisher and Chang, 1985), whereas in other studies (Kraft *et al.*, 1968) no abnormal localized EMG findings were observed. One recent investigation has provided evidence, based on ultrasound scanning, that localized effects of muscle spasm are measurable (Hides *et al.*, 1994). Muscle wasting and 'rounded' muscles (measured by a shape ratio index computed from the muscle cross-sectional area) were present at symptomatic sites.

It is safe to say on the basis of reviewing the research literature that muscle spasm has still eluded proper characterization and further basic science investigations are needed to help clarify its presence and role in LBP. Research is particularly lacking in determining the effects of therapeutic modalities, such as exercise, heat, and spinal mobilization techniques, or the effects of medications, such as muscle relaxants or analgesics, on muscle spasm associated with LBP. The influence of ergonomic approaches to redesigning a work-site or modifying a job task has not been evaluated in terms of reducing muscle spasm or modifying abnormal muscle functioning. The current lack of research in these areas is truly remarkable considering that numerous treatment regimens currently in vogue are based on relieving symptoms associated with muscle spasm. In a recent workshop among a diverse group of LBP experts, the need to address these topics was recognized. In the report emanating from this workshop, the participants suggested that the first research step should be towards defining and documenting muscle spasm (Garrett *et al.*, 1989a). Of note, is the observation that surface EMG techniques were specifically mentioned as the most likely resources for accomplishing this important objective.

9.4 EMG STUDIES OF PARASPINAL MUSCLE FUNCTION DURING DYNAMIC TASKS

Although there is little agreement among researchers identifying which dynamic tasks are most important when evaluating back muscle function, most of the surface EMG studies described in the literature include a trunk flexion task, usually initiated from a standing posture. The interest in this task probably originated with the work of Floyd and Silver (1951, 1955) who studied the phenomenon of 'flexion–relaxation'. Flexion–relaxation refers to the observation that during forward flexion of the trunk, when paraspinal muscle activity is typically increasing in magnitude (as measured by the EMG signal), there is an eventual 'silent period', or period of significantly reduced paraspinal muscle activity, during the latter part of trunk flexion. It was hypothesized that this silent period results from reflex inhibition. However, most others have explained this phenomenon as being due to the passive resistance of either stretched muscles, ligaments, fascia, and/or facet joints that relieve the muscle of the need to contract actively (Farfan and Lamy, 1975; Gracovetsky *et al.*, 1977; Kippers and Parker, 1984; Schultz *et al.*, 1985). The interest in this phenomenon is related primarily to its implications for muscle functioning during lifting and other work activities that require a flexed posture.

The question of whether this phenomenon is absent in patients with LBP has prompted a number of surface EMG investigations among patients with LBP. Among the earliest of these investigations, Golding (1952) studied forward flexion in 120 patients and reported that 86 achieved the expected flexion–relaxation. Later studies by Floyd and Silver (1955) and Yashimoto *et al.*, (1978) confirmed these findings in a similar population of LBP patients. Contradictory findings have also been reported concerning the presence of a relaxation response (Chapman and Troup, 1970; Troup and Chapman, 1972). Wolf and Basmajian (1977) in their study of nine patients with chronic LBP, reported that muscle activity was lower than or the same as that in a healthy control population studied under static and dynamic conditions, including trunk flexion (Wolf and Basmajian, 1977; Wolf *et al.*, 1979). The results of this study have been questioned because no tests for significant differences were reported and some of the patients included in the study had a history of back surgery which could have altered the EMG signal (Miller, 1985). Nouwen *et al.* (1987) reported that all of the 20 patients they studied with LBP had significantly higher paraspinal muscle activity near full flexion than control subjects. Their subjects were tested during trunk flexion, extension, lateral bending, and rotation. They also reported no evidence of muscle imbalances on the basis of comparisons between left and right paraspinal EMG amplitudes. Others have explored the possibility of the flexion–relaxation phenomena as a useful clinical measure for LBP (Triano and Schultz, 1987). Results of the flexion–relaxation test from LBP patients and controls were compared to a disability rating scale. A positive relationship between the degree of disability and the loss of the flexion–relaxation phenomena was reported. All of the control subjects tested in this study exhibited the flexion–relaxation phenomena, however, this phenomena was absent in half of the patients tested. According to the results of this study, the test had excellent specificity but poor sensitivity.

Surface EMG studies that assess functional tasks have included a few evaluations of manual lifting, however they have for the most part been limited to normal healthy subjects rather than patients with LBP. Studies in normal healthy subjects have generally investigated the relationship between either lifting moments, posture, or load and EMG activity of paraspinal muscles (Seroussi and Pope, 1987; Scholtz, 1992). A number of other studies have evaluated different lifting strategies and its effect on muscle activity or coordination (Grieve, 1974; Andersson *et al.*, 1976a,b; Freivalds *et al.*, 1984; Scholtz, 1992) while others were undertaken to help validate biomechanical models to predict muscle load sharing (Bejjani *et al.*, 1984; Freivalds *et al.*, 1984; Jäger and Luttman, 1989). Many of these studies were limited to static or quasi-static conditions in which lifting was limited to a single plane of movement or to isometric conditions. This line of research is in its very early stages of development since real-life lifting tasks rarely involve just one axis of motion and are dynamic rather than static.

9.5 THE USE OF EMG SPECTRAL PARAMETERS TO DESCRIBE PARASPINAL MUSCLE FUNCTION

9.5.1 Muscle Fatigue and LBP

Studies in which surface-detected EMG signals from paraspinal muscles are analyzed to extract spectral parameters, such as the median or mean frequency, have contributed to our understanding of the role of muscle fatigue in LBP. Preliminary research findings indicate that muscle performance, as measured by spectral estimates of the EMG signal, may provide a more objective measure than purely mechanical indices (Biedermann *et al.*, 1991; Klein *et al.*, 1991; De Luca, 1993a,b; Roy *et al.*, 1995). The EMG spectral technique measures the shift in the EMG power spectrum associated with the biochemical events that occur during a sustained contraction. Spectral parameters of the

EMG signal are influenced by metabolic fatigue processes not cognitively perceived or voluntarily regulated by the subject when performing a sustained contraction, particularly when numerous muscle groups are being monitored (Basmajian and De Luca, 1985; De Luca, 1985). This aspect of the technique appears to provide the user with a more objective measure of muscle performance capability than techniques that rely on mechanical indices that can be volitionally regulated, such as torque or force measurements from back dynamometers. The second possible advantage of the technique is that it can enable the user to obtain localized measurements specific to a particular segment of the superficial paraspinal musculature. In this way, information describing the interaction between muscles associated with a work task can be measured, rather than modeling the back as a single extensor, as is the case for back dynamometer measurements. A third advantage is that the median frequency parameter is more reliable than the amplitude of the EMG signal (De Luca, 1985).

The current approach is based on the concept that by simultaneously monitoring the median frequency from multiple electrode sites, it is possible to evaluate the relative contributions of individual paraspinal muscle groups during a sustained extension of the trunk (De Luca, 1985, 1993b). This concept has been reviewed in detail in a recent position paper (De Luca, 1993b). Based upon the accepted notion that muscle dysfunction may follow injury, pain, or disuse, it is reasonable to expect that some muscles would compensate for these deficits, resulting in a relative alteration in their EMG activity during induced localized muscle fatigue. This concept is depicted in Figure 9.1.

The earliest applications of EMG spectral techniques to back muscles were limited by the use of only a few EMG electrodes, the failure to isolate the trunk extensor muscles properly, and the reliance upon cumbersome methods of spectral analyses (Morioka, 1964; De Vries, 1968a,b; Jayasinghe *et al.*, 1978). Many of these initial limitations have been resolved by technical and methodological improvements (Roy *et al.*, 1989; Biedermann *et al.*, 1990, 1991; Klein *et al.*, 1991; Roy, 1992; De Luca, 1993b). The earliest studies of back muscle fatigue investigated the behavior of EMG spectral parameters during static, constant-force contractions of paraspinal muscles. Initial studies by Morioka (1964), Okada (1970), and Okada *et al.* (1970) reported an increase in the low frequency components of the EMG power spectrum while subjects performed static lifts of incremental loads. In one of these studies, subjects were asked to sustain the contractions until the point of muscle pain, at which time it was observed that the EMG signal characteristics changed. This was the first demonstration that back pain and fatigue resulted in a consistent change in muscle activation.

Andersson *et al.* (1979) contributed a major advancement to the application of EMG signal techniques in assessing paraspinal muscle function by (1) monitoring more paraspinal muscle sites (bilaterally in the thoracic and lumbar region), (2) carefully restraining the subject's posture during EMG signal

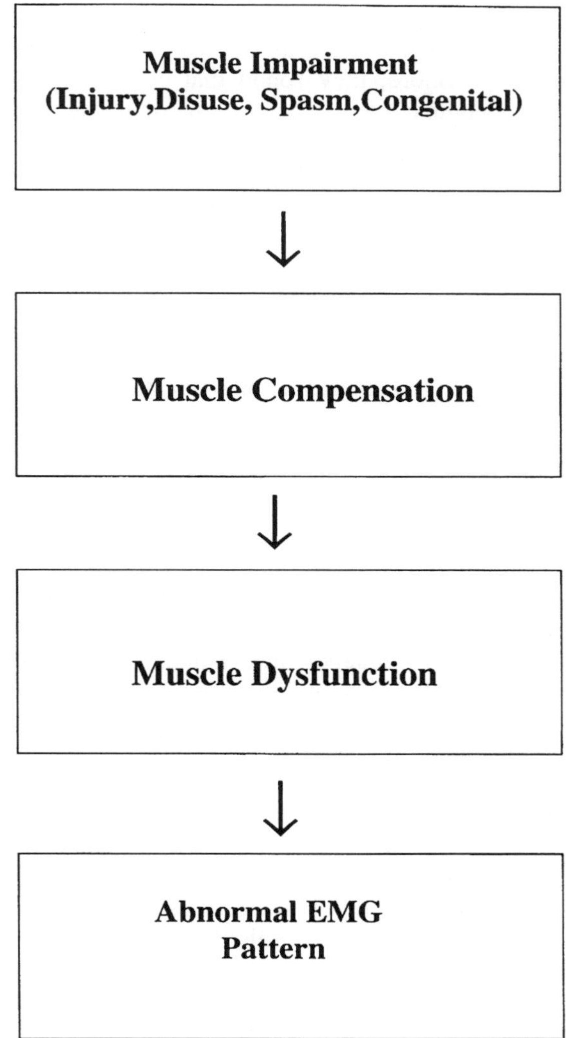

Figure 9.1 A diagram representing a proposed model of back muscle impairment that accounts for the changes to the surface EMG signals observed from lumbar paraspinal muscles. Muscle impairment from various types of LBP disorders result in the back muscles compensating for this impairment during a sustained contraction. Muscle compensation disturbs the normal functioning of back muscles which is measured as a change in EMG signal variables (either in amplitude or median frequency) detected from the lower back.

detection, and (3) measuring changes in EMG amplitude and spectral parameters concurrently during sustained contractions at different degrees of trunk flexion (from 10 to 50°). They observed significant EMG spectral shifts towards lower frequencies that were associated with increases in the EMG signal amplitude. Furthermore, they found that an increased level of EMG activity was always accompanied by a greater rate of decay of the EMG power spectrum.

Among the first to apply surface EMG techniques to compare back muscle fatigue in patients with LBP and controls, DeVries (1968a,b) showed that subjects who developed pain during a sustained trunk extension had a corresponding increase in EMG signal amplitude from the paraspinal muscles. In those subjects without pain, the EMG signal amplitude decreased. It was concluded by the authors that these differences in the characteristics of the EMG signal were indicative of the weakness and fatigue associated with LBP. Jayasinghe *et al.* (1978) conducted a similar study and confirmed these findings. Others have reported that patients with LBP develop more fatigue in their back muscles than controls, according to mechanical measures of endurance capacity (Jorgensen and Nicolaisen, 1987) and according to surface EMG spectral measurement (Roy *et al.*, 1989; Biedermann *et al.*, 1991; Klein *et al.*, 1991; De Luca, 1993b). Our group studied whether differences in fatigability between chronic LBP and control subjects were influenced by the force level of a sustained contraction and the muscle recording site (Roy *et al.*, 1989). Twelve patients with chronic LBP were compared to an equal number of control subjects. Median frequency measurements from six bilateral lumbar paraspinal muscles (the longissimus thoraces, iliocostalis lumborum and multifidus muscles at the L1, L2 and L5 interspinous lumbar levels) were analyzed during isometric trunk extension sustained at 40, 60, and 80% of MVC for a maximum of 1 min duration each. The median frequency slope (MF slope), a measure of the rate of change of the median frequency and an index of muscle fatigue (De Luca, 1985), was calculated for each EMG recording site using a linear regression interpolation procedure. The results (Figure 9.2) demonstrated that the MF slope was significantly higher (i.e. more negative) for patients compared to controls, but only for recordings from the multifidus and iliocostalis muscles corresponding to the 80% MVC contractions.

These findings demonstrated that median frequency measurements from back muscles are muscle specific and load dependent. Furthermore, they provided evidence that, to characterize paraspinal muscle function during sustained isometric tasks properly, several muscle sites are necessary. This recommendation may explain the poor reliability and conflicting data that have characterized previous attempts at studying back muscle function using only one or two electrode sites. The interpretation of these results from a physiological or biomechanical perspective was, however, less definitive. The higher fatigue rates in the LBP group may be explained by (1) either a greater proportion of type II muscle fibers (Jorgensen and Nicolaisen, 1987; Roy

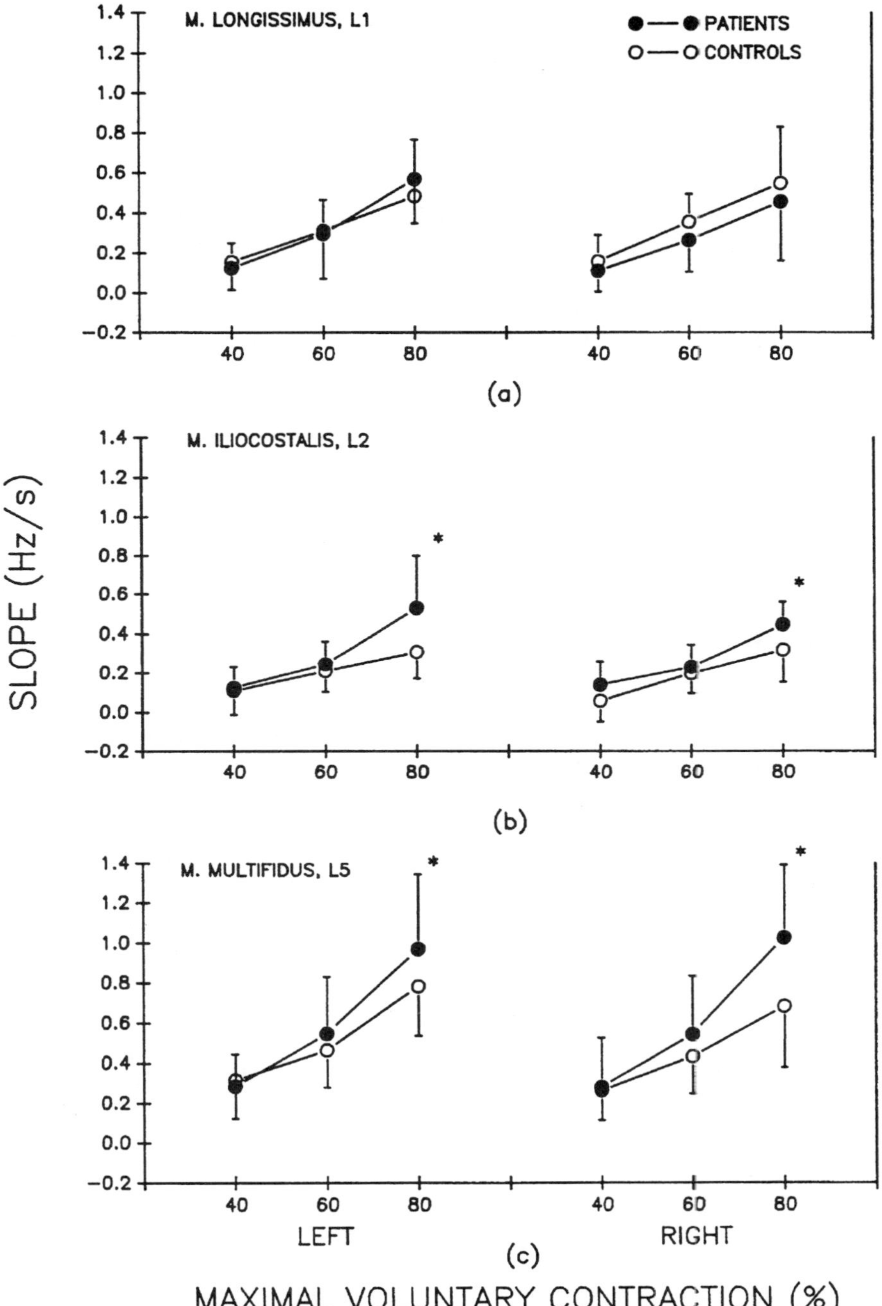

Figure 9.2 Mean values of the median frequency slope (Hz s^{-1}) for 12 patients with LBP (•) and 12 control subjects (o) tested at 40, 60, and 80% of MVC sustained for 60 s. Data is plotted separately for (a) the longissimus thoracis, (b) the iliocostalis lumborum, and (c) the multifidus muscles. In each figure, the results from the left and right side of the back are presented separately. $^{*}p < 0.05$ (with permission from Roy *et al*., 1989).

et al., 1989, 1995; Biedermann *et al.*, 1991; Thompson *et al.*, 1992; Thompson and Biedermann, 1993), (2) a greater pre-contraction metabolite level resulting from persistent spasm (Armstrong, 1984; Roy *et al.*, 1989; Biedermann *et al.*, 1991), or (3) a redistribution of the loads among the various paraspinal muscles causing some muscles to fatigue more readily than normal (Roy *et al.*, 1989). The increase in the negative slope of the median frequency with increased force levels from 40 to 80% MVC is consistent with numerous similar studies from limb muscles which have argued that this effect results from an increased recruitment of type II fibers that are more glycolytic, and therefore contribute more metabolites to the muscle membrane environment (Basmajian and De Luca, 1985; De Luca, 1985).

Different strategies for utilizing EMG spectral measurements to evaluate paraspinal muscles were evaluated by Kondraske *et al.* (1987). Although their study did not include subjects with LBP, the issues raised are relevant to this application. They carried out a two-phase study in which phase I identified the most appropriate EMG measures of fatigue and phase II tested different normalization procedures for accommodating the effects of the absolute force level of the contraction on the median frequency. A test frame provided pelvic and lower limb stabilization and a visual feedback display was used to help produce isometric extension torques during standing. A number of relevant findings were reported and are summarized: (1) mean and median frequency results were comparable, (2) rate of spectral shift, although exponential, could be accurately measured by the slope of a linear regression line for a portion of the data, (3) maximum stabilization of the subjects posture and provision of force-feedback was highly recommended, and (4) target force levels based on percent body weight rather than % MVC was a more suitable method for normalizing muscle loads. Their last finding in this list raises an important question of whether it is appropriate to normalize contractions with respect to MVC when assessing paraspinal muscles. This question arises because the presence of LBP or fear of reinjury to the back might negatively influence the voluntary effort needed to produce a 'maximal' contraction. Biedermann (1990) has directly addressed this issue by implementing a technique in which subjects generate a constant load to the lower back by holding a specific absolute weight at arm's length for a prescribed duration. A reference frame was used to standardize the position of the subject's feet, pelvis, and spine. The reliability of the technique was evaluated in 31 subjects by having them perform the weight-holding test for 45 s and then repeating the test 5 days later (Thompson and Biedermann, 1993). The reliability of the EMG parameters from four bilateral lumbar sites was within an acceptable range ($r = 0.85$) according to calculations of a Pearson's correlation coefficient for the median frequency of the EMG signal. Similar reliability estimates have been reported for EMG parameters derived from contractions specified as a percentage of the MVC (Roy *et al.*, 1989). A further discussion of this issue is included later in this chapter where the influence of the MVC on classification techniques for muscle impairment is described.

9.6 APPLICATION OF EMG SPECTRAL TECHNIQUES TO LBP ASSESSMENT

9.6.1 Classification of Muscle Impairments

It has been suggested by several research groups that the analysis of paraspinal muscles using surface EMG spectral techniques may provide the clinician or occupational health professional with the following two requirements for a LBP evaluation procedure: (1) identification and classification of paraspinal muscle impairment and (2) monitoring changes of muscle impairment associated with treatment progression. With respect to the first requirement, it is often important to know more than if the back extensor muscles are functioning normally. Additional information is often needed to identify a specific type of impairment as well as the probability that the classification is correct. These additional features are important in maximizing the usefulness of the technique to planning treatment or modifying the work environment. Sometimes researchers overlook the requirement that a musculoskeletal assessment procedure should be of practical value for the intended user rather than just providing an abundance of objective data. One criterion of practicality is that the assessment results are interpretable to the user and are relevant to their particular area of interest. For example, a physical therapist should be able to use the EMG results to assist in the formulation of a back exercise program or to modify the physical demands of a particular work-site or job specification. Validated guidelines are still needed to bridge this gap between the data and how it should be interpreted to accomplish a particular objective. Techniques that produce volumes of objective data without suggesting a treatment plan are not ready for general use and should be considered only as investigative tools.

Although still in the early stages of development, there is evidence that surface EMG analysis techniques are effective in discriminating between LBP patients and control subjects. This indication is based in part on a series of four studies conducted using a device and technique referred to as a Back Analysis System (BAS®) (Roy *et al.*, 1989; Klein *et al.*, 1991; Roy, 1992; De Luca, 1993b), as well as from independent studies that have reached similar findings (Biedermann *et al.*, 1991; Thompson *et al.*, 1992). Details of the BAS system have been described previously (De Luca, 1993b); however the basic functional elements are depicted in Figure 9.3 and consist of (1) custom surface EMG electrodes having a specific architecture and electrical properties designed for the detection of EMG signals suitable for spectral analysis (De Luca *et al.*, 1979), (2) a muscle fatigue monitor which processes the EMG signals to obtain the EMG signal variables (Gilmore and De Luca, 1985, 1987), (3) a postural restraint apparatus which constrains the posture of the subject so that the measured force is related, as much as possible, to the force generated by the muscles being monitored, and (4) a software package to provide system calibration, monitor signal-quality, specify test configurations, organize the database, and analyze and present the results. The electrodes are

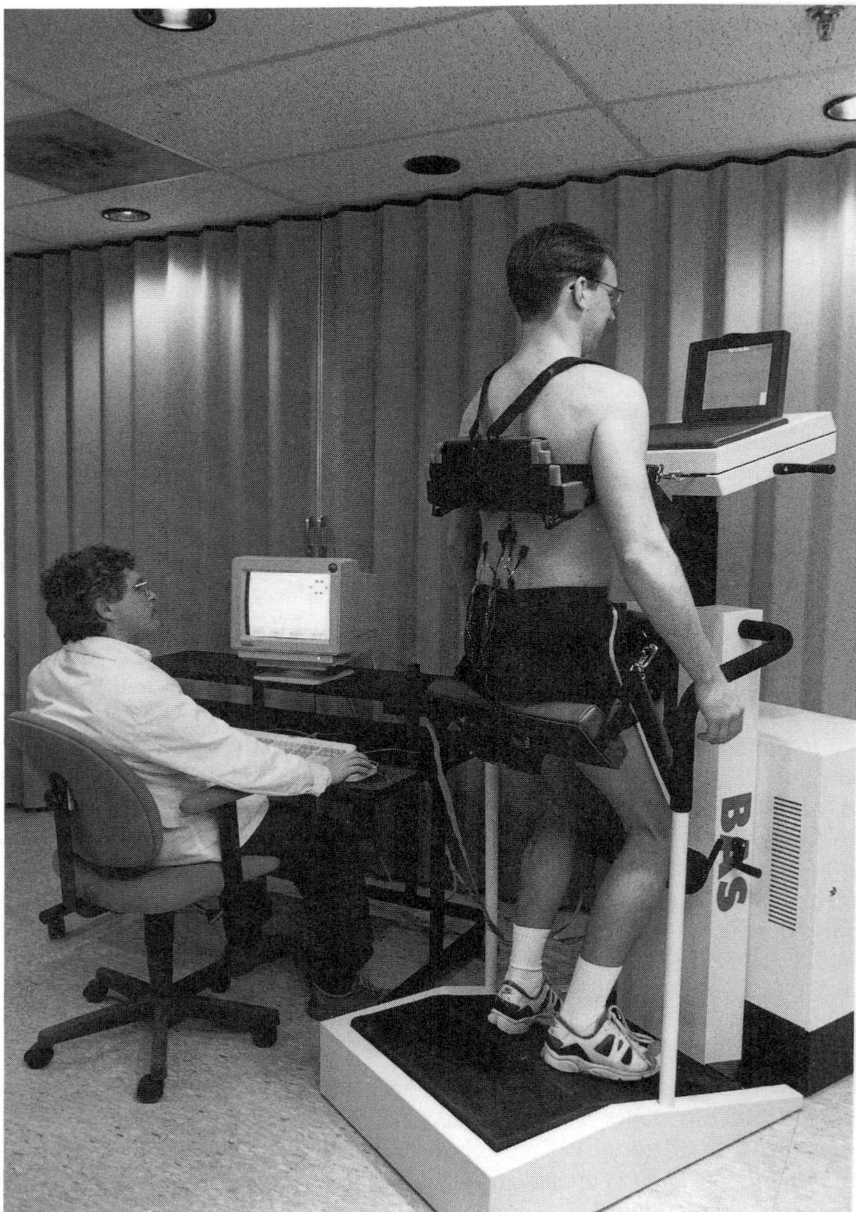

Figure 9.3 A photograph of a prototype Back Analysis System (BAS) used to acquire and process surface EMG signals and isometric trunk extension forces during various protocols designed to fatigue the lumbar paraspinal muscles. The subject is shown performing isometric trunk extension in an erect posture. The screen in front of the subject provides a visual display for feedback of the target force and the force they are producing during the task.

placed at six lumbar sites corresponding to specific superficial paraspinal muscle groups and the torso is immobilized to limit the extension task to an isometric contraction and to provide a means for standardizing the posture of the subject during the test. The restraint device also provides visual force feedback to limit contractions to constant force levels. These are important features to consider when conducting EMG spectral analysis because constant muscle length and tension are specified to insure signal stationarity (i.e. constant mean and variance) (Basmajian and De Luca, 1985; Merletti *et al.*, 1990). Studies in which EMG spectral parameters are derived from variable-force or dynamic contractions may be considered as methodologically flawed until more definitive analysis methods are available to identify stationary epochs within the data. Tests for EMG signal stationarity are under development (Bilodeau *et al.*, 1991; Franklin, 1993).

Most of our studies conducted to date on back muscles share similar protocols. A maximal voluntary contraction (MVC) is first obtained by selecting the highest force value from several attempts at producing a maximal trunk extension. One or more fatigue-inducing contractions are then sustained at a specified force level (typically 40, 60, or 80% MVC) and for a fixed time duration (typically 30 or 60 s). At least two median frequency parameters are calculated from each sustained contraction: (1) the median frequency slope (MF slope), defined as the coefficient of a linear regression fit by a least-squares procedure and (2) the initial median frequency (IMF), defined as the *y*-intercept of the linear regression used to derive MF slope. In some instances, the fatigue-inducing contraction is followed at exactly 1 min by a contraction of shorter duration (typically 10 s) to derive a parameter to measure the amount of recovery of the median frequency (REC). Classification of subjects into 'LBP' and 'normal' groups are obtained by the use of a multivariate statistical procedure referred to as discriminant analysis (Zar, 1974). The procedure consists of developing a rule or 'discriminant function' based on the EMG median frequency measurements that best separates the subjects into their respective groups (i.e. the analysis maximizes the 'between-group variation' while minimizing the 'within-group variation'). Measurements from the LBP and control subjects (whose classifications are known *a priori*) comprise a 'learning set' to map the median frequency parameters from the six muscle sites into a single discriminant parameter which can then be used to make classification predictions. As a result, EMG measurements from future individuals whose LBP classification is not known *a priori*, can be predicted on the basis of the discriminant function from the learning set. The function is developed using a stepwise regression procedure to identify an optimal set of EMG parameters for discriminating 'normal' from 'LBP' groups. The stepwise regression procedure selects only those EMG variables that add to the discriminating ability of the function. Therefore, it is quite possible to start out with a dozen or more variables which can be reduced to a few discriminant variables for making the selection or classification. Another possible advantage of this technique is that in addition to providing a classification, the 'distance' of the discriminant function value from the cut-off point between one classification group and the other can also be plotted as Fisher *z*-scores. These scores can provide a measure of the strength of the classification prediction. Plots of *z*-scores are shown later in this section for clinical research results. A more detailed explanation of this procedure can be obtained in statistical textbooks that include multivariate regression analysis (Zar, 1974; Kleinbaum and Kupper, 1978).

The first of a series of case–control studies performed by our group evaluated male chronic LBP patients ($n = 12$; average duration of LBP = 5.2 years) and control subjects ($n = 12$) matched for age and height who had never experienced debilitating LBP (Roy *et al.*, 1989). Patients were not complaining of pain at the time of testing (i.e. they were in remission) and none had previous back surgery or radiological evidence of structural disorders of the spine. Tests were conducted at 40, 60, and 80% MVC. On the basis of the EMG parameters from the discriminant analysis functions, the test was able to identify the control subjects with similar high levels of accuracy (approximately 85%) for each of the three contraction levels (Table 9.1). The EMG variables selected by the discriminant procedure were primarily the initial median frequency from the L1 spinal level and the median frequency slope from the L2 and L5 spinal levels. These were the same parameters as those whose mean values were found to be significantly different between the LBP patients and controls (Figure 9.2). This study also demonstrated a higher

Table 9.1 Results from discriminant analyses – all lumbar levels

	Percent correct classification		
Contractile level (% MVC)	NLBP (n = 12)	LBP (n = 12)	Variables used in classification (in order)
40	92	82	(R)IMF, L1 (R)SLOPE, L2 (L)SLOPE, L2 (L)SLOPE, L1 (L)IMF, L2 (L)IMF, L1
60	67	75	(R)IMF, L1 (L)SLOPE, L1
80	84	91	(L)IMF, L1 (R)SLOPE, L5 (L)SLOPE, L5 (L)SLOPE, L2

(R), (L), right side, left side; L1, L2, L5, lumbar spinal levels; LBP, with low back pain; NLBP, without low back pain; IMF, initial median frequency; SLOPE, median frequency slope.

percentage of correct classifications for analyses in which data from all electrode sites were included in the discriminant analysis rather than by treating muscle groups separately according to lumbar level. The results further support the recommendation that arrays of electrodes, rather than a few bilateral sites, provide the most accurate results for classification. The limited success and conflicting results of previous EMG studies may have been due in part to these factors. Earlier, three reasons were postulated for the altered fatigue characteristics of the lower back muscles.The results of this study excludes one of them – the one postulating a greater pre-contraction metabolite level resulting from persistent spasm.

The successful implementation of the EMG signal procedures for chronic LBP subjects led us to the second in a series of investigations which utilized the technique to identify individuals with LBP ($n = 6$) within a population of elite athletes ($n = 24$) (Roy *et al.*, 1990). We were interested in determining whether athletes with LBP might still have muscular impairment associated with their disorder despite the fact that they were highly conditioned. The study addressed the larger question of whether different kinds of muscular disorders are present in LBP and whether they require separate discriminant functions. We speculated that there may be recognizable patterns of EMG signal disturbances among the various muscles being monitored for specific kinds of LBP disorders (e.g. acute versus chronic) or within specific populations of individuals (e.g. athletic versus sedentary). Twenty-three members of a men's collegiate varsity crew team ($n = 13$ port and $n = 10$ starboard rowers) were tested at 80% MVC × 30 s using the Back Analysis System. Recovery from fatigue was also included in the protocol and was analyzed as the percent recovery of median frequency at 1 min following the sustained contraction. This parameter, referred to as the median frequency recovery (REC), expressed the degree to which the median frequency returned to its baseline, pre-fatigued level as a result of the 1 min rest period. We speculated that the median frequency recovery parameter represented the ability of a muscle to recover from metabolite accumulation following strenuous exercise. The discriminant analysis resulted in 100% correct classification of the LBP rowers and 93% correct classification of non-LBP rowers (one false-positive classification). EMG variables selected by the classification function were primarily the median frequency recovery parameter from the L5 and L1 electrode locations. Interestingly, none of the parameters used to classify LBP and controls from our previous study on sedentary chronic LBP patients were selected by the discriminant analysis procedure to classify the rowers with LBP. This finding indicates that there are different muscle disorders associated with LBP for the sedentary and athletic populations studied. In the former group, the impairment was likely representative of deconditioning whereas in the latter, deconditioning and muscle disuse would be unlikely since these subjects continued their training and competitive rowing despite their LBP symptoms. The question remains, however, as to the nature of the muscle disorder represented by the EMG findings in the rower study. We have speculated that the differences in the median frequency recovery parameter may be related to the unique energy requirements of the sport of rowing. During competition, rowers quickly achieve a marked anaerobic response and must tolerate high levels of excessive lactate throughout the remainder of the race (Hagerman *et al.*, 1979). It is possible that the inefficient physiological removal of these high levels of lactate is a sequela of LBP in rowers. We also conducted a discriminant analysis between the port and starboard rowers in this study who load their back muscles asymmetrically by use of a single oar placed on either the port or starboard side of the hull. The test identified 100% of the port and starboard rowers correctly. The ability to identify muscle imbalances may have implications for ergonomic applications where chronic asymmetrical loading might also cause muscle imbalances.

A third study was conducted to compare EMG spectral methods of classifying LBP to standard clinical assessment procedures (Klein *et al.*, 1991). Twenty-five freshman sweep rowers were recruited for this study, eight of whom had a history of LBP (four chronic and four acute). None of the subjects reported pain during the test. Protocols similar to those described for previous studies using the Back Analysis System were conducted on all subjects. In addition, each subject was evaluated for trunk and spinal mobility. The clinical measurements consisted of maximal trunk extension strength and trunk range of motion (ROM) for forward, backward, and lateral bending, as well as trunk rotation. The discriminant analysis identified 88% (seven out of eight) of the patients with LBP and 100% (nine) of the subjects without LBP. In contrast, the conventional clinical tests were less accurate: identifying 57% (four out of seven) of the patients with LBP and 67% (10 out 15) of the subjects without LBP. The high sensitivity and specificity of the EMG procedure suggests that the technique may be superior to conventional clinical measures of trunk mobility and strength for LBP screening or identification of muscle impairment. No comparisons of this kind have been reported for populations of workers and therefore we cannot extrapolate these results to other populations.

The most recently published work from this series of investigations addressed many of the limitations of previous studies (Roy *et al.*, 1995). Most of the prior studies had been limited to evaluating relatively young, 'white-collar' males with non-specific spinal disorders. The yet to be confirmed applicability of the technique to the population at large, particularly among 'blue-collar' manual laborers at risk of LBP needed to be addressed. In addition, until this particular study, classification results had only been reported for subjects whose EMG data was included in the original formulation of the discriminant function. A total of 92 patients with chronic LBP and 52 control subjects with no history of debilitating LBP were studied. Many of the clinical characteristics of the patient population in this study were different from previous studies. Most of the patients had either a verified herniated disc or spinal surgery, conditions which were excluded in our previous studies. Secondly, we did not limit our selection of subjects in the present study to patients in remission from pain, as was also previously specified in other studies (Roy *et al.*, 1989). All of the patients tested described pain localized to the lumbar region and most were unable to produce a normal MVC. Patients were tested immediately prior to their entry into an occupational rehabilitation center where they participated in a full-time, multidisciplinary work hardening program for a 30 day period. Subjects were divided into a 'learning sample' and a 'holdout sample' of LBP and normal groups (Table 9.2). The learning sample was used to formulate the LBP discriminant function and then this function was used to classify the holdout sample. The discriminant function selected the initial median frequency and median frequency slope parameters from the L1 electrode site, and the initial median frequency from the three left lumbar sites to classify 85% of patients with LBP and 86% of control subjects for the learning set correctly (Figure 9.4).

Fisher discriminant function values or *z*-scores, representing the distance from the group classification cut-off point, are displayed for each of the LBP

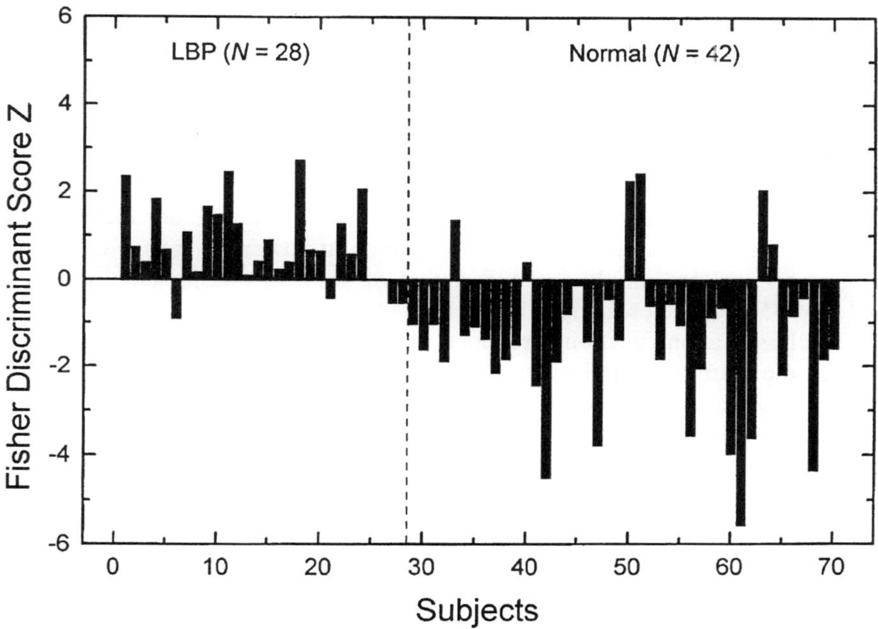

Figure 9.4 Fisher discriminant function scores (z-scores), a measure of the distance from the classification cutoff point, are presented for each of the LBP ($n = 28$) and control ($n = 42$) subjects that formed the 'learning set' for the discriminant function. A positive score indicates a LBP classification, a negative score indicates a normal classification. Subjects are divided along the x-axis according to those that are known *a priori* to be LBP or normal. Incorrect classifications are identified in the lower left quadrant for LBP subjects (four false-negative scores) and the upper right quadrant for normal subjects (six false-positive scores) (with permission from Roy *et al.*, 1995).

Table 9.2 Characteristics of subjects – mean values (SD)

	LBP (n = 28)	Normal (n = 42)	LBP (holdout) (n = 57)	Normal (holdout) (n = 6)
Age (years)	35.3 (8.9)	26.7 (5.2)	37.1 (8.9)	23.8 (2.5)
Height (m)	1.8 (0.1)	1.8 (0.1)	1.8 (0.1)	1.8 (0.1)
BMI (kg m^{-2}	27.0 (4.6)	23.0 (2.5)	27.4 (5.0)	25.7 (3.7)
Weight (kg)	84.2 (16.2)	70.5 (9.7)	86.4 (19.0)	81.4 (11.1)
MVC (lbs)	140.7 (57.0)	184.8 (73.0)	120.5 (70.5)	241.3 (77.5)
Surgery (%)	43	–	1	–
HNP (%)	75	–	27	–
Duration LBP (months)	26.3 (31.4)	–	15.2 (12.2)	–

LBP, low back pain; HNP, herniated disc; BMI, body mass index; MVC, maximal voluntary contraction.

control subjects. The four LBP subjects identified in the figure with a negative z-score represent false-negative classifications whereas the six normal subjects with a positive z-score represent false-positive classifications. The classification results were independent of the subject's ability to exert a maximal extension because the stepwise discriminant analysis procedure rejected an attempt to include MVC into the classification function. In other words, forcing MVC into the function did not change the accuracy of the classification. This implies that the discriminating power of the EMG spectral parameters was not simply a manifestation of the fact that patients with LBP had lower MVC values than the normal population.

The classification function performed as well among the holdout sample as it did for the learning sample with 88% of LBP patients and 100% of normal subjects correctly classified (Figure 9.5). The relatively high levels of correct classifications in this study cannot be explained as a consequence of the stepwise regression overfitting the model. The size of our sample population should have been adequate to avoid overfitting the data because it was approximately 14 times the number of EMG parameters used (Zar, 1974). The favorable classification results among the holdout sample further dispels the likelihood of overfitting and demonstrates that the test was highly sensitive and specific for a population that was not included in the initial learning set. Unfortunately, few if any procedures used to classify muscle function have demonstrated the sensitivity and specificity of their measures beyond those of the sample comprising the learning set (Newton and Waddell, 1993). Although there is no way of verifying this interpretation at this time, it has been suggested that the discriminant parameters based on median frequency measures differed between patients and controls as a result of a different pattern of muscle activation and load sharing between the different paraspinal muscles in response to pain or fear of reinjury (Roy *et al.*, 1989; Thompson *et al.*, 1992; De Luca, 1993b).

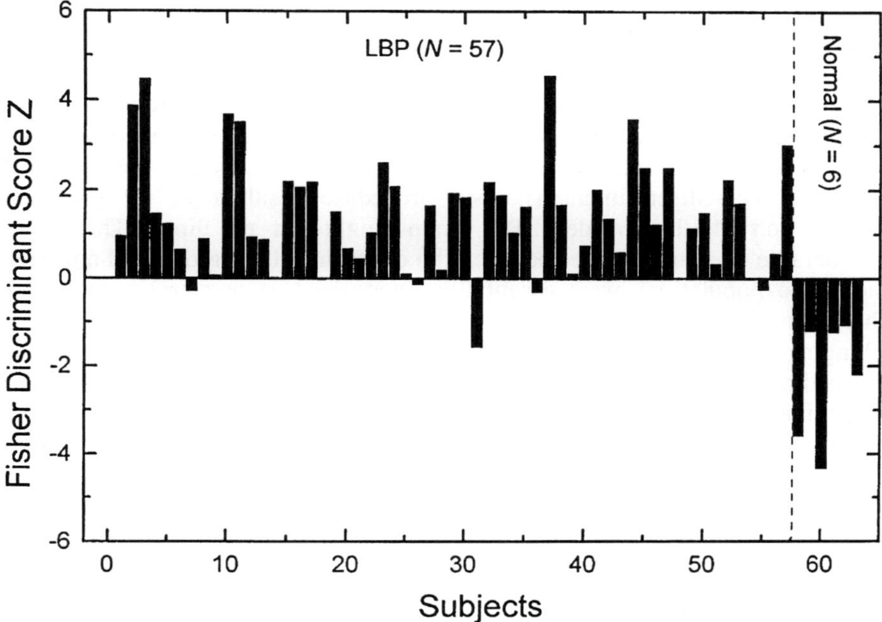

Figure 9.5 Fisher discriminant function scores (z-scores), a measure of the distance from the classification cut-off point, are presented for each of the LBP ($n = 57$) and control ($n = 6$) subjects that formed the 'holdout sample' for the discriminant function. A positive score indicates a LBP classification, a negative score indicates a normal classification. Subjects are divided along the x-axis according to those that are known *a priori* to be LBP or normal. Incorrect classifications are identified in the lower left quadrant for LBP subjects (five false-negative scores) and the upper right quadrant for normal subjects (zero false-positive scores) (with permission from Roy *et al.*, 1995).

Although few in number, other relevant reports in the literature describe the ability of EMG spectral techniques to identify and classify back muscle disorders. Biedermann *et al.* (1991) tested 22 healthy subjects and 24 patients with chronic LBP. Of particular interest was the method used to fatigue the back extensor muscles and the ability of median frequency parameters to distinguish between 'pain behaviors' in the LBP group. The technique employed a weightlifting procedure to produce a paraspinal contraction of constant isometric force and avoid the need to rely on an MVC determination. Each subject was asked to hold a barbell, which had two 5 lb weights attached symmetrically about its center, with both arms outstretched at 90° degrees of shoulder flexion. The subjects held the weight for a 45 s period while standing in a test reference frame which reliably positioned the feet, pelvis, and spine. Four surface electrodes were used to monitor EMG signals bilaterally from the iliocostalis lumborum muscle at the L2–L3 interspinous level and the multifidus muscle at the L4–L5 interspinous level. In addition to the 'normal' category, LBP subjects were divided into a physically-active or 'confronter' group, and a physically-passive 'avoider' group on the basis of their response to a pain behavior checklist (Zarkowska, 1981). It was postulated that these categories reflected the clinical observation that some patients remained very active despite their reported back problems, whereas others tended to avoid physical and social activities as much as possible to protect their painful condition. The discriminant analysis procedure resulted in 89% correct classification of the avoider LBP group (eight out of nine). There was considerable overlap however between the confronter LBP group and normals. These two populations were essentially categorized as one group. Only 8% (three out of 37) of the normals and confronters were misclassified as belonging to the avoider LBP group. Median frequency parameters similar to the median frequency slope and initial median frequency were selected as discriminant parameters, but only from multifidus muscle sites. Generally, the avoiders LBP group had greater spectral changes toward lower frequencies which were interpreted as an indication that physically passive LBP patients had more fatigable paraspinal muscles. This finding is consistent with earlier results on chronic LBP patients who had similar strength but higher fatigue as compared to controls (Jorgensen and Nicolaisen, 1987; Roy *et al.*, 1989).

9.6.2 Monitoring Treatment Progression

Independent reports by others lend support to the suggestion that the surface EMG technique can be useful as an objective, non-invasive measure of LBP treatment outcome. The physiological basis for this application of the technique is based in part on the effect that muscle adaptation has on muscle fiber conduction velocity and, hence, EMG power spectral parameters. Although the sensitivity of the technique to paraspinal muscle adaptations has not been demonstrated in a well-designed prospective study, there is sufficient experimental support to consider this likelihood. This evidence has been related to the following two mechanisms which have been shown to be sensitive to changes in muscle use: (1) conduction velocity and median frequency parameters are related to muscle fiber diameter, particularly in fast fibers (De Luca, 1985; Roy *et al.*, 1994) and (2) the conduction velocity and median frequency are related to intramuscular H^+ (De Luca, 1985; Brody *et al.*, 1991; Roy, 1993b) and extracellular K^+ (Juel, 1986), two metabolites that can change with muscle adaptation (Booth, 1987). Directed studies have also related muscle training and/or differences in muscle fiber type to conduction velocity and EMG spectral parameters (Sadoyama *et al.*, 1988; Thompson, *et al.*, 1992, Thompson and Biedermann, 1993; Roy *et al.*, 1994).

Retrospective and prospective EMG studies have been conducted to evaluate the influence of activity and training on back muscle function in normal and LBP subjects. Power spectrum analysis has been used in normal (pain-free) subjects to examine differences between those that are physically active from those that are sedentary (Biedermann *et al.*, 1991). Active subjects had a reduced fatigue rate, as measured by the change in the median frequency from the iliocostalis muscle during a weightlifting task. These same subjects had greater left–right side variability of the initial median frequency as well. However, the retrospective study design could not determine the sensitivity of the EMG spectral parameters to training or even whether these differences were the result of training or the result of pre-existing characteristics. A more recent prospective study has been reported (Moffroid *et al.*, 1993) examining the sensitivity of the EMG power spectrum analysis to the effects of a training program for paraspinal muscles. Healthy female subjects ($n = 28$) were randomly assigned to groups which (1) maintained their previous lifestyle or (2) completed a series of isometric back exercises twice daily, for 6 weeks. Subjects were tested before and after the intervention using a Sorenson test in which the endurance time and EMG median frequency from the L3 lumbar level were measured while the subject maintained a prone trunk extension with only the lower limbs and pelvis supported. The results indicated that while trained subjects showed a significant improvement in the endurance time, they did not change on any of the spectral parameters (initial median frequency and median frequency slope). A number of methodological limitations have been pointed out by others (Thompson *et al.*, 1992; Thompson and Biedermann, 1993) concerning the acquisition of the EMG signals in this study. Most of the criticism relates to the placement of the EMG electrodes which did not correspond to any specific muscle and did not take into consideration muscle fiber orientation. The interelectrode spacing of 25 mm was questioned as being too large to provide proper muscle localization. The authors of the study provided their own explanation for the lack of EMG sensitivity to training. They suggested that the lack of change in the initial median frequency and the median frequency slope could have been caused by the training program being too brief and not specific to reducing muscle fatigue. The training may have also improved the endurance of the hamstring and glutei muscles, which contribute significantly to back extension tasks such as the Sorenson test (Jones *et al.*, 1988; Pollock *et al.*, 1989).

In another training study, the EMG median frequency of the multifidus and iliocostalis muscles were compared to adaptive changes associated with physical training in normal subjects using a test–retest experimental design (Thompson *et al.*, 1992). Sedentary women were randomly assigned to a control group ($n = 24$) or a 1 h fitness class, three to five times per week for 12 weeks (exercise group, $n = 22$). In addition to monitoring the EMG median frequency from four lumbar paraspinal muscles during a 45 s weight-holding task, the investigators also measured several physical fitness parameters for aerobic capacity, back muscle strength, and back flexibility. Various anthropometric measures were also included. Only two EMG parameters meeting an acceptable level of reliability were included in the analysis. They were the initial median frequency (IMF) defined previously in this chapter and an adjusted fatigue parameter (FTG) defined as the decline in the median frequency over the trial, adjusted by arm length, because the fatigue of the back muscles during a constant force contraction is known to be affected by the distance of the weight from the body (Biedermann, 1990). Parameters first considered but eventually excluded from further analysis because of poor reliability were parameters measuring recovery of the initial median frequency after a second trial, and the absolute left–right difference in the initial median frequency from the contralateral back muscles. The results described significant improvements in aerobic capacity, back strength, and flexibility following training with no comparable changes in the control group. EMG power spectrum parameters were also responsive to training. For the exercise group, the FTG showed a significant reduction (approximately 35%) from pre- to post-test, being significantly lower than that of the control group at post-test. Interestingly, the same parameter from the iliocostalis muscle did not change with training.

The authors explained the differential response between muscle groups as possibly representing the different demands placed on these muscle groups by the specific exercises of the fitness class. The changes in the FTG associated with training were explained by improvement in the oxidative capacity and oxygenation of the muscle, which were confirmed indirectly. The analysis did not find a significant change in the initial median frequency with training, however the initial median frequency from the multifidus muscle decreased at a near significant level ($p = 0.06$). The observed change in the initial median frequency was explained on the basis of disuse atrophy producing smaller diameter muscle fibers and therefore correspondingly lower values of conduction velocity and median frequency. This finding and interpretation are consistent with previous results of our studies which compared patients with chronic LBP with controls (Roy *et al.*, 1989). In addition, the data were analyzed further to determine if there was a relation between changes in fitness parameters and changes in EMG median frequency parameters since both were responsive to training. However, the results did not identify a significant overall relationship between these two sets of parameters probably because both measures are indirect and the underlying physiological adaptations were specific to the muscles involved in the test.

In a similar study, without controls, a group of LBP patients participating in a back-care exercise and education program were assessed (Thompson *et al.*, 1992). The subjects were middle aged, non-obese males and females, employed or retired, with chronic LBP with episodic remissions and recurrences, sometimes necessitating time off from work. They completed EMG and fitness assessment as described above for the training/control study, before and after participating in 20 1 h classes over a 10 week period. The exercises consisted of progressive extension and flexion strengthening and flexibility exercises and low-intensity aerobic activities (mostly walking) at home. The results were a significant improvement in back flexibility but no significant improvement in aerobic capacity or lifting strength. Concomitantly, there was a substantial reduction in the median frequency fatigue parameter

(FTG) in both muscle groups following training. In fact these differences in the median frequency resulted in a reclassification of patients from 'inactive LBP patient' to 'normal control' on the basis of the discriminant function from a previous diagnostic study (Biedermann *et al.*, 1991). There was also a small, but non-significant, increase in the multifidus initial median frequency. Because of the small sample size, no analysis was conducted on the possible relationship between fitness and median frequency changes with training. These results confirm that EMG spectral measurements of back muscles are sensitive to training effects, not only in healthy subjects, but also in patients with LBP.

Mayer *et al.* (1989) used EMG spectral analysis for fatigue assessment in normal subjects and deconditioned patients. Ten industrial workers undergoing functional restoration for chronic disabling spinal disorders and 11 healthy volunteers were tested by monitoring the changes in the EMG mean frequency during a Roman Chair exercise. This exercise is similar to the Sorenson test described earlier in this chapter. Subjects were required to complete two successive sessions of ten trials where each trial consisted of 15 s of unsupported trunk extension followed by 10 s of rest. The slope of the decline in mean frequency was significantly greater (i.e. more negative) in LBP subjects at initial testing than either normal subjects or back pain subjects after reconditioning. These findings indicated that the EMG spectrum shifted further and recovered more slowly for LBP patients compared to normals, and that muscle reconditioning can result in a more normal fatigue response as measured by the EMG spectral parameter. The results of this study were less successful in demonstrating consistency between the spectral parameter estimate of fatigability and actual endurance. This unexplained discrepancy, as well as other issues raised by the authors relating to the validity of the EMG fatigue index and its reliability, were discussed in terms of the applicability of the technique as a clinical assessment procedure (Standridge *et al.*, 1988). The problems identified by these investigators could be related to the protocol they used. Repeated duty cycle contractions can be notoriously unstable in terms of EMG spectral estimates because of poor signal stationarity resulting from length and force changes in the muscle and significant disturbances to muscle blood flow (De Luca, 1985). During static isometric contractions constrained to a constant force level, the EMG signal is more stable. Although more work is needed, the relationship between changes in EMG spectral parameters and the loss of force-producing capability in muscles have been studied in lower back muscles. A recent study by Mannion and Dolan (1994) reported significant correlations between median frequency parameters and trunk extension endurance during static paraspinal muscle contractions.

Other studies have evaluated the usefulness of surface EMG procedures for evaluating work-related LBP in patients undergoing multidisciplinary rehabilitation (Roy *et al.*, 1995). Twenty eight-patients (24 males and four females) with a history of chronic LBP and clinical evidence of mechanical strain or sprain were tested using the Back Analysis System just prior to and immediately following their participation in a multidisciplinary functional restoration program for work-related back injuries. Twelve of the patients with LBP had prior back surgery for a decompression laminectomy, the average duration of LBP syndrome was 26.3 months, and all complained of debilitating LBP in the lumbar region. Subjects participated in the rehabilitation program for an average of 40 h per week according to a fixed schedule that included physical therapy, occupational therapy, circuit weight training, work-hardening, back pain school, and psychological counseling. Baseline EMG testing was conducted during trunk extension sustained at 40 and 80% MVC for 30 s. Although the MVC value was reassessed at follow-up, the sustained contractions requested at follow-up were specified at 40 and 80% of the baseline MVC. In this way, any EMG changes in fatigability would be independent of differences associated with muscle force. Data acquired before and after the rehabilitation program were analyzed using a repeated-measures, four-way ANOVA to study the influence of the following main effects on the initial median frequency and median frequency slope parameters: time (pre-and post-rehabilitation), % MVC (40 and 80%), lumbar level (L1, L2, and L5), and lumbar side (left and right). The ANOVA results indicated that the median frequency slope significantly changed ($p<0.05$) in a manner consistent with improved muscle fatigability (i.e. the median frequency slope was less negative) in patients with LBP undergoing rehabilitation. Because no pain was reported by the patients and no change in the amplitude of the signal was observed following training, it was speculated that training resulted in a change in the muscle's production and/or elimination of metabolites during the contraction rather than a redistribution of muscle forces among the different muscle groups. Changes in the median frequency slope following rehabilitation were not significantly influenced by test protocol factors such as electrode location or the % MVC of the trial. The only main effect from the ANOVA analysis of the initial median frequency and median frequency recovery was between lumbar levels ($p<0.05$). Similar findings have been explained on the basis of possible fiber type differences between the different erector spinae muscles (Biedermann *et al.*, 1991; Thompson *et al.*, 1992). Because the MF slope differed significantly between patients and normals ($p<0.05$) as well as when comparing baseline to post-rehabilitation trials ($p<0.05$), the ANOVA analysis for the MF slope was repeated with the baseline MVC as a co-variate. The results demonstrated a significant interactive effect for time (pre- and post-rehabilitation) and MVC ($p=0.03$). This interactive effect is plotted in Figure 9.6 and demonstrates that patients with a low baseline MVC had the most 'improvement' in the median frequency slope. Interestingly, it was also found in this study that patients with a history of previous back surgery did not 'improve' their median frequency slopes as much as those without previous surgery. The surgical history in this instance was decompression laminectomies for herniated discs, all from posterior approaches in which muscle fibers could have been damaged.

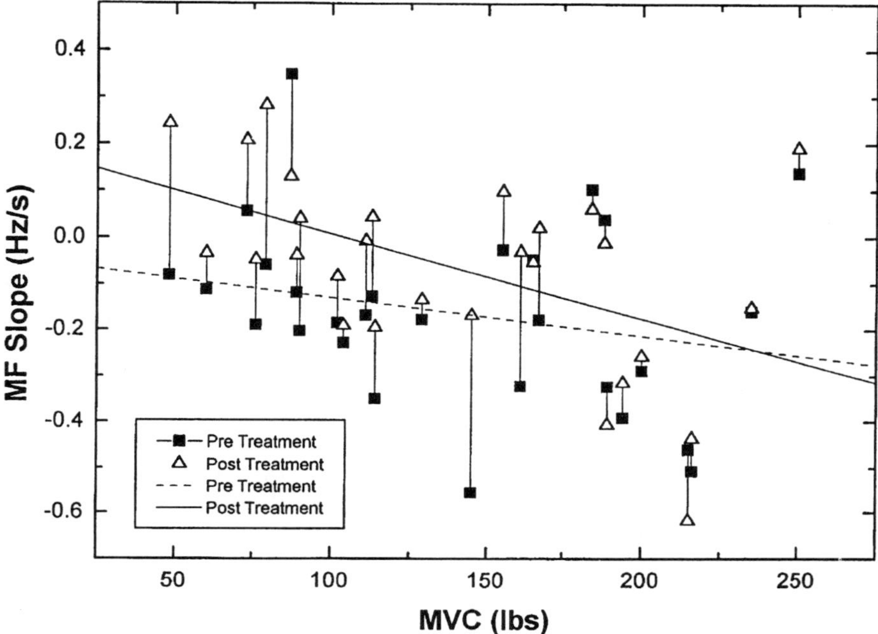

Figure 9.6 The interactive effect of baseline MVC and time (pre- and post-rehabilitation) on MF slope is depicted. The mean MF slope (all electrode sites and % MVC trials) for pre-rehabilitation [■] and post-rehabilitation [△] are connected for each individual patient. The MF slope is plotted according to the baseline MVC for each patient. Least-squares linear regressions are displayed separately for pre- and post-rehabilitation data points (with permission from Roy *et al.*, 1995).

Adaptive change following exercise, producing such effects as muscle hypertrophy and changes in muscle bioenergetics, have been associated with median frequency parameters (Thompson *et al.*, 1992; Roy, 1993a; Thompson and Biedermann, 1993), in back muscles as well as other muscle groups. Other than the study by Mayer *et al.* (1989) already mentioned, only one other independent group has reported changes in EMG spectral parameters in back muscles following exercise. Thompson *et al.* (1992) compared sedentary women without LBP participating in a 12 week fitness class to a second group consisting of LBP patients participating in a 10 week back-care exercise program. The results in both study groups indicated that the median frequency decreased by a lesser amount during an isotonic, isometric task when trials conducted at baseline were compared to similar trials at the end of training. The reduced decrement of median frequency during the fatigue-inducing task was evident in both the multifidus and iliocostalis muscles for the LBP group, whereas this effect was only present for the multifidus muscles in the non-LBP group. LBP patients also presented with a significant increase in the initial median frequency parameter following the training period. These results are fully consistent with our own results reported earlier (Roy *et al.*, 1995). Although this study documented that the intervention resulted in concurrent improvements in muscle strength, endurance and flexibility, we are still left without definitive evidence identifying a specific physiological adaptation with the changes in the EMG parameters. For this to be clarified, *in vitro* models will need to be conducted to clarify these points.

Although it can be appreciated from this brief overview of surface EMG signal analysis of paraspinal muscles that much has been gained by this relatively new science, much remains to be developed before we can fully realize the potential benefits of the described technique. Specialists in the field of ergonomics and their clinical colleagues in occupational and physical therapy have begun to incorporate some of these techniques in their research and/or clinical practice. Many more have likely considered its use but are hesitant because of their unfamiliarity with the technical and procedural aspects of the technique. The routine use of surface EMG techniques in clinical practice will be achieved when technological advances are combined with the guidelines set forth by clinical research.

This work was supported by Liberty Mutual Insurance Company and the Rehabilitation and Development Service of the Department of Veterans Affairs.

References

AHERN, D.K., FOLLICK, M.J. COUNCIL, J.R., LASER-WOLSTON, N. & LITCHMAN, H. (1988) Comparison of lumbar paravertebral EMG patterns in chronic low back pain patients and non-patient controls. *Pain*, **34**, 153–60.

ANDERSSON, G.B.J. & SCHULTZ, A.B. (1979) Transmission of moments across the elbow joint and the lumbar spine. *J. Biomech.*, **12**, 747–55.

ANDERSSON, G.B.J., HERBERTS, P. & ÖRTENGREN, R. (1976a) Myoelectric back muscle activity in standardized lifting postures. In *Biomechanics 5-A* (Ed. P.V. KOMI), pp. 520–9. University Park Press, Baltimore, M.D.

ANDERSSON, G.B.J., ÖRTENGREN, R. & NACHEMSON, A. (1976b) Quantitative studies of back loads in lifting. *Spine*, **1**, 178–85.

ANDERSSON, G.B.J., ÖRTENGREN, R. & HERBERTS, P. (1979) Quantitative electromyographic studies of back muscle activity related to posture and loading. *Orthoped. Clin. in N. Am.*, **8**, 85–96.

ANDERSSON, G.B.J., BOGDUK, N. & DE LUCA, C.J. (1989) Muscle: clinical perspective. In *New Perspective on Low Back Pain* (Ed. J.W. FRYMOYER & S.L. GORDON, pp. 293–334. American Academy of Orthopedic Surgeons, Park Ridge, Ill.

ARENA, J.G., SHERMAN, R.A., BRUNO, G.M. & YOUNG, T.R. (1989) Electromyographic recordings of 5 types of low back pain subjects and non-pain controls in different positions *Pain*, **37**, 57–65.

ARMSTRONG, R.B. (1984) Mechanisms of exercise-induced delayed onset muscular soreness: a brief review. *Med. Sci. Sports Exercise*, **6**(13), 529–38.

ARROYO, P. (1966) Electromyography in the evaluation of the reflex muscle spasm. *J. Florida Med. Assoc.*, **53**, 29–31.

BASMAJIAN, J.V. & DE LUCA, C.J. (1985) *Muscles Alive: Their Functions Revealed by Electromyography*. Williams & Wilkins, Baltimore, MD.

BEJJANI, F.J., GROSS, C.M. & PUGH, J.W. (1984) Model for static lifting: relationship of loads on the spine and the knee. *J. Biomech.*, **17**(4), 281–6.

BIEDERMANN, H.J. (1990) Weight lifting in a postural restraining device: a reliable method to generate paraspinal constant force contractions. *Clin. Biomech.*, 180–2.

BIEDERMANN, H.J., SHANKS, G.L. & INGLIS, J. (1990) Median frequency estimates of paraspinal muscles: reliability analysis. *EMG Clin. Neurophysiol.*, **30**, 83–8.

BIEDERMANN, H.J., SHANKS, G.L., FORREST, W. & INGLIS, J. (1991) Power spectrum analyses of electromyographic activity: discriminators in the differential assessment of patients with chronic low back pain. *Spine*, **16**, 1179–84.

BIGOS, S.J. & BATTIE, M.C. (1987) Surveillance of back problems in industry. In *Clinical Concepts in Regional Musculoskeletal Illness* (Ed. N.M. HADLER), pp. 99–315. Grune and Stratton, Orlando.

BILODEAU, M., ARSENAULT, A.B., GRAVEL, D. & BOURBONNAIS, D. (1991) EMG power spectra of elbow extensors during ramp and step isometric contractions. *Eur. J. Appl. Physiol.*, **63**, 24–8.

BOOTH, F.W. (1987) Physiologic and biochemical effects of immobilization on muscle. *Clin. Orthoped. Related Res.*, **219**, 15–20.

BREWER, B.J. (1960) Mechanism of injury to the musculotendinous unit. In *American Academy of Orthopaedic Surgeons Instructional Course Lectures, XVII* (Ed. F.C. REYNOLD), pp. 354–8. Mosby, St. Louis.

BRODY, L.R., POLLOCK, M., ROY, S.H., DE LUCA, C.J. & CELLI, B. (1991) pH-induced effects on median frequency and conduction velocity of the myoelectric signal. *J. Appl. Physiol.*, **71**, 1878–85.

CASSISI, J.E., ROBINSON, M.E., O'CONNER, P. & MACMILLAN, M. (1993) Trunk strength and lumbar paraspinal muscle activity during isometric exercise in chronic low-back pain patients and controls. *Spine*, **18**(2), 245–51.

CHAFFIN, D.B. & PARK, K.S. 1973, A longitudinal study of low-back pain as associated with occupational weight lifting factors. *Am. Indust. Hyg. Assoc. J.*, **34**, 513–25.

CHAPMAN, A.E. & TROUP, J.D.G. (1970) Prolonged activity of lumbar erectores spinae: an electromyographic and dynamometric study of the effect of training. *Ann. Phys. Med.*, **10**, 262–9.

COHEN, M.J., SWANSON, G.A. & NALIBOFF, B.D. (1986) Comparison of electromyographic response patterns during posture and stress tasks in chronic low back pain patterns and control *J. Psychosom. Res.*, **30**, 135–41.

COLLINS, G.A., COHEN, M.J. & NALIBOFF, B.D. (1982) Comparative analysis of paraspinal and frontalis EMG, heart rate and skin conductance in chronic low back pain patients and normals to various postures and stress. *Scand. J. Rehabil. Med.*, **14**, 39–46.

DARLING, D. (1993) In search of the perfect treatment. In *Back Pain Rehabilitation* (Ed. B. D'ORAZIO), pp. 3–31. Andover Medical Publishers, Boston).

DE ANDRADE, J.R., GRANT, C. & DIXON, A.J. (1965) Joint distension and reflex muscle inhibition in the knee. *J. Bone Joint Surgery*, **47A**, 313–22.

DE LUCA, C.J. (1979) Physiology and mathematics of myoelectric signals. *IEEE Trans. Biomed. Engng*, **26**, 313–25.

DE LUCA, C.J. (1985) Myoelectric manifestation of localized muscular fatigue in humans. *CRC Crit. Rev. Biomed. Engng*, **11**, 251–79.

DE LUCA, C.J. (1993a) Neuromuscular fatigue. In *Spectral Compression of the EMG Signal as an Index of Muscle Fatigue*, (Ed. A.J. SARGEANT & D. KERNELL), pp. 44–51. North-Holland, Amsterdam.

DE LUCA, C.J. (1993b) The use of the surface EMG signal for performance evaluation of back muscles. *Muscle Nerve*, **16**, 210–16.

DE LUCA, C.J., LE FEVER, R.S. & STULEN, F.B. (1979) Pasteless electrode for clinical use. *Med. Biol. Engng Comput.*, **17**, 387–90.

DE VRIES, H.A. (1968a) Method for evaluation of muscle fatigue and endurance from electromyographic fatigue curve. *Am. J. Phys. Med.*, **47**, 125–35.

DE VRIES, H.A. (1968b) EMG fatigue curves in postural muscles. A possible etiology for idiopathic low back pain. *Am. J. Phys. Med.*, **47**, 175–81.

DENSLOW, J.S. & CLOUGH, G.H. (1941) Reflex activity in the spinal extensors. *J. Neurophysiol.*, **4**, 430–7.

DEYO, R.A. (1987) Reducing work absenteeism and diagnostic costs for backache. In *Clinical Concepts in Regional Musculoskeletal Illness*, (Ed. N.M. HADLER), pp. 25–50. Grune & Stratton, Orlando.

DEYO, R.A. & TSUI WU, Y.J. (1987) Descriptive epidemiology of low-back pain and its related medical care in the United States. *Spine*, **12**, 264–8.

DEYO, R.A., CHERKIN, D., CONRAD, D. & VOLINN, E. (1991) Cost, controversy and crisis: low back pain and the health of the public. *Am. Rev. Public Health*, **12**, 141–56.

D'ORAZIO, B. (1993) Exercise prescription in low back pain. In *Back Pain Rehabilitation* (Ed. B. O'RAZIO), pp. 32–71. Andover Medical Publishers, Boston.

EDWARDS, R.H.T. (1981) Human muscle function. In *Human Muscle Fatigue: Physiological Mechanisms* (Ed. R. PORTER & J. WHALEN), pp. 1–18. Pitman Medical, London.

EKSTRAND, J. & GILLQUIST, J. (1983) The avoidability of soccer injuries. *Int. J. Sports Med.*, **4**, 124–8.

ELLIOT, F.A. (1944) Tender muscles in sciatica: electromyographic studies. *Lancet*, **1**, 47–9.

ENGLAND, R.W. & DELBERT, P.W. (1972) Electromygraphic studies: part I. Consideration in the evaluation of osteopathic therapy. *Am. J. Osteopathy Assoc.*, **72**, 221–3.

FARFAN, H.L. & LAMY, C. (1975) *Human Spine in the Performance of Dead Lift*. St. Mary's Hospital, Montreal.

FISHER, A.A. & CHANG, C.H. (1985) Electromyographic evidence of paraspinal muscle spasm during sleep in patients with low back pain. *Clin. J. Pain*, **1**, 147–54.

FLOYD, W.F. & SILVER, P.H.S. (1951) Function of erectores spinae in flexion of the trunk. *Lancet*, **1**, 133–4.

FLOYD, W.F. & SILVER, P.H.S. (1955) The function of the erectores spinae muscles in certain movements and postures in man. *J. Physiol. (London)*, **129**, 184–203.

FRANKLIN, D.M. (1993) *Analysis of nonstationarities in surface myoelectric signals.* (MSc thesis, Boston University.

FREIVALDS, A., CHAFFIN, D.B., GARG, A. & LEE, K.S. (1984) A dynamic biomechanical evaluation of lifting maximum acceptable loads. *J. Biomech.*, **17**(4), 251–64.

FRYMOYER, J.W. (1989) Epidemiology. In *New Perspectives on Low Back Pain* (Ed. J.W. FRYMOYER & S.L. GORDON), pp. 19–33. American Academy of Orthopedic Surgeons Symposium.

FRYMOYER, J.W., POPE, M. & CLEMENTS, J.H. (1983) Risk factors in low-back pain: an epidemiological survey. *J. Bone Joint Surgery*, **65A**, 213–8.

GARRETT, W.E., JR, SAFRAN, M.R. & SEABER, A.V. (1987) Biomechanical comparison of stimulated and nonstimulated skeletal muscle pulled to failure. *Am. J. Sports Med.*, **15**, 452.

GARRET, W., ANDERSSON, G. & RICHARDSON, W. (1989a) Muscle: future directions. In *New Perspectives on Low Back Pain* (Ed. J.W. FRYMOYERS & S.L. GORDON), pp. 373–80. American Academy of Orthopedic Surgeons Symposium, Park Ridge, Ill.

GARRET, W., BRADLEY, W., BYRD, S., EDGERTON, V.R. & GOLLNICK, P. (1989b) Muscle: basic science perspective. In *New Perspectives on Low Back Pain* (Ed. J.W. FRYMOYER & S.L. GORDON), pp. 335–72. American Academy of Orthopedic Surgeons Symposium).

GILMORE, L.D. & DE LUCA, C.J. (1985) Muscle fatigue monitor: second generation, *IEEE Trans. Biomed. Engng BME*, **32**, 75–8.

GILMORE, L.D. & DE LUCA, C.J. (1987) Muscle fatigue monitor (MFM): an IBM-PC based measurement system. In *Proceedings of the Ninth Annual Meeting of the IEEE-Engineering in Medicine and Biology Society*. pp. 239–40.

GOLDING, J.S.R. (1952) Electromyography of the erector spinae in low back pain. *J. Postgraduate Med.*, **28**, 401–6.

GRABEL, J.A. (1973) Electromyographic study of low back muscle tension in subjects with and without chronic low back pain. *Dissertat. Abstr. Int.*, **34(B)**, 2929–30.

GRACOVETSKY, S., FARFAN, H.F. & LAMY, C. (1977) A mathematical model of the lumber spine using an optimized system to control muscles and ligaments. *Orthoped. Clin. N. Am.*, **8**, 135–53.

GRIEVE, D.W. (1974) Dynamic characteristics of man during crouch and stoop lifting. In *Biomechanics* (Ed. R.C. NELSON & C.A. MOREHOUSE), pp. 19–29. University Park Press, Baltimore.

HAGERMAN, F.C., HAGERMAN, G.R. & MICKELSON, T.C. (1979) Physiological profiles of elite rowers. *Physician Sports Med.*, **7**, 76–83.

HIDES, J.A., STOKES, M.J., SAIDE, M., JULL, G.A. & COOPER, D.J.H. (1994) Evidence of lumbar multifidus muscle wasting ipsilateral to symptoms in patients with acute/subacute low back pain. *Spine*, **19**(2), 165–72.

HOLBROOK, T.L., GRAZIER, K. & KELSEY, J.L. (1984) *The Frequency of Occurrence, Impact and Cost of Selected Musculoskeletal Conditions in the United States*. American Academy of Orthopaedic Surgeons.

HOYT, W.H., HUNT, H.H. & DE PAUW, M.A. (1981) Electromyographic assessment of chronic low back pain syndrome. *J. Am. Osteopath. Assoc.*, **80**, 728–30.

ISERNHAGEN, S.J. (1993) Advancements in functional capacity evaluation. In *Back Pain Rehabilitation* (Ed. B. D'ORAZIO), pp. 180–204. Andover Medical Publishers, Boston.

JACOBSON, E. (1944) *Progressive Relaxation; a Physiological and Clinical*

Investigation of Muscular States and their Significance in Psychology and Medical Practice. University of Chicago Press, Chicago, IL.

JÄGER, M. & LUTTMAN, A. (1989) Biomechanical analysis and assessment of lumbar stress during load lifting using a dynamic 19-segment human model. *Ergonomics*, **32**, 93–112.

JAYASINGHE, W.J., HARDING, R.H., ANDERSON, J.A.D. & SWEETMAN, B.J. (1978) An electromyographic investigation of postural fatigue in low back pain – a preliminary study. *ECG Clin. Neurophysiol.*, **18**, 191–98.

JONES, A., POLLACK, M., GRAVES, J., FULTON, M., JONES, W., MACMILLAN, M. *et al.* (1988) *Safe, Specific Testing and Rehabilitative Exercise for the Muscles of the Lumbar Spine*. Sequoia Communications, Santa Barbara, CA.

JORGENSEN, K. & NICOLAISEN, T. (1987) Trunk extensor endurance: determination and relation to low-back trouble. *Ergonomics*, **30**, 259–67.

JUEL, C. (1986) Potassium and sodium shifts during *in vitro* isometric muscle contraction, and the time course of the ionic-gradient recovery. *Pflügers Arch.*, **406**, 458–63.

KIPPERS, V. & PARKER, A.W. (1984) Posture related to myoelectric silence of erectores spinae during trunk flexion *Spine*, **9**, 740–45.

KLEIN, A.B., SNYDER-MACKLER, L., ROY, S.H. & DE LUCA, C.J. (1991) Comparison of spinal mobility and isometric trunk extensor strength to EMG spectral analysis in identifying low back pain. *Phys. Ther.*, **71**, 445–54.

KLEINBAUM, D.G. & KUPPER, L.L. (1978) *Applied Regression Analysis and Other Multivariable Methods*. Duxbury Press, Belmont, California.

KONDRASKE, G.V., DEIVANAYAGAM, S., CARMICHAEL, T., MAYER, T.G. & MOONEY, V. (1987) Myoelectric spectral analysis and strategies for quantifying trunk muscular fatigue. *Arch. Phys. Med. Rehabil.*, **68**, 103–10.

KRAFT, G.H., JOHNSON, E.W. & LABAN, M.M. (1968) The fibrositis syndrome. *Arch. Phys. Med. Rehabil.*, **49**, 155–62.

KRAUS, H. & RABB, W. (1961) *Hypokinetic Disease*. C.C.Thomas, Springfield, IL.

KRAVITZ, E., MOORE, M.E. & GLAROS, A. (1981) Paralumbar muscle activity in chronic low back pain. *Arch. Phys. Med. and Rehabil.*, **62**, 172–76.

LUND, J.P., REVERS, D., WIDMER, C.G. & STOHLER, C.S. (1991) The pain-adaptation model: a discussion of the relationship between chronic musculoskeletal pain and motor activity. *Can. J. Physiol. Pharmocol.*, **69**, 683–94.

MAGORA, A. (1970) Investigation of the relation between low back pain and occupation: 1. Age, sex, community, education, and other factors. *Indust. Med. Surgery*, **39**, 465–71.

MANNION, A.F. & DOLAN, P. (1994) Electromyographic median frequency changes during isometric contraction of the back extensors to fatigue. *Spine*, **19**(11), 1223–9.

MAYER, T.G., GATCHEL, R., KISHINO, N., KEELEY, J., MAYER, H., CAPRA, P. & MOONEY, V. (1986) A prospective short-term study of chronic low back pain patients utilizing novel objective functional measurement. *Pain*, **25**, 53–68.

MAYER, T.G., GATCHEL, R., MAYER, H., KISHINO, N., KEELEY, J. & MOONEY, V. (1987) A prospective two-year study of functional restoration in industrial low back injury: an objective assessment procedure. *J. Am. Med. Assoc.*, **258**, 1763–67.

MAYER, T.G., KONDRASKE, G., MOONEY, V., CARMICHAEL, T. & BUTSCH, R. (1989) Lumbar myoelectric spectral analysis for endurance assessment: a comparison of normals with deconditioned patients. *Spine*, **14**(9), 986–91.

MERLETTI, R., KNAFLITZ, M. & DE LUCA, (1990) Myoelectric manifestations of fatigue in voluntary and electrically elicited contractions. *J. Appl. Physiol.*, **69**, 1810–20.

MILLER, D.J. (1985) Comparison of electromyographic activity in the lumbar paraspinal muscles of subjects with and without chronic low back pain. *Phys. Ther.*, **65**(9), 1347–54.

MILLER, J.A., SCHULTZ, A.B. & ANDERSSON, G.B. (1987) Load-displacement behavior of sacroiliac joints. *J. Orthoped. Res.*, **5**, 92–101.

MOFFROID, M.T, HAUGH, L.D., HAIG, A., HENRY, S. & POPE, M. (1993) Endurance training of trunk extensor muscles. *Phys. Ther.*, **73**, 10–17.

MORIOKA, M. (1964) Some physiologic responses to the static muscular exercises. *Rep. Inst. Sci. Labour*, **63**, 6–24.

NEWTON, M. & WADDELL, G. (1993) Trunk strength testing with iso-machines, Part I: review of a decade of scientific evidence. *Spine*, **18**(7), 801–11.

NICOLAISEN, T. & JORGENSEN, K. (1985) Trunk strength, back muscle endurance and low-back trouble. *Scand. J. Rehabil. Med.*, **17**, 121–7.

NOUWEN, A. & BUSH, C. (1984) The relationship between paraspinal EMG and chronic low back pain. *Pain*, **20**, 109–23.

NOUWEN, A., VAN AKKERVEEKEN, P.F. & VERSLOOT, J.M. (1987) Patterns of muscular activity during movement on patients with chronic low-back pain. *Spine*, **12**, 561–5.

O'DONOGHUE, D.H. (1984) Injuries to the muscle-tendon unit. In *Treatment of Injuries to Athletes*, pp. 57–63. W.B. Saunders, Philadelphia.

OKADA, M. (1970) Electromyographic assessment of muscular load in forward bending postures. *J. Faculty Sci. (Tokyo)*, **8**, 311.

OKADA, M., KOGI, K. & ISHII, M. (1970) Enduring capacity of the erectores spinae in static work. *J. Anthropol. Soc. Nippon*, **78**, 99–110.

PERRY, J. & BEKEY, G.A. (1981) EMG–force relationships in selected muscle. *CRC Crit. Rev. Biomed. Engng*, **7**, 1–22.

POLLOCK, M.L., LEGGETT, S.H., GRAVES, J.E., JONES, A., FULTON, M. & CIRULLI, J. (1989) Effect of resistance training on lumbar extension strength. *Am. J. Sports Med.*, **17(5)**, 624–29.

PRICE, J.P., CLARE, M.H. & EWERHARDT, R.H. (1948) Studies in backache with persistent muscle spasm. *Arch. Phys. Med. Rehabil.*, **29**, 703–9.

ROLAND, M.D. (1986) A critical review of the evidence for a pain–spasm–pain cycle in spinal disorders. *Clin. Biomech.*, **1**, 102–9.

ROY, S.H. (1992) Instrumented back testing. *Phys. Ther. Practice*, **1**, 32–42.

ROY, S.H. (1993a) The role of muscle fatigue in low back pain. In *Back Pain Rehabilitation* (Ed. B. D'ORAZIO), pp. 149–79. Andover Medical Publishers, Boston, Massachusetts.

ROY, S.H. (1993b) Combined surface EMG and ^{31}P-NMR spectroscopy for the study of muscle disorders. *Phys. Ther.*, **73**, 892–901.

ROY, S.H., DE LUCA, C.J. & CASAVANT, D.A. (1989) Lumbar muscle fatigue and chronic lower back pain. *Spine*, **14**, 992–1001.

ROY, S.H., DE LUCA, C.J., SNYDER-MACKLER, L., EMLEY, M.S., CRENSHOW, R.L. & LYONS, J.P. (1990) Fatigue, recovery and low back pain in varsity rowers. *Med. Sci. Sports Exercise*, **22**(4), 463–9.

ROY, S.H., KUPA, E.J., KANDARIAN, S.C. & DE LUCA, C.J. (1994) Effects of muscle fiber type and size on EMG median frequency. In *Proceedings of the 10th*

Congress of the International Society of Electrophysiology and Kinesiology, (Ed. R. SHIAVA & S. WOLF), pp. 32–3. Butterworth-Heinemann Publishing, Oxford, England.

ROY, S.H., DE LUCA C.J., EMLEY, M. & BUIJS, R.J.C. (1995) Spectral EMG assessment of back muscles in patients with LBP undergoing rehabilitation. *Spine*, **20**(1), 38–48.

SADOYAMA, T., MASUDA, T., MIYATA, H. & KATSUTA, S. (1988) Fibre conduction velocity and fibre composition in human vastus lateralis. *Eur. J. Appl. Physiol.*, **57**, 767–71.

SAFRAN, M.R., GARRETT, W.E,. JR, & SEABER, A.V. (1988) The role of warm-up in muscular injury prevention. *Am. J. Sports Med.*, **16**, 123–9.

SCHOLTZ, J.P. (1992) Low back injury and manual lifting: Review and new perspectives. *Phys. Ther. Practice*, **1**(3), 20–31.

SCHULTZ, A.B., HADERSPECK-GRIB, K. & SINKORA, G. (1985) Quantitative studies of the flexion–relaxation phenomenon in the back muscles. *J. Orthoped. Res.*, **3**, 189–97.

SEIDEL, H., BEYER, H. & BRAUER, D. (1987) Electromyographic evaluation of back muscle fatigue with repeated sustained contraction of different strengths. *Eur. J. Appl. Physiol.*, **56**, 592–602.

SERROUSSI, R.E. & POPE, M.H. (1987) The relationship between trunk muscle electromyography and lifting moments in the sagital and frontal planes. *J. Biomech.*, **20**, 135–46.

SHERMAN, R.A. (1985) Relationships between strengh of low back muscle contraction and reported intensity of chronic low back pain. *Am. J. Phys. Med.*, **64**, 190–200.

SODERBERG, G.L. & BARR, J.O. (1983) Muscular function in chronic low-back dysfunction. *Spine*, **8**, 79–85.

STANDRIDGE, R., KONDRASKE, G., MOONEY, V., CARMICHAEL, T. & MAYER, T. (1988) Temporal characterization of myoelectric spectral moment changes: analysis of common parameters. *IEEE Trans. Biomed. Engng*, **35**, 789–97.

TEUFEL, R. & TRAUE, H. (1989) Myogenic factors in low back pain. In *Clinical Perspectives on Headache and Low Back Pain* (Ed. H. BISCHOFF, H. TRAUE & H. ZENZ), pp. 38–54. Hogrefe & Huber Publishers, Toronto.

THOMPSON, D.A. & BIEDERMANN, H.J. (1993) Electromyographic power spectrum analysis of the paraspinal muscles: long-term reliability. *Spine*, **18**, 2310–23.

THOMPSON, D.A., BIEDERMANN, H.J., STEVENSON, J.M. & MACLEAN, A.W. (1992) Changes in paraspinal EMG spectral analysis with exercise: two studies. *J. Electromyogr. Kinesiol.*, **2**, 179–86.

TIDBALL, J.G. (1984) Myotendinous junction: morphological changes and mechanical failure associated with muscle cell atrophy. *Exp. Mol. Pathol.*, **40**, 1–12.

TRAVELL, J., RINZLER, S. & HERMAN, M. (1942) Pain and disability of the shoulder and arm. *J. Am. Med. Assoc.*, **120**, 417–22.

TRIANO, J.J. & SCHULTZ, A.B. (1987) Correlation of objective measure of trunk motion and muscle function with low-back disability ratings. *Spine*, **12**, 561–65.

TROUP, J.D.G. & CHAPMAN, A.E. (1972) Changes in the waveform of the electromyogram during fatiguing activity in the muscles of the spine and hips: the analysis of postural stress. *EMG Clin. Neurophysiol.*, **12**, 347–65.

WHITE, A.A., III & GORDON, S.L. (1982) Synopsis: workshop on idiopathic low-back pain. *Spine*, **7**, 141–9.

WIKTORSSON-MOLLER, M., OBERG, R. & EKSTRAND, J. (1983) Effects of warm-up, massage, and stretching on range of motion and muscle strength in the lower extremity. *Am. J. Sports Med.*, **4**, 124–8.

WILDER, D.G., WOODWORTH, B.B. & FRYMOYER, J.W. (1982) Vibration and the human spine. *Spine*, **7**, 243–54.

WOLF, S.L. & BASMAJIAN, J.V. (1977) Assessment of paraspinal electromyographic activity in normal subjects and in chronic back patients using a muscle biofeedback device. In *Biomechanics VI-B* (Ed. E. ASMUSSEN & K. JORGENSEN), pp. 319–24. University Park Press, Baltimore.

WOLF, S.L., BASMAJIAN, J.V. & RUSSE, C.T. (1979) Normative data on low back mobility and activity levels: implications for neuromuscular reeducation. *Am. J. Phys. Med.*, **58**, 217–29.

YASHIMOTO, K., ITAMI, I. & YAMAMOTA, M. (1978) Electromyographic study of low back pain. *Jap. J. Rehabil. Med.*, **15**, 252–.

ZAR, J.H. (1974) *Biostatistical Analysis*. Prentice Hall, Englewood Cliffs, NJ.

ZARKOWSKA, A.W. (1981) The relationship between subjective and behavioral aspects of pain in people suffering from lower back pain. MPhil thesis, University of London.

CHAPTER TEN

Application of EMG in ergonomics: a clinical perspective

RUBIN M. FELDMAN

University of Alberta, Edmonton

10.1 DEFINITION

The use of electromyography in the clinical diagnosis of problems related to the motor unit (anterior horn cell, peripheral nerve, neuromuscular junction, and peripheral muscle) and the principles of its use in diagnosis are well established. Over the past 10 years, there has been a gradual increase in the use of this diagnostic method in determining the way in which muscles are used in such areas as upper extremity function in the work-place, analysis of muscle use in ambulation, and the determination of the presence of fatigue with repetitive movement. The effect of muscle training and muscle use in sport activities, attempts at defining more appropriate, non-fatiguing methods of performing muscle function in disease processes, and evaluation of muscle function for use in prosthetic devices, have also been studied by electromyography (EMG).

In addition, EMG biofeedback has been used as a means of reducing muscle spasm in muscles that have been overused, as well as a means of instructing individuals on the best way to use their muscles after disuse. In muscles that have been subjected to disuse with the subsequent decrease in proprioceptive input to the individual when attempts are made to start using the muscles again, retraining with the use of EMG biofeedback has been an effective treatment method.

This method of evaluation of muscle function is now being applied with increasing frequency in the study of ergonomics and the evaluation of muscle use and the appearance of muscle fatigue and pain. It provides guidance when attempts are made to improve the method of function in the normal situation and to correct abnormalities in relation to abnormal muscle function because of repetitive activities in the work-place.

If we define ergonomics as the study of all the factors involved in achieving the most efficient use of human energy, then the evaluation of muscle function by EMG becomes an integral part of this evaluation, particularly since clinical abnormalities can appear if abnormal and inefficient muscle function are perpetuated.

This chapter will review the methods of evaluation of normal and abnormal muscle activity, as well as the way in which they can be used as a guide for treatment of these abnormalities, so as to achieve a more appropriate and less stressful use of muscle.

10.2 EVALUATION

The use of surface electrodes with electromyographical equipment that allows for greater signal amplification and reduction of noise has introduced to EMG evaluation, a greater amount of comfort and accuracy than had previously been possible with needle electrode methods (Lynn *et al.*, 1978; De Luca, 1993; Barkhaus and Nandedkar, 1994). While in the experimental animal, needles embedded into muscle, remaining on site for long periods of time, achieve an understanding of muscle function which may be more intense than that which is possible with the use of surface electrodes, the precision in the use of skin electrodes has improved to the point where excellent studies have been performed. Using skin electrodes, the measurement of the amplitude and frequency of the EMG signal and the determination of power spectrum analysis have permitted a broader study of muscle fatigue and the identification of the exact moment that fatigue actually occurs.

In ergometric studies, this method of using skin electrodes provides an excellent opportunity of studying muscle function in the dynamic state such as the study of repetitive strain injuries during overuse of the upper extremities in the work-place (Lundervold, 1951; Volz *et al.*, 1980; Brumfield and Champoux, 1984; Sigholm, 1984; Fernstrom *et al.*, 1994; Zijdewind and Kernell, 1994).

10.2.1 Gait Analysis

One of the first areas in which this methodology was put to the test was in the development of methods of gait analysis (Gurth *et al.*, 1979; Eke-Okoro, 1993). Within a short time, there was a profusion of gait laboratories, applicable to the clinical setting in large centers. However, these proved to be rather cumbersome, requiring a great deal of sophisticated and expensive technology, so that gait laboratories are now relegated to use in research laboratories, almost exclusively. One of the initial reasons for the use of gait laboratories was for the diagnosis of abnormal gait patterns. However, it soon became apparent that the accuracy which these analyses could achieve was equivalent to that of a well-trained observer who could identify any gait abnormalities that were present in the patient population.

10.2.1.1 Previous Injury

Nevertheless, EMG is still used as a means of determining the type of previous injury that might have occurred which would have had a negative effect on an otherwise, normal gait pattern. Electromyography is effective in identifying the presence of any neuropathic or myopathic processes.

10.2.1.2 Other Disease Processes

In addition, although in the developed countries, poliomyelitis is an historical oddity, it is still a major health problem in Third World countries. The presence of a late complication of poliomyelitis, post polio syndrome, has attracted a great deal of attention because of its symptomatology which is reminiscent of the weakness and paralysis that was present in the initial poliomyelitis episode. EMG needle electrode studies have been able to identify methods of differentiating between weakness in post-polio syndrome caused by incomplete recovery from acute polio and weakness caused by the changes in the motor units that are known to be the cause of new weaknesses appearing 25–40 years after the acute episode (Feldman, 1989a).

10.2.1.3 Teaching Aid

The use of surface EMG to determine the amount of potential muscle function that is present in the previously paralysed muscle, as well as the use of biofeedback to teach the individuals how to improve function in the muscles, has become an accepted method of management within the treatment protocol of these clinical syndromes.

The decreased use of some muscles in ambulation frequently results in overuse of other muscles or early development of fatigue in the muscles because of lack of efficiency. The use of surface electromyography has been found helpful in teaching individuals how to improve efficiency and, thereby, reduce fatigue in affected muscles. This method is also helpful in determining when overuse of some muscles occurs, as well as in identifying for the patient, methods of reducing the use of these muscles so as to avoid overuse and its complications (Feldman, 1988, 1989b, 1992).

10.2.2 Upper Extremity Muscle Training

These methods of analysis of muscle function are certainly not limited to the lower extremities. Needle electrode EMG studies are crucial in being able to determine if any disease processes are causing abnormal muscle function in the upper extremities. When these types of abnormalities are identified, accommodation to the abnormalities is frequently possible. Surface electromyography can then be of help in determining the best way to achieve non-fatiguing and more efficient muscle function in the upper extremities (Zijdewind and Kernell, 1994). A dramatic example is the training of individuals who have spasticity of the upper extremities after such medical problems as stroke with subsequent improvement in hand function as greater control is achieved and greater efficiency takes place (Zijdewind and Kernell, 1994).

In both the upper and the lower extremities, surface electromyography can be used as a guide to determine the need for electrical stimulation in spinal cord-injured individuals with electrical stimulation being used as a means of stimulating muscle function and achieving some improvement in either wheelchair function or short-distance ambulation. The increased sophistication of the electromyographic components and their portability provide opportunities which could not have been attempted just a few years ago.

10.2.3 EMG Evaluation in Prosthetics

This sophistication has also been extended to the area of myoelectric prosthesis (Northmore-Ball *et al.*, 1980). Electromyographic evaluation with needle and surface electrodes helps to determine the amount of EMG signal available which could frequently then be amplified to provide a method of initiating the function of small motors used in prosthetics, particularly of the upper extremities to achieve more appropriate upper extremity hand function. Unfortunately, while this is a very well-accepted method of providing another method of prosthetic use, the slowness of movement of the prosthetic components and the length of the reaction time between the thought of wanting to move the prosthetic fingers and the actual achievement of this movement are negative factors in the acceptance of this type of prosthesis. However, there are many individuals who have adapted to this method of prosthetic use and are functioning very well with myoelectric prostheses.

10.2.4 Evaluation of Fatigue

The electromyographic evaluation of fatigue with the exactness of being able to determine the exact moment when muscle fatigue occurs, has been very instrumental in being able to determine when there is a compromise of efficiency and muscle function. The observation that the EMG signal increases while the force of contraction is reduced is one indication of fatigue when EMG and force measurements are performed simultaneously. The improved study of the EMG signal with measurement of the frequency and amplitude of the signal and the determination of the power spectrum analysis have achieved the degree of accuracy in fatigue studies which provides an excellent method of determining how much fatigue occurs and the exact moment that muscle fatigue takes place (Lindstrom *et al.*, 1970; Komi and Tesch, 1979; Bigland-Ritchie *et al.*, 1981; Mills, 1982; Bigland-Ritchie and Woods, 1984; Babier *et al.*, 1993; Zijdewind and Kernell, 1994).

This evaluation has also been helpful in being able to train individuals to avoid fatigue by teaching them how to monitor the sensory changes within the muscle, which occur just prior to the appearance of EMG changes indicative of fatigue. This method of ergomonic evaluation to avoid fatigue, coupled with the training of individuals in methods of avoiding fatigue is very helpful in dealing with individuals who are susceptible to abnormalities related to such problems as repetitive strain injuries. This method has also been helpful in being able to teach individuals how to avoid muscle fatigue in syndromes in which fatigue must be avoided during exercise, such as multiple sclerosis and post-polio syndrome.

The combination of training the individual to perceive the sensory changes in muscle, combined with the additional information obtained from surface electrode EMG studies, provides an excellent method by which avoidance of fatigue and therefore, improvement in efficiency of muscle function can take place (Feldman, 1988, 1992).

10.2.5 Evaluation of Repetitive Strain Injuries

Finally, the evaluation of individuals with repetitive strain injuries resulting in painful muscle, tendon, and joint abnormalities because of overuse, particularly in the work-place, has been greatly facilitated by the use particularly of surface electromyography of the potentially affected muscles (Lundervold, 1951; Komi and Tesch, 1979; Sigholm, 1984; Fernstrom *et al.*, 1994; Zijdewind and Kernell, 1994). When these electrodes are placed on muscles that are known to suffer from overuse, evaluation of the intensity of the EMG signal, comparison of one limb with the contralateral limb if there is concern about unilateral repetitive strain, and the study of electromyographic changes relative to changes in force when simultaneous force measurements are done, provide very powerful tools in determining the mechanism by which repetitive strain injuries occur, as well as methods of their prevention. These studies performed either in a work-simulated or in the actual work situation provide timed hard copy results which can be studied and repeated as changes are made in the work environment. The achievement of ergonomic improvement in muscle function, particularly in the upper extremities in the work-place is greatly facilitated by the use of these electromyographic studies.

While changes in intensity of the EMG signal can be studied, fatigue studies of the corresponding muscles can also take place, providing yet another method of determination of abnormal changes in muscle in the work-place. The universal use of computers involving concentration on a monitor which may be poorly placed relative to the position of the worker as well as the use of the keyboard, have resulted in a tremendous increase in the number of repetitive strain injuries that are reported every year. The monitor position can easily have an abnormal effect on shoulder musculature, particularly in individuals who wear bifocals and are, therefore, required to frequently move their heads into different positions, so as to accommodate for their visual difficulties. EMG surface electrode studies of shoulder musculature provides a way of determining the effect of the abnormally positioned monitor on how these muscles are being used (Sigholm, 1984). If it is known that the individual develops a pain syndrome in these areas, evaluation with electromyography provides a method of determining to what degree overuse of these muscles has occurred.

In addition, forearm and upper arm muscles and, in particular, wrist extensors and biceps muscles and occasionally the triceps and deltoid muscles can be studied so as to determine the effect of the use of particular keyboards on the function of these muscles (Lundervold, 1951; Brumfield and Champoux, 1984; De Luca, 1993; Fernstrom *et al.*, 1994; Zijdewind and Kernell, 1994). When ergonomic changes take place in the work-place after evaluation, the effect of these changes can also be studied by repeat electromyographic studies, while attempting to achieve the greatest possible ergonomic correction for the worker involved in these areas.

10.3 TREATMENT

10.3.1 EMG Biofeedback

In the presence of pain produced by abnormal muscle function, regardless of whether this pain is secondary to abnormal muscle activity or whether it is joint pain as a result of decreased protection from overlying muscles, there is a tendency on the part of these muscles to reduce their function. If this process continues, there is a gradual reduction in the strength of contraction as well as the efficiency of movement, to the point where eventually, the use of this muscle or muscle group becomes very ineffective.

To this is added the problem of decreased proprioceptive input from the abnormal muscle as its contraction becomes less, so that, eventually, even if attempts are made to strengthen the muscle, the patient is unable to participate in a strengthening exercise program because the proprioceptive input required to know when a contraction is occurring, is now lacking.

A method which is commonly used to replace the decrease in proprioceptive input of muscle contraction is the use of electromyographic feedback (EMG biofeedback). This method of muscle training provides the affected individual with EMG signals which correspond to the intensity of the diminished but still existing muscle contraction. The EMG signal from this muscle, as identified by a surface electrode on the muscle, is amplified and is delivered to the patient, either as an auditory or as a visual signal or a combination of both. The patient is then asked to increase the signal which is either heard or visualized by improving and increasing the amount of muscle contraction. By responding to the need for increasing the visual and/or auditory input, the muscle is required to increase its intensity of contraction. Eventually, proprioceptive input returns and once that occurs, the EMG biofeedback can be discontinued. However, until such time as this happens, biofeedback by this method is a valid way of training muscles that have been subjected to disuse, either because of their involvement in a pain syndrome or because of other pathological reasons such as paralysis with only partial return of function.

In addition, individuals who suffer from back injury and have slowly reduced the use of some muscles, can learn to use these muscles again by this method (De Luca, 1993). Osteoarthritic changes particularly in the hip and shoulder girdles frequently result in secondary muscle disuse. Because these muscles are virtually normal, they can be trained to begin to function again with the use of biofeedback and, later on, in response to returned normal proprioceptive feedback.

Muscle disuse that relates to changes in muscle function as a result of limb injuries can also be treated in this way. Individuals who have had poliomyelitis and who have post-polio syndrome can experience an increase in the amount of muscle function that occurs in response to this method of treatment. Biofeedback is not limited only to the use of muscles that have become weak because of disuse. In the presence of overuse of muscles, resulting in muscle spasm, biofeedback can be used to reduce the intensity of muscle contraction. Individuals who cannot reduce the intensity of contraction adequately themselves, can be taught to do this by responding to the auditory and/or visual signals that have been mentioned. In this case, training consists of attempting to reduce the signals which are presented as a result of muscle contraction. As they reduce the signal, this reduction corresponds to a decrease in the intensity of muscle contraction, thereby achieving greater efficiency of muscle function with less pain. Since pain and spasm reinforce each other, a reduction in spasm results in a reduction in pain.

When muscles have been subjected to overuse in the work situation, as a result of spasm, the pain of the repetitive strain injury becomes apparent. The use of EMG biofeedback to reduce intense or prolonged muscle contraction, i.e. spasm, thereby reducing pain, can be used in the work situation to provide the worker with more comfort.

Similarly, individuals who have been involved in industrial accidents which result in muscle spasm or automobile accidents resulting in whiplash injuries can also benefit from the use of biofeedback to reduce the secondary muscle spasm.

These soft tissue injuries occurring so frequently in industry and in automobile accidents result in severe inefficiency in the use of muscles with subsequent fatigue of muscle as well as generalized fatigue. This method of reducing spasm can go a long way towards achieving improved efficiency in muscle function.

EMG biofeedback is also an effective method of reducing spasticity in central nervous system (upper motor neurone) dysfunction. The same methodology is used as was described in the treatment of muscle spasm, the requirement in this case being the need for cognitive and learning skills in a sensory intact patient.

10.4 SUMMARY

This chapter has dealt with the use of electromyography with the use of needle or surface electrodes and an amplified EMG signal, either as a method of evaluation of normal and abnormal muscle function or as a method of treatment.

The need for evaluation in many situations including work-related and recreational activities means that paradigms must be developed which will provide a portable method of evaluation of problems related to abnormal muscle function. An accurate, reproducible, and comfortable method of analysis of muscle function can be achieved by the use of electromyography using surface electrodes. This method of evaluation not only provides a better understanding as to the reasons for the abnormal and inefficient muscle function but it also provides guidance to help professionals in the formulation of treatment protocols which will reduce the adverse effects of the abnormal muscle function.

As can be seen by the information in this chapter, not only can electromography be used in the evaluation process but, in many cases, it can also be used as a treatment modality to achieve greater effectiveness and efficiency of muscle function where abnormal function was present before.

It becomes essential, therefore, that health care professionals become acquainted with the methodology by which EMG evaluation takes place as well as the interpretation of the results of this evaluation, so that they can be guided in the development of treatment protocols and ergonomic changes in the work-place after appropriate interpretation of EMG results has occurred. The combination of evaluation and interpretation is essential in our evolving and increasingly technical environment. This not only permits a more efficient use of the musculoskeletal system as the demands on this system change, but it also provides access to the work-place by a larger number of members of the disabled population.

References

BABIER, M., GUILLOT, C., LAGIER-TESSONIER, F., BURNET, H., & JAMMES, Y. (1993) EMG power spectrum of respiratory and skeletal muscles during static contraction in healthy man. *Muscle Nerve*, **16**, 601–9.

BARKHAUS, P.E. & NANDEDKAR, S.D. (1994) Recording characteristics of the surface EMG electrodes. *Muscle Nerve*, **17**, 1317–23.

BIGLAND-RITCHIE, B., DONOVAN, E.F. & ROUSSOS, C.S. (1981) Conduction velocity and EMG power spectrum changes in fatigue of sustained maximal efforts. *J. Appl. Physiol.*, **51**, 1300–5.

BIGLAND-RITCHIE, B. & WOODS, J.J. (1984) Changes in muscle contractile properties and neural control during human muscular fatigue. *Muscle Nerve*, **7**, 691–9.

BRUMFIELD, R.H. & CHAMPOUX, J.A. (1984) A biomechanical study of normal functional wrist motion. *Clin. Orthop.*, **187**, 23.

DE LUCA, C.J. (1993) Use of surface EMG signal for performance evaluation of back muscles. *Muscle Nerve*, **16**, 210–16.

EKE-OKORO, S. (1993) An electromyographic examination of the effect of load on human gait. *Clin. Biomech.*, **3**, 118–20.

FELDMAN, R.M. (1988) The use of EMG in the differential diagnosis of muscle weakness in post-polio syndrome. *EMG Clin. Neurophysiol.*, **28**, 269–72.

FELDMAN, R.M. (1989a) *The Determination of Fatigue Changes by Electromyography in Post-Polio Muscles.* American Association of Electromyography and Electrodiagnosis, San Diego, CA.

FELDMAN, R.M. (1989b) *EMG Evidence of Fatigue in Exercising Post-Polio Muscles.* Royal College of Physicians and Surgeons of Canada, Canadian Association of Physical Medicine and Rehabilitation, Toronto.

FELDMAN, R.M. (1992) *EMG Evidence of Muscle Fatigue in Post-Polio Syndrome – Its Use in Determining Treatment (in Peripheral Muscles and the Diaphragm).* International Association of EMG (CN 1X), Jerusalem, Israel. Poster Presentation.

FERNSTROM, E., ERICSON, M. & MALKER, H. (1994) EMG activity during typewriting and keyboard use. *Ergonomics*, **37** (3), 477–84.

GURTH, V., ABBINK, F. & THEYSOHN, H. (1979) Electromyographic investigation of gait. Methods and applications in orthopaedics *EMG Clin. Neurophysiol.*, **19**, 305–23.

KOMI, P.V. & TESCH, P. (1979) EMG frequency spectrum, muscle structure and fatigue during dynamic contractions in man. *Eur. J. Appl. Physiol.*, **41**, 41–50.

LINDSTROM, L., MAGNUSSON, R. & PETERSON, I. (1970) Muscular fatigue and action potential conduction velocity changes studies with frequency analysis of EMG signals. *Electromyography*, **10**, 341–55.

LUNDERVOLD, A. (1951) EMG investigation during sedentary work, especially typewriting. *Br. J. Phys. Med.*1, **14**, 32–26.

LYNN, P.A., BETTLES, W.D., HUGHES, A.D. & JOHNSON, S.W. (1978) Influences of electrode geometry on bi-polar recordings of the surface electrode. *Med. Biol. Engng Comput.*, **16**, 651–60.

MILLS, K.R. (1982) Power spectral analysis of electromyogram and compound muscle action potential during muscle fatigue and recovery. *J. Physiol. (London)*, **326**, 401–9.

NORTHMORE-BALL, M.D., HEGER, H. & HUNTER, G.A. (1980) The below-elbow myoelectric prosthesis. *J. Bone Joint Surg.*, **62B**, 363–7.

SIGHOLM, G. (1984) EMG analysis of shoulder muscle load. *J. Orthop. Res.*, **1**, 379–86.

VOLZ, R.G., LIET, M. & BENJAMIN, J. (1980) Biomechanics of the wrist. *Clin. Orthop.*, **149**, 112–7.

ZIJDEWIND, I. & KERNELL, D. (1994) Fatigue associated EMG behaviour of the first dorsal interosseous and adductor pollicis muscles in different groups of subjects. *Muscle Nerve* **17**, 1044–54.

Index

The following abbreviations have been used:
EMG electromyography
LBP low back pain